Table of Contents

Dedication

This book is dedicated to my parents, Palma and Roland Leveille, my son Eric, and my husband Kim, with honorable mention to my beloved cat, Holstein.

Preface

"There are some remedies worse than the disease"

Publius Syrus

It goes without saying that one of the incentives for me to write this QUICK LOOK *Clinical Pharmacology* book was a strong interest in the field of pharmacology. However, the real force that drove this project to completion was my passion for the practice of veterinary medicine. For the last several years I have worked as a small animal medicine internist at a secondary referral center. This means that I routinely see patients with complex medical problems who have a long history of pharmacologic manipulation of their disease that usually require continued polypharmaceutical therapy. This affords me the opportunity to see not only the best of what pharmacologic agents can do, but also instances where the "remedy truly is worse than the disease." I witness the cure of life-threatening infections with broad spectrum antibiotics, and at the same time see animals on chronic antibiotic therapy die from pneumonia acquired from their own enterically-derived multiple drug resistant bacteria. I see lymphoma patients in remission due to the administration of multi-drug chemotherapy regimes die from myelosuppression caused by the very same agents. I see patients with life-threatening cardiac arrhythmias die suddenly after starting anti-arrhythmic therapy and wonder if it was their cardiac disease or the pro-arrhythmic potential of the drug which was responsible for their death.

These experiences have convinced me that the responsible use of pharmacologic agents in our patients requires that we develop and maintain a rudimentary knowledge of basic pharmacologic principles. I have attempted to write this book with that thought in mind. Although one does not need to become proficient in the derivation of formulas that describe drug absorption, distribution, and excretion, as veterinarians we should have a basic understanding of how these mathematical manipulations affect our everyday decisions on drug use. It is not difficult for us to remember a drug's mechanism of action. This information is practical and directly relevant to why we prescribe the drug. Each time we use the drug, we witness the beneficial effect in our patients. It is not as easy to recall the drug's potential side effects, its interaction with other drugs, or the pathophysiologic states in which use of the drug is contraindicated. I suspect that well over 99% of the drugs we prescribe are used successfully; i.e., they control the disease condition and do not harm the patient. As a referral clinician where adverse drug reactions/interactions tend to concentrate, I realize that a 99% success rate may not be good enough. We need to appreciate how important our understanding and recognition of adverse drug reactions and interactions is to our patient's well being.

As course director for the second year course in veterinary clinical pharmacology, the book arose intially from my course syllabus. It is primarily intended for veterinary students enrolled in such a course, but should also be useful for interns and residents studying for board preparation and as an update for practicing veterinarians. The book is not meant to be an authoritative resource on veterinary pharmacology, but a brief, easy-to-follow review of major concepts. The format developed by my publishers minimizes the amount of text, so I have had to be concise. Each chapter is accompanied by diagrams, charts, or tables that visually re-inforce the important concepts.

This is the second book in Teton's series of Quick Look veterinary titles and my first venture into primary authorship of a book. I will admit that the book is slanted toward small animal applications because, after all, I am a small animal clinician. Both the publishers and I would appreciate any feedback on the style and content that might be incorporated into the second edition.

Cynthia R. L. Webster

Acknowledgments

I would like to acknowledge all the lecturers in my clinical pharmacology course at Tufts University School of Veterinary Medicine—Dr. Don Brown, Dr. Antony Moore, Dr. Andrew Hoffman, Dr. Linda Ross, Dr. Sawkat Anwer, Dr. Mary Rose Paradis, Deborah Phillips, and Dr. Mary Labato—all of whom have taught me so much about pharmacologic agents used in their area of specialty.

QUICK LOOK SERIES in Veterinary Medicine

CLINICAL PHARMACOLOGY

1 Basic Pharmacology: Drug Absorption

A Drug Absorption and Distribution

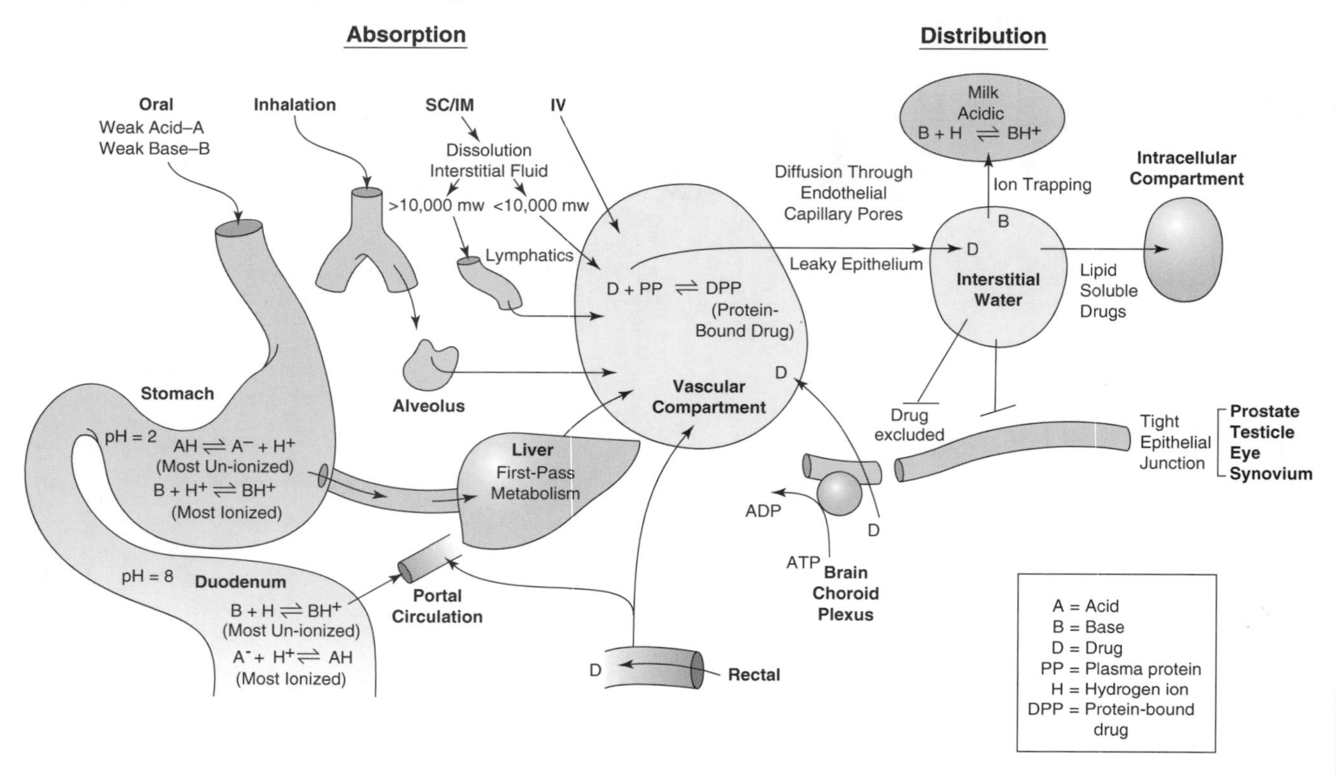

B Routes of Drug Administration

Route of Administration	Advantages	Disadvantages	Factors Affecting Absorption
IV	Complete absorption Accurate titration dose Can give large volumes Can give tissue irritants	Increased risk of adverse reaction Requires sterile technique Requires vascular access	Completely absorbed
SC/IM	Almost complete absorption Repository formulations for slow release and prolonged action	Cannot give large volume May cause tissue irritation/pain ↑ Muscle enzymes	Tissue vascularity Size of drug
Oral	Easy Economical Safe	Variable absorption Relies on owner/patient compliance	Stability of the drug in gastric acids and digestive enzymes Food in stomach Hepatic first-pass metabolism Lipid solubility: degree of ionization Gastrointestinal motility
Rectal	Easy to administer Good in vomiting, or comatose patients	Variable absorption Rectal mucosal irritation	50% of drug absorbed directly into circulation and bypass hepatic metabolism
Inhalation	Rapid absorption Avoidance of systemic side effects of some drugs	Patient compliance Regulation of dose administered	Solubility of drug in blood Size of drug

In order for a drug to produce its characteristic pharmaceutical effect, it must be present at the site of action in an appropriate concentration. Many factors influence the concentration of drug at the site of action. These include the rate and extent of absorption, distribution, metabolism, and excretion.

Drug Absorption

Absorption describes the rate and extent at which a drug leaves its site of administration. The route of administration and the physiochemical properties of the drug influence the rate of absorption. In order to exert cellular effects, drugs must penetrate cell membranes. Some drugs move across membranes by active transport or facilitated diffusion, utilizing membrane transport proteins present for endogenous compounds. This type of uptake is most important in epithelial tissues that normally transport solutes, such as the kidneys, liver, and choroid plexus. Most drugs cross cellular membranes by passive diffusion. A number of factors control the ability of a drug to diffuse across cell membranes. These include molecular size and shape, lipophilicity, degree of ionization, and solubility at the site of absorption. Since cell membranes are composed of a phospholipid bilayer, the more lipid soluble the drug is, the better the degree of cellular penetration. Highly polar, water-soluble substances are excluded. Most drugs are either weak acids or bases that are present in solution in both ionized and nonionized forms. Ionized drug is highly polar and cannot cross membranes, whereas the nonionized form is more lipid soluble and can diffuse through membranes. Therefore, the transmembrane distribution of the weak acids and bases is determined primarily by their degree of ionization. The degree of ionization is determined by the drug's pK_a and the pH of the medium in which it is dissolved. The pK_a is the pH at which 50% of the drug is in the ionized form. The ratio of ionized to nonionized drug is easily calculated using the Henderson-Hasselbach equation for weak acids (1) or weak bases (2):

1. pK_a = pH + log concentration of nonionized acid/concentration of ionized acid
2. pK_a = pH + log concentration of ionized base/concentration of nonionized base

Weak bases are poorly absorbed in the acidic environment of the stomach since they will be protonated and thus poorly lipid soluble. In the small intestine where the contents are alkaline, weak bases are not protonated and thus readily absorbed by passive diffusion through the intestinal epithelium (**Part A**).

Route of Administration

Drugs may be administered parenterally (by injection or inhalation), orally, topically, or rectally (**Part A**). The major routes of parenteral administration are intravenous (IV), subcutaneous (SC), and intramuscular (IM). Since IV injection bypasses all barriers to absorption, the desired blood concentration is obtained accurately and immediately. Since the rate of absorption is determined by the rate of administration, IV administration permits accurate titration of drug dose. The IV route permits administration of large quantities of drug at one time and the infusion of tissue irritants. Since IV administration results in high serum and tissue drug levels, it is associated with the highest risk of adverse drug effects. Other disadvantages of IV administration are the needs to establish venous access and to adhere to a strict aseptic technique (**Part B**).

Tissue vascularity and the solubility of the drug in interstitial fluids determine the rate of absorption of drugs after SC or IM injection (*see* **Part A**). Drugs given IM or SC must dissolve in the interstitial fluid and then move into the vascular compartment through large pores in the capillary endothelium. Lipid-soluble drugs passively diffuse through the capillary endothelial cells. Drugs with a molecular weight >10,000 are excluded from the endothelial pores and gain access to the circulation via lymphatic channels. This results in a considerable slowing of absorption. The rate of absorption after SC or IM injection can be slowed by altering the physiochemical properties of the drug either by dissolving it in a nonaqueous base or by attaching fatty acid esters. The local action of drugs administered IM or SC can be prolonged by concurrent administration of a vasoconstricting agent. This is commonly exploited by combining a local anesthetic agent with a vasoconstricting agent like epinephrine. The major disadvantage of the IM or SC route is that drug absorption cannot be terminated (*see* **Part B**). In addition, large volumes of drugs cannot be administered by this route and drugs associated with tissue irritation cannot be given. Pain or hemorrhage, or both, may occur at the site of administration. IM injections may lead to increases in serum concentrations of muscle leakage enzymes.

Oral administration of drugs is safe, convenient, and economical, but has the potential for the most variable absorption pattern. Since orally administered drugs must pass through the gastrointestinal mucosa to reach the bloodstream, the more lipid soluble the drug is, the better the absorption. A number of mechanical factors may interfere with absorption. Some drugs may irritate the gastric mucosa, resulting in emesis. Others may be destroyed or inactivated by gastric acid, digestive enzymes, or the intestinal microflora. Gastrointestinal motility disorders may also delay drug absorption. The presence of food in the stomach often delays absorption for a variety of reasons, including 1) formation of a dilutional/physical barrier to reaching the mucosal absorptive surface, 2) delays in gastric emptying (since most absorption occurs in the duodenum, anything that slows gastric emptying slows drug absorption), and 3) binding of drug to food components. Once absorbed across the gastrointestinal epithelium, drugs must enter the portal circulation and pass through the liver prior to entry into the systemic circulation. Some drugs undergo extensive hepatic metabolism during their first pass through the liver so that limited amounts of active drug reach the systemic circulation (first-pass effect) (*see* **Part A**). Another major disadvantage to the oral route is owner and patient compliance (*see* **Part B**).

Rectal administration of drugs is useful in vomiting or unconscious patients. Approximately 50% of any drug administered rectally bypasses hepatic circulation, so that the negative effects of hepatic first-pass metabolism can be minimized. Rectal drug absorption, however, is highly variable. In addition, many substances cause rectal mucosal irritation.

Gaseous and volatile drugs may be administered by inhalation and absorbed through the pulmonary epithelium and mucous membranes of the respiratory tract. Typically drugs are absorbed rapidly by this route, owing to the large surface area of the respiratory epithelium. Some nonvolatile drugs can also be aerosolized into small droplets and given by the inhalation route. While inhalational administration of volatile anesthetics to patients under anesthesia can be carefully controlled, major disadvantages of inhalational therapy in conscious patients are the difficulties inherent in patient compliance and the poor ability to regulate the dose administered.

2 Basic Pharmacology: Drug Distribution

A Drug Absorption and Distribution

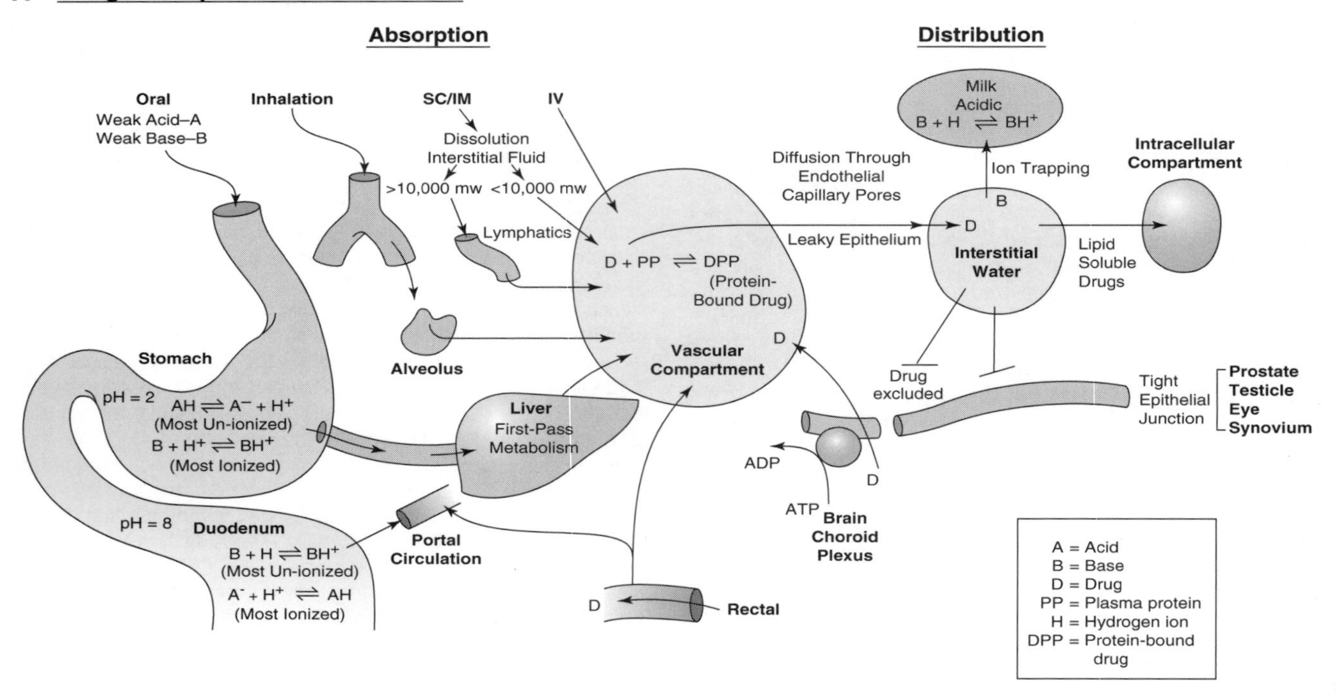

B Factors Affecting Drug Distribution

Factors	Consequences
Physiochemical properties	
pK_A	Ion Trapping Nonionized drug is more lipid soluble.
Lipid solubility	Lipid soluble drugs penetrate intracellularly. Highly polar (lipid insoluble) drugs confined to interstitial space.
Molecular weight	Drugs with molecular weight< 60 kd freely move through capillary endothelium.
Blood flow to tissue	Better perfused tissues achieve greater drug concentrations. Poorly perfused tissues (adipose tissue) have low drug concentrations that can serve as a reservoir of drug or as a site of redistribution.
Binding to macromolecules	
Plasma proteins	Only free (unprotein bound) drug can distribute outside of vascular space.
Tissues	Tissue binding can concentrate a drug at site of action and/or toxicity.
Anatomic Barriers	
Tight capillary endothelium CNS Eye Testicle	Tight junctions between capillary endothelium limits bulk flow of macromolecules. Drug penetration is now highly dependent on lipid solubility.
Fibrous capsules	Limited vascularity impedes drug entry in abscesses.

Drug distribution is the reversible transfer of drugs from one body compartment to another. It is governed by four factors: 1) the physiochemical properties of the drug (pK_a, lipid solubility, molecular weight); 2) blood flow to the tissue; 3) binding of the drug to plasma proteins and tissue macromolecules and; 4) the presence of anatomic barriers. (**Part B**).

Distribution of a drug begins as soon as it enters the bloodstream. The initial phase of drug distribution reflects cardiac output and regional blood flow so that the liver, brain, and kidneys receive most of the drug while distribution to more poorly perfused tissues such as fat, skin, and most viscera is slower. A second phase of drug distribution reflects movement of the drug into interstitial spaces or cellular compartments (**Part A**). The rate at which a drug leaves the circulation is proportional to the concentration gradient of the free drug from plasma to the extracellular water. Small molecules less than 60,000 daltons (most drugs) can freely move into the interstitial spaces by diffusion through the relatively large pores in the capillary endothelium. The degree of lipid solubility of a drug is generally not a limiting factor in the distribution of the agent to the interstitial space. The intracellular distribution of drugs is, however, governed by their lipid solubility. Lipid-soluble drugs are free to penetrate cellular membranes, whereas hydrophilic agents are excluded.

Many drugs bind to plasma proteins, which limits their distribution. Acidic drugs bind to albumin and basic drugs to α1-acid glycoprotein. The percentage of drug bound affects distribution, since protein-bound drugs cannot escape the capillary endothelium and gain access to cellular sites of action.

Drug distribution is also influenced by the presence of anatomical barriers to drug penetration. One of the most important barriers is present in the capillaries in the central nervous system (CNS). The capillary endothelial cells in the brain are held together by tight junctions that limit the bulk flow of macromolecules. Thus, drug penetration into the CNS is limited to highly lipid-soluble compounds. Drugs that do gain access to the brain, especially organic ions, can be actively extruded from the cerebrospinal fluid into the blood at the choroid plexus by specific membrane transport proteins. Other capillary beds with tight junctions include those in the testicles, eyes, and synovia. In many instances, disease processes may disrupt these capillary barriers and permit the entry of drugs normally excluded in the nondiseased state.

Drugs may accumulate in certain tissues as a result of pH gradients. In body compartments that are more acidic or basic than plasma, basic or acidic drugs may become trapped. For example, breast milk is acidic relative to the plasma, so weak bases that diffuse into milk in the nonionized form will become protonated (ionized) in milk (*see* **Part A**). Since the ionized drug is not lipid soluble, its movement back into plasma is prevented.

Tissue distribution may also be altered by drug binding to intracellular components. For example, tetracyclines bind to bone and aminoglycosides bind to the renal tubular epithelium.

Drug redistribution can contribute to termination of a drug's therapeutic action. Highly lipid-soluble drugs such as the barbiturates slowly redistribute to fat stores. This lowers the concentration of the drug at the site of action and diminishes the drug's action. Since fat has a low vascular supply, fat stores can serve as a stable reservoir of the drug.

3 Basic Pharmacology: Drug Elimination and Metabolism

A Drug Metabolism

Phase I Reactions

Aliphatic and Aromatic Hydroxylation

R-benzene → R-benzene-OH

$RCH_2CH_3 \longrightarrow RCH(OH)CH_3$

Oxidative or N-Dealkylation

$RNHCH_3 \longrightarrow RNH_2 + CH_2O$

$ROCH_3 \longrightarrow ROH + CH_2O$

Reduction

$RN = NR' \longrightarrow RNH_2 + R'NH_2$

$RNO_2 \longrightarrow RNH_2$

Hydrolysis of Esters + Amides

$R_1C(=O)OR_2 \longrightarrow R_1COOH + R_2OH$

$RC(=O)NR \longrightarrow RCOOH + R_1NH_2$

Deamination

$R_2CHNH_2 \longrightarrow R_2CO_3 + NH_3$

Desulfuration

$R_2CS \longrightarrow R_2CO$

Sulfoxide Formation

$RSR' \longrightarrow RS(=O)R'$

Phase II Reactions

Conjugation of OH, COOH, NH_2, or SH Group
–Glucuronidation
–Acetylation
–Sulfate Conjugation
–Glycine Conjugation
–O, S, and N Methylation
– Glutathione

R = Side Chain
BPD = Biotransformed Polar Drug
PP = Plasma Protein
D = Drug
ND = Nonpolar Drug
PD = Polar Drug

B Factors Affecting Drug-Metabolizing Enzymes

Individual Genetic Diversity

Species Variation
- Cats ↓ Glucuronyl transferases
- Dogs ↓ Acetyltransferases
- Pigs ↓ Sulfate conjugation

Activation by Drugs
- Barbiturates
- Phenylbutazone

Inhibition by Drugs
- Cimetidine
- Chloramphenicol
- Quinidine
- Organophosphates
- Monoamine oxidase inhibitors

↓ Activity in Neonatal and Geriatric Patients
↓ Activity in Chronic Hepatic Failure

C Drug Elimination

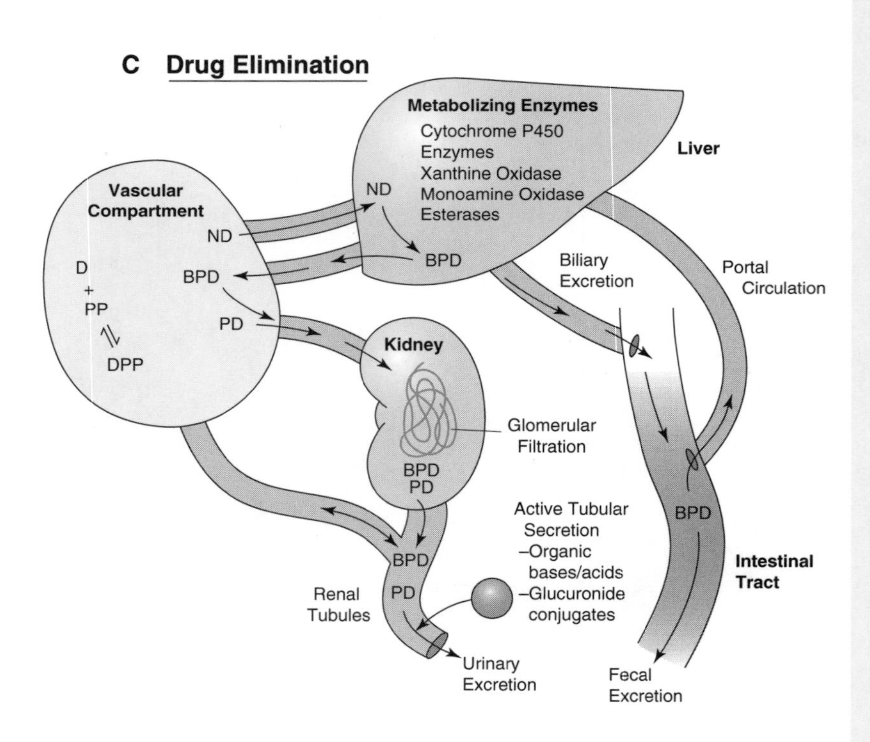

The kidneys are the major route by which drugs are excreted from the body. Non-protein-bound drugs are filtered at the glomerulus and enter into the tubular lumen. Polar drugs do not undergo tubular reabsorption and are excreted unchanged in the urine. On the contrary, lipid-soluble drugs present in the tubular fluid passively diffuse through the renal tubular cells back into the plasma compartment. In order to be excreted, these drugs must first undergo biotransformation to a more polar form.

Biotransformation

The enzymes responsible for drug biotransformation are primarily found in the smooth endoplasmic reticulum of the liver and to a limited extent, in the gastrointestinal tract, kidneys, and lungs. Chemical reactions involved in drug biotransformation are classified as phase I and phase II reactions. Phase I reactions convert the drug to a more polar metabolite by oxidation, reduction, or hydrolysis. Phase II reactions, which are called *conjugation reactions*, couple the more polar metabolite with glucuronide, sulfate, acetate, glutathione, or an amino acid.

Phase I Reactions

Phase I reactions are depicted in **Part A**. Oxidation is quantitatively the most important phase I reaction. The majority of oxidation reactions are catalyzed by a large family of microsomal enzymes known as the cytochrome P450 (CP450) enzymes. Additional nonmicrosomally mediated oxidation reactions are catalyzed by monoamine oxidases (MAOs), xanthine oxidase, and aldehyde dehydrogenase. Phase I hydrolysis reactions are carried out by nonspecific esterases present in the liver, plasma, and gastrointestinal tract. Phase I reduction reactions occur primarily in liver microsomes. The products of phase I reactions are most often inactive metabolites, although in some instances phase I metabolism results in the production of an active metabolite. In this case, the parent compound is called a *prodrug*.

Phase II Reactions

Phase I metabolites or drugs that contain a polar side group such as a hydroxyl, carboxyl, amino, or sulfhydryl group can be substrates for phase II conjugation reactions (**Part A**). The products of phase II reactions are highly hydrophilic and more rapidly undergo elimination in the urine. In addition, some phase II conjugates, particularly glucuronides, may serve as substrates for specific transporters in the renal tubule or biliary canalicular membrane. The products of phase II reactions are almost always inactive.

Phase II reactions require the activity of specific enzymes. Glucuronyl transferases are a family of microsomal enzymes found primarily in the liver and kidneys that catalyze the addition of glucuronide in phase II reactions. Cats have relatively low levels of certain members of this enzyme family and consequently metabolize drugs requiring this conjugation reaction very slowly. Cats appear to be able to conjugate endogenous substrates such as bilirubin, thyroxine, and corticosteroids with glucuronide, but are deficient in glucuronyl transferases for phenols and aromatic amines. Acetyltransferases, which are present in the cells of the reticuloendothelial system, are responsible for the addition of acetyl groups to drug metabolites. Dogs are relatively deficient in the acetyltransferase that catalyzes acetylation of aromatic amines. Pigs have a relative deficiency in the enzymes necessary for sulfate conjugation, while in cats sulfate-conjugating systems are easily saturated.

Factors Influencing Drug Biotransformation

A large number of variables can influence drug metabolism. These include genetic, environmental, and physiological factors (**Part B**). In humans (and most likely animals), considerable genetic polymorphism exists in the expression of the various members of the CP450 superfamily of enzymes. A number of environmental agents and drugs can inhibit or activate the CP450 enzymes. Drugs that inhibit the CP450 system include cimetidine, chloramphenicol, quinidine, and organophosphates. Drugs that induce CP450 activity include the barbiturates and phenylbutazone. In general, older individuals and neonates have lower activity of all hepatic biotransforming enzymes. Severe liver disease may compromise the activity of these enzymatic pathways. Since hepatic biotransformation pathways are saturable with limited capacity, when the amount of drug presented to the liver exceeds enzyme capacity, drug elimination will be slowed. Alternatively, if 2 drugs compete for the same enzyme system, elimination will also be slowed.

The enzyme γ-glutamyl transpeptidase catalyzes the conjugation of xenobiotics to glutathione. Although glutathione conjugation is not an important player in drug biotransformation, it is quite important in inactivating unstable and potentially toxic intermediates. In some instances, the products of phase I oxidation reactions yield reactive intermediates that are cytotoxic. Rapid conjugation of these intermediates to glutathione limits toxicity. When conjugation reactions temporarily fail to keep up with the generation of reactive intermediates, such as may occur with genetic lack of a specific enzyme or toxic overdosage, cytotoxicity may occur. This is the basis behind the unique sensitivity of cats to acetaminophen. Acetaminophen is normally metabolized by conjugation with glucuronide or sulfate. Since these conjugation reactions are quickly saturated in cats, acetaminophen is metabolized by the CP450 system to a reactive cytotoxic intermediate. Normally, the reactive intermediate is detoxified by conjugation with glutathione. Cats, however, have a limited amount of hepatic glutathione. When feline glutathione stores are depleted, oxidant damage to red blood cells and hepatocytes by the reactive intermediate occurs.

Drug Elimination

Renal excretion is the most common mechanism for the elimination of polar drugs or their metabolites (**Part C**). For most drugs or drug metabolites, renal elimination involves filtration of non-protein-bound drugs at the glomerulus and subsequent tubular excretion. Some drugs are actively secreted into the tubular lumen by renal proximal tubule membrane transport proteins that normally transport endogenous substrates. For example, organic acids (penicillins, furosemide) and glucuronide drug metabolites are excreted via the uric acid transport system. Organic bases (procainamide, trimethoprim, dopamine) are transported by a carrier that normally excretes histamine and choline. The extent of reabsorption of nonionized forms of weak acids and bases by the renal tubules can be modified by changing urine pH. When urine pH is rendered alkaline, weak acids become more ionized and are excreted more rapidly. The converse is true for weak bases. The presence of renal disease can decrease the rate of drug elimination. If drugs are excreted unchanged in the urine, then renal disease can significantly slow the rate of elimination of the active drug and result in prolongation of drug action.

Other routes of drug elimination include biliary and fecal excretion. The biliary canalicular membrane contains several transporters for low-molecular-weight organic anions and cations. Drugs excreted in bile enter the intestine, where they are either excreted in the feces or passively reabsorbed. Glucuronide conjugates of drugs that enter the intestinal tract may undergo hydrolysis by intestinal β-glucuronidases and the lipid-soluble parent drug reabsorbed. This process is called *enterohepatic cycling* and may significantly prolong the action of some drugs. Species differences in the extent of biliary excretion exist. Dogs, cats, and humans are considered good, fair, and poor biliary excretors, respectively.

4 Basic Pharmacology: Pharmacokinetics I

A Plasma Concentration–Time Curve Following Oral Drug Administration

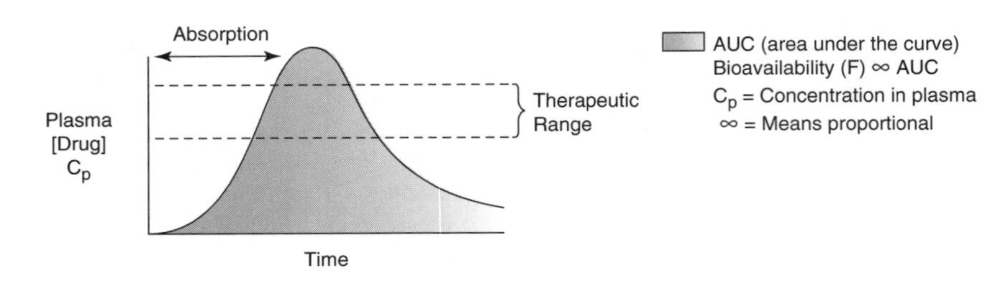

B One Compartment Modeling

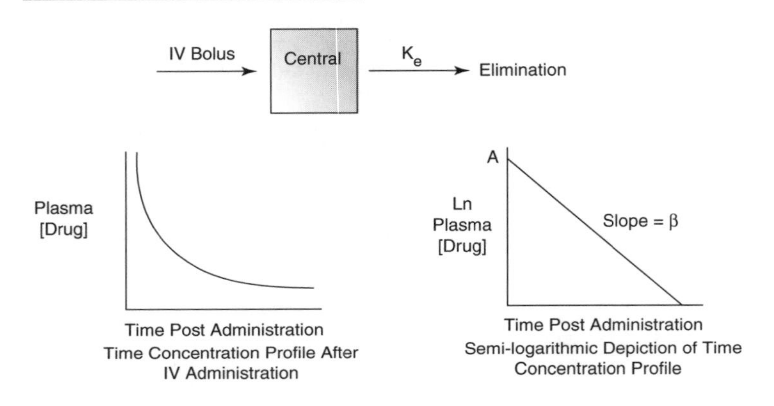

C Two Compartment Modeling

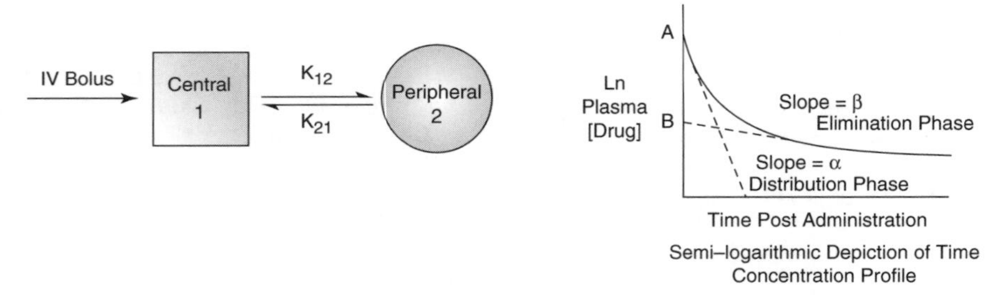

D Multiple Dosing Regimes

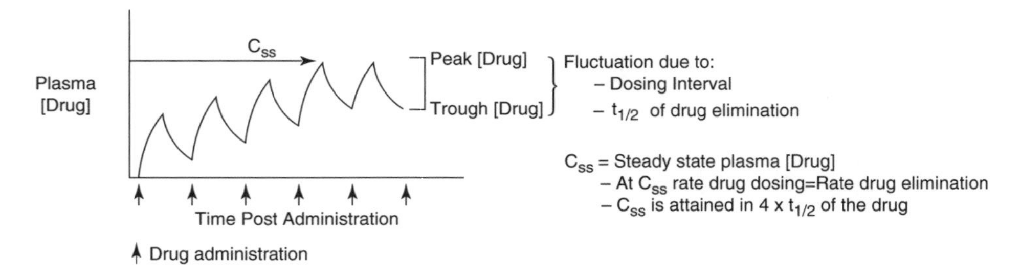

Pharmacokinetics is a mathematical description of the time course of drug absorption, distribution, metabolism, and elimination. The goal of pharmacokinetic modeling of drugs is to establish therapeutic drug doses and dosing regimens. It also forms the basis for establishing drug withdrawal times for meat and milk in food animals. The four most important pharmacokinetic parameters are bioavailability, clearance, serum half-life ($t^{1}/_{2}$), and volume of distribution (Vd).

Bioavailability

Bioavailability is a term used to describe the rate and extent to which a drug reaches its site of action. Drugs given IV are essentially 100% bioavailable. The bioavailability of drugs administered by other routes is the fraction of the dose that reaches the arterial blood circulation in an active form. Any factor that affects drug absorption (*see* Chapter 1) influences drug bioavailability. Intravenously administered drugs are 100% bioavailable. Drugs administered by other parenteral routes (IM or SC) have greater bioavailability than those given orally. For orally administered drugs, the extent of bioavailability depends on the extent of gastrointestinal absorption and the extent of hepatic and gastrointestinal metabolism. The oral bioavailability of some drugs can be severely limited by extensive first-pass hepatic metabolism. In general, orally administered drugs should have a relatively high bioavailability, but in reality the reproducibility of absorption is more clinically important. Mathematically, bioavailability is represented by the area under the curve (AUC) of a drug's plasma concentration–time curve (*see* **Part A**).

Volume of Distribution

The *Vd* is a quantitative estimate of the extent of drug distribution (**Part A**). It represents a theoretical volume in which the total amount of drug would have to be uniformly distributed to give the observed plasma concentration. The Vd is a proportionality constant that relates the amount of drug in the body to the concentration of drug in the blood:

Vd = total amount of drug in the body/ serum drug concentration at steady state (Css)

If the Vd of a compound equals plasma volume, then the compound is confined to the plasma and essentially is not useful as a drug. If Vd exceeds plasma volume, the compound is localized somewhere other than in the plasma. Highly lipid-soluble drugs that penetrate cell membranes and distribute intracellularly will have a larger Vd than drugs confined to interstitial fluids.

The Vd reflects a drug's affinity for plasma and tissue binding. High levels of plasma protein binding limit a drug's Vd to the plasma volume. Drug binding to tissues increases a drug's Vd. A highly lipid-soluble drug slowly partitions into adipose tissue, which increases its Vd. Since adipose tissue has a very low blood supply, drug deposition in fat results in persistent accumulation of a drug in the body. Obesity may actually increase the Vd of highly lipophilic drugs.

Clearance

Clearance is the volume of blood cleared of a drug per unit of time. The total clearance of a drug is the sum of clearances from each organ of elimination such as the liver, kidneys, and lungs.

Clearance = rate of drug elimination/serum drug concentration

Clearance is important in determining dosage regimens required to produce a desired steady-state serum drug concentration. Steady-state serum concentrations of a drug are established when clearance equals the rate of drug administration.

Dosing rate = clearance/steady-state serum concentration of drug

Thus, if the steady-state or therapeutic concentration of the drug is known, the drug's clearance will determine the dosing rate. Clearance is constant over a range of drug doses as long as the systems for drug uptake and elimination are not saturated (i.e., follow first-order kinetics where a constant fraction of the drug is eliminated per unit of time). When any component of drug uptake or elimination becomes saturable, zero-order kinetics prevail where a constant amount of drug is eliminated per unit of time. Under these circumstances, clearance becomes variable.

Clearance may be influenced by plasma protein binding, organ perfusion, drug-metabolizing enzyme ability, and the efficiency of renal excretion. A drug bound to plasma proteins is not filtered at the glomerulus, and if drug excretion by filtration is the major mode of elimination from the body, changes in the extent of protein binding can affect the rate of renal clearance. Since the degree of protein binding does not usually limit the rate of hepatic biotransformation, plasma protein binding does not alter the clearance of drugs that are metabolized by the liver and excreted in the bile. Clearance will vary depending on the drug hepatic extraction ratio. The hepatic extraction ratio is the fraction of drug presented to the liver that is cleared after a single pass through the liver. If $>70\%$ of a drug is extracted by the liver in one pass, the drug is said to have a high extraction ratio (propranolol, diltiazem, chlorpromazine). Clearance of this drug is described as blood flow limited and any factor that alters hepatic blood flow will affect drug clearance. Changes in the ability of the liver to metabolize the drug, or in the amount of free drug in the plasma, have little effect on clearance. If the hepatic extraction ratio of a drug is $<30\%$ (most drugs), then clearance is closely tied to free drug concentration. Now changes in plasma protein binding or in the intrinsic ability of the organ to metabolize the drug will have a greater effect on total clearance.

Plasma Half-Life ($t^{1}/_{2}$)

The plasma half life ($t^{1}/_{2}$) is defined as the time it takes for the total amount of drug in the body to decrease by 50%. It gives an estimate of the duration of drug effects in the body. The $t^{1}/_{2}$ is a derived parameter dependent on both clearance and Vd.

$$t^{1}/_{2} = 0.693\ \text{Vd/clearance}$$

Since disease states can alter clearance and Vd independently, $t^{1}/_{2}$ is an unreliable indicator of elimination. Some compounds may have similar clearances, but very different $t^{1}/_{2}$ values. For example, ampicillin and digoxin have similar clearances, but significant differences in Vd result in different $t^{1}/_{2}$ values of 48 and 1,680 min, respectively. Although it is not a good indicator of drug elimination, $t^{1}/_{2}$ is a useful parameter for the determination of the amount of time required to reach steady-state drug concentrations (about $4 \times t^{1}/_{2}$) and the time it takes for a drug to be removed from the body (50% should be eliminated in each $t^{1}/_{2}$), and provides a means to estimate dosing interval.

5 Basic Pharmacology: Pharmacokinetics II

A Plasma Concentration–Time Curve Following Oral Drug Administration

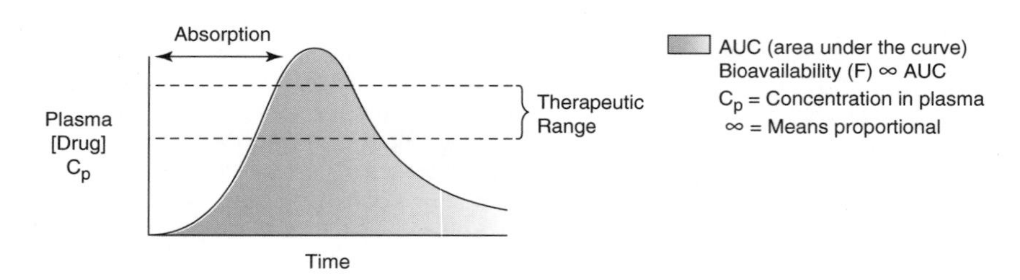

B One Compartment Modeling

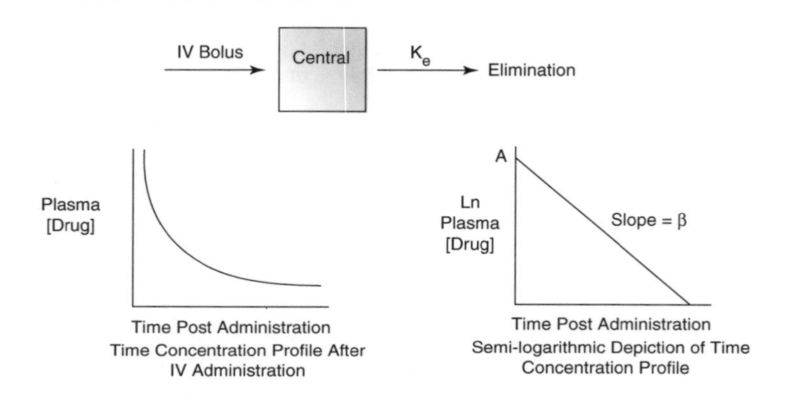

C Two Compartment Modeling

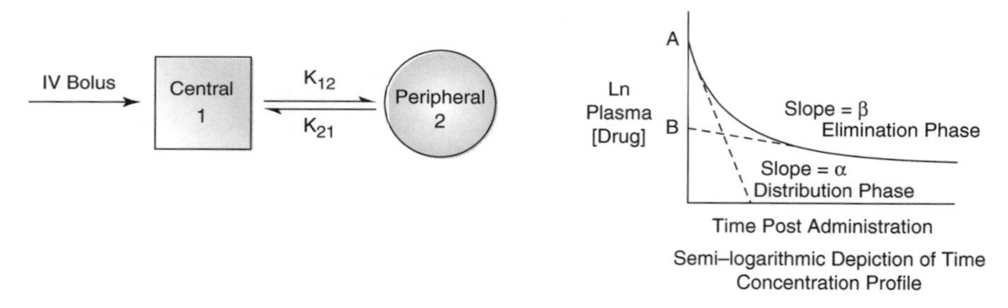

D Multiple Dosing Regimes

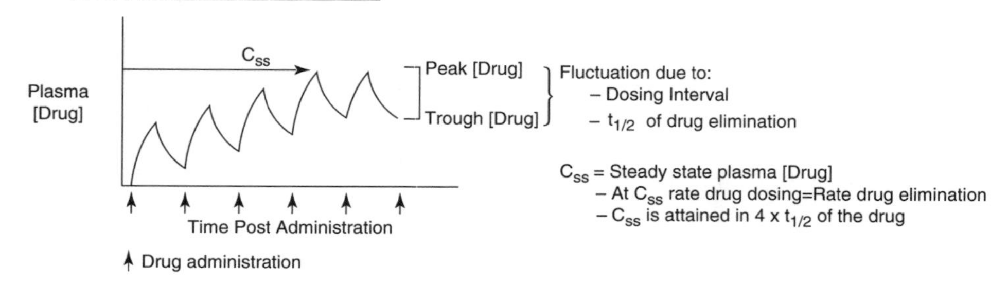

Clinical Dosage Regimen Design

Pharmacokinetic analysis uses the classic linear compartmental design. This system models the body as a series of interconnected compartments in which drugs are distributed and eliminated. Most drugs used in veterinary medicine can be described by a 1- or 2-compartment model (**Parts B** and **C**). The construction of a concentration-time curve for the drug after IV administration allows the determination of clinically relevant pharmacokinetic parameters. A basic assumption in analyzing pharmacokinetic data is that the concentration of any drug at its site of action is related to its concentration in the blood. In most cases, this is a valid assumption.

One Compartment Modeling

The simplest model is the one compartment model in which the body is envisioned as a single compartment into which drugs are distributed and eliminated. A plasma drug concentration-time curve for a single dose of an IV administered drug is depicted in **Part B**. In this graph the plasma drug concentration at any time (C_t) decreases exponentially (i.e., by a constant percentage per unit time). The rate of change in C_t can be described as:

$$dC_t/dt = -k_e(C_t) \text{ [or upon rearrangement } dC_t/Ct = -k_e dt] \quad (1)$$

in which k_e is the rate constant representing the fraction of drug eliminated from the body per unit of time. Integration of this differential equation yields:

$$C_t = C_o e^{-kt} \quad (2)$$

where C_o is the concentration of drug at time 0. A natural logarithm transformation of this equation results in the linear equation:

$$\ln Ct = \ln C_o - kt \quad (3)$$

A graphical depiction of this log transformation is shown in **Part B**. In this graph, the y intercept of this line (A) is equal to the C_o and k_e rate constant determined by the slope of the line, β. If the slope (β) and the intercept A of the line are known, all of the relevant pharmacokinetic parameters can be determined.

The volume of distribution, the proportionality constant relating the serum drug concentration to the total amount of body area is defined as:

$$Vd = Dose/A \quad (4)$$

The clearance of drug from the body, the amount of drug removed from the plasma for a given time interval is expressed as:

$$dxt/dt = Cl \times C_t$$

and since $dxt/dt = keVd\ Ct$

$$Cl = \beta Vd\ Ct/Ct \quad \text{or} \quad Cl = \beta \times Vd \quad (5)$$

The $t^1/_2$ can be calculated as the time it takes the initial drug plasma concentration, C_o, to decrease by $^1/_2$. This can be expressed as:

$$\ln {}^1/_2\ C_o/C_o = -\beta\ t^1/_2$$

and after rearrangement

$$t^1/_2 = \ln 2/\beta \quad \text{or} \quad t^1/_2 = 0.693/\beta \quad (6)$$

Two Compartment Modeling

In the 2 compartment model, which more accurately describes the pharmacokinetic behavior of most drugs, it is assumed that the body consists of a central and peripheral compartment between which reversible distribution and elimination occur (*see* **Part C**). In this model elimination and administration occur only from the central compartment. The rate of change in drug concentration in the central compartment equals:

$$dC_1/dt = -(k_{12} + k)\ C_1 + k_{21}C_2 \quad (7)$$

where C_1 is the concentration of drug in the central compartment, C_2 the concentration of drug in the peripheral compartment, k_{12} the rate constant for drug elimination from the central compartment, k_{21}, the rate constant for drug removal from the peripheral compartment (which equals drug addition to the central compartment) and k is the elimination rate constant.

Equation 7 can be integrated and expressed as the sum of 2 exponentials

$$Cp = Ae^{-\alpha} + Be^{-\beta}$$

where A and B are the intercepts and α and β the slope of the distribution and elimination phases, respectively as depicted in a plot of the log transformation of the time concentration curve (*see* **Part C**). The first phase of this curve is the distribution phase in which the drug is transferred to the plasma. The slope of this line is $-\alpha$ and the intercept A. The second phase is the elimination phase during which the drug is leaving the body. The slope of this line is $-\beta$ and the intercept is B.

The Vd, Cl, and $t^1/_2$ for a drug using the two compartment model can then be calculated:

$$Vd = Dose/AUC \times \beta$$

in which the AUC represents the area under the curve of the concentration-time profile from t = 0 to t = infinity and is calculated graphically using the trapezoidal method.

Clearance is calculated as:

$$Cl = Dose/AUC$$

The $t^1/_2$ is calculated with the terminal slope of the line, β:

$$t^1/_2 = 0.693/\beta$$

Multiple Dose Administration

In clinical medicine, multiple drug doses are administered separated by a time interval in an attempt to maintain a steady state plasma drug concentration (Css) within the drugs therapeutic range. **Part D** shows a typical serum concentration time profile after multiple dose oral drug administration. There is a rise in drug concentration that results immediately after drug administration followed by a first order exponential decay that predominates until the next dose is given. Drug accumulation occurs when the entire administered dose is not eliminated in a single dose interval. Peak and trough concentrations gradually rise with repeat administration of the drug until the amount of drug eliminated is equal to the amount of drug administered, i.e., Css has been reached. At steady state the amount of drug in the plasma will vary consistently between a peak and trough concentration. The amount of drug that accumulates is related to the dosing interval and the elimination of $t^1/_2$. When the dosing interval is less than $t^1/_2$, the drug accumulates and when it is longer than $t^1/_2$ the drug does not accumulate. The average Css is equal to the dose when the dosing interval equals $1.44 \times t^1/_2$.

When the average steady state concentration of the drug (Css) necessary to produce the pharmacological response is known the dosing interval can be calculated as

$$Css = Dose/Dosing\ interval \times Cl$$

or

$$Dosing\ interval = Dose/Css \times Cl$$

Another consideration in attaining a therapeutic response with any given drug is the time necessary to reach Css. Typically this occurs after 4 $t^1/_2$'s. In instances where steady state concentrations must be reached faster, a loading dose of the drug is administered. A loading dose (LD) is calculated as:

$$LD = Css \times Vd$$

Determination of Drug Bioavailability

The plasma drug concentration-time curve is generally constructed after IV drug administration. Drugs that are given intravenously are 100% bioavailable. The bioavailability of drugs administered orally or by alternative parenteral routes must be determined before clinical dosing regimes are designed for these routes of administration. Bioavailability (F) is to be determined by comparing the concentration time curves after oral and IV administration.

$$F = AUC^{oral} \times dose^{IV} \times \beta^{oral}/AUC^{IV} \times dose^{oral} \times \beta^{IV}$$

Taking bioavailability into account, the equation for determining dosing interval becomes:

$$Dosing\ interval = F \times Dose/Css \times Cl$$

6 Basic Pharmacology: Pharmacodynamics

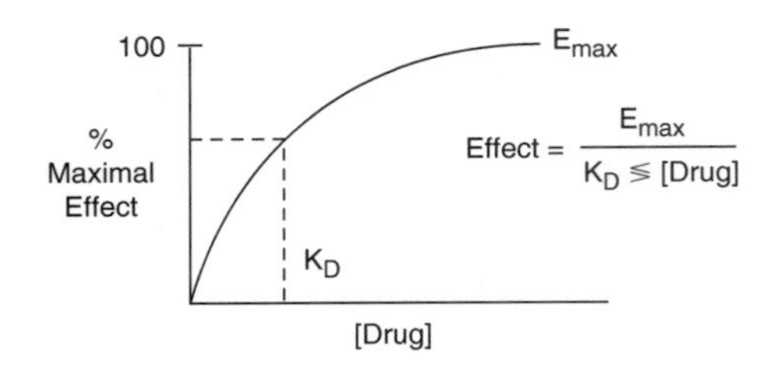

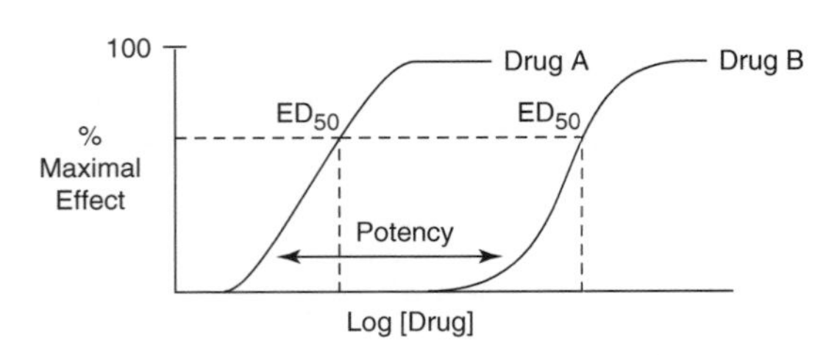

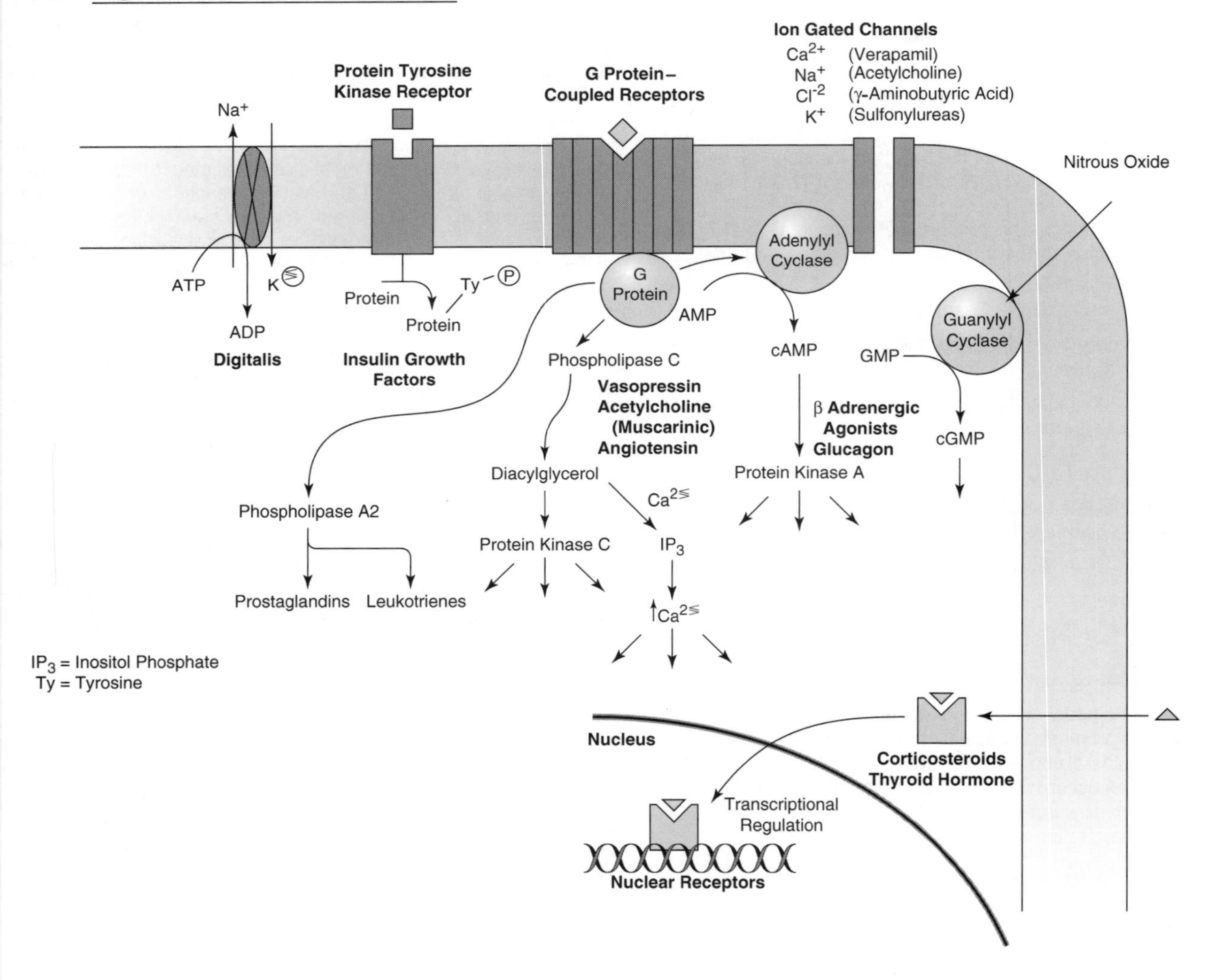

Pharmacodynamics is the study of the biochemical and physiological effects and mechanisms of action of drugs. The majority of drugs exert their effects by binding to receptors. This binding may be to physiological receptors that normally mediate hormone or neurotransmitter action or proteins involved in transport processes. Some drugs act independently of receptor binding by interfering with or acting as false substrates for important enzymatic pathways (e.g., pyrimidine and purine analogues) or by inhibiting enzymes (e.g., carbonic anhydrase or monoamine oxidase inhibitors). A few drugs act by virtue of their unique physiochemical properties, for example; 1) general anesthetics associate with lipid membranes in the CNS, leading to decreased excitability; 2) antacids work by neutralizing gastric acid; and 3) osmotic diuretics or cathartics stimulate increased volumes of urine or feces by drawing water into the tubular or gastrointestinal lumen.

Drug-Receptor Interactions

The nature of the drug-receptor interaction can be defined graphically by a dose-response curve (**Part A**). The effect the drug produces is proportional to the number of receptors occupied up to a maximal effect (E_{max}) at which all the receptors are occupied. This curve can be mathematically described by the Michaelis-Menten equation that is used to describe the interaction of enzymes with substrate so that:

$$\text{Drug response} = E_{max}/K_D + [\text{dose}]$$

where K_D is the dissociation constant for the drug-receptor complex. A plot of the response on a logarithmic scale (**Part B**) simplifies graphical interpretation of drug-receptor interactions. The sigmoidal curve defines the potency of the drug, which is dependent on the affinity of the drug for the receptor. The curve also defines E_{max}, which is an indicator of the drug's efficacy. In the depiction in **Part B**, drug A is more potent than drug B, but both drugs have the same efficacy. A useful parameter for comparing drug efficacy is ED_{50}, which is the dose of drug that produces the desired effect in 50% of the population. Often ED_{50} is compared to LD_{50}, which is the dose of drug that is lethal in 50% of the population. The resultant ratio (LD_{50}/ED_{50}) is known as the therapeutic index (TI). The higher the TI, the safer the drug.

The binding of a drug to its receptor is governed by its chemical structure. Small changes in chemical structure, such as changes in stereoisomerism, can have major effects on the nature of the drug-receptor interaction. In many cases, drugs have been developed by minor chemical modification of physiological ligands. Endogenous ligands have been modified to prolong the compound's $t^1/_2$, to increase affinity with which the compound binds to the receptor, or to eliminate the ability of the compound to elicit postreceptor effects.

Drugs that bind at receptors and mimic the action of the endogenous ligand are called *agonists*, whereas drugs that bind and do not activate the receptor are called *antagonists*. If inhibition by an antagonist can be overcome by increasing the concentration of the agonist, the antagonist is considered competitive. The antagonist is noncompetitive if it prevents the agonist from having an effect at any concentration.

Physiological receptors can be divided into categories based on structural homology and similarity between the postreceptor events they trigger. These postreceptor events are also called *signal transduction pathways* (**Part C**). Receptors can be classified as; 1) agonist or voltage-gated ion channels, 2) agonist-activated receptors that engage enzyme systems such as protein kinase C or protein tyrosine kinases, 3) G protein-linked receptors, and 4) nuclear receptors.

Nuclear receptors are present for corticosteroids, thyroid hormone, vitamin D, and retinoids. These receptors are soluble DNA-binding proteins that regulate the transcription of specific genes. Receptors for peptide hormones that regulate growth and differentiation are frequently transmembrane enzymes that phosphorylate target proteins on tyrosine residues. These include the receptors for insulin and various growth factors. Receptors for other ligands are tied to activation of enzymes involved in the production of cGMP or cAMP. Receptors for several neurotransmitters are ligand- or voltage-gated ion channels. These are the nicotinic acetylcholine (ACh) receptor (a Na^+ channel) and γ-aminobutyric acid (GABA) receptor (a Cl^- channel). A large number of cell membrane receptors work by coupling binding to activation of G proteins to downstream signal transduction events. Some G protein-coupled receptors activate adenylyl cyclase, leading to the production of cAMP and subsequent activation of protein kinase A. Other G protein-coupled receptors result in activation of phospholipase C, leading to the generation of increases in intracellular calcium and activation of protein kinase C. G protein receptors may also be coupled to activation of phospholipase A_2 or to the activation of ion-specific channels.

Receptors themselves can be regulated. Continued stimulation of a receptor may lead to its desensitization or downregulation. Downregulation may occur by modification of receptor structure, relocalization of the receptor intracellularly, or destruction of the receptor. Receptor desensitization has been documented after long-term treatment of asthmatics with β_2 agonists to promote bronchodilation. Conversely, long-term use of receptor antagonists may lead to upregulation of the receptors and hypersensitivity.

Drug Interactions

Drug interactions occur when one drug alters the intensity or duration of the pharmacological effect of a concurrently administered drug. The interaction can lead to an increased or decreased effect. Drug interactions may occur at several levels. Drugs may be chemically or physically incompatible in solution prior to administration, as is the case with penicillins and aminoglycosides. Drugs may interact with each other to alter gastrointestinal absorption. Anticholinergics or other drugs that delay gastric emptying slow the absorption of many drugs, since the majority of drug absorption occurs in the duodenum. Some drugs may bind to each other, such as occurs when fluoroquinolones or tetracyclines are administered with antacids. Drugs that alter or raise the gastric pH, such as antacids or H2 receptor antagonists, may decrease the absorbance or activation of drugs that require a low gastric pH such as ketoconazole or sucralfate. Highly protein-bound drugs can compete with each other for binding to plasma proteins. If drug A has a higher affinity for protein binding than drug B, it will displace drug B. This results in higher concentrations of the free drug B and an enhanced pharmacological effect. For example, the anticoagulant effect of the highly protein-bound drug warfarin is enhanced when it is displaced from its binding site on albumin by a nonsteroidal anti-inflammatory drug (NSAID). In general, drug interactions are significant only when the displaced drug is $>$90% bound to plasma proteins.

Drugs may interact at the receptor level. For example, aminoglycosides and clindamycin interfere with ACh release and potentiate the action of other drugs that act at the ACh receptor on the neuromuscular junction. Drugs that inhibit (cimetidine) or induce (phenobarbital) the hepatic microsomal enzyme systems can alter the elimination of drugs that require hepatic biotransformation for excretion. Drugs may also compete for clearance mechanisms. For instance, probenecid and penicillin compete for tubular secretion by the same anionic transporter in the renal tubules. Coadministration results in higher penicillin concentration and prolongs penicillin's $t^1/_2$. Another clinically relevant interaction in the renal tubules is the inhibition of digoxin excretion by quinidine and verapamil.

7 Autonomic Nervous System: Introduction

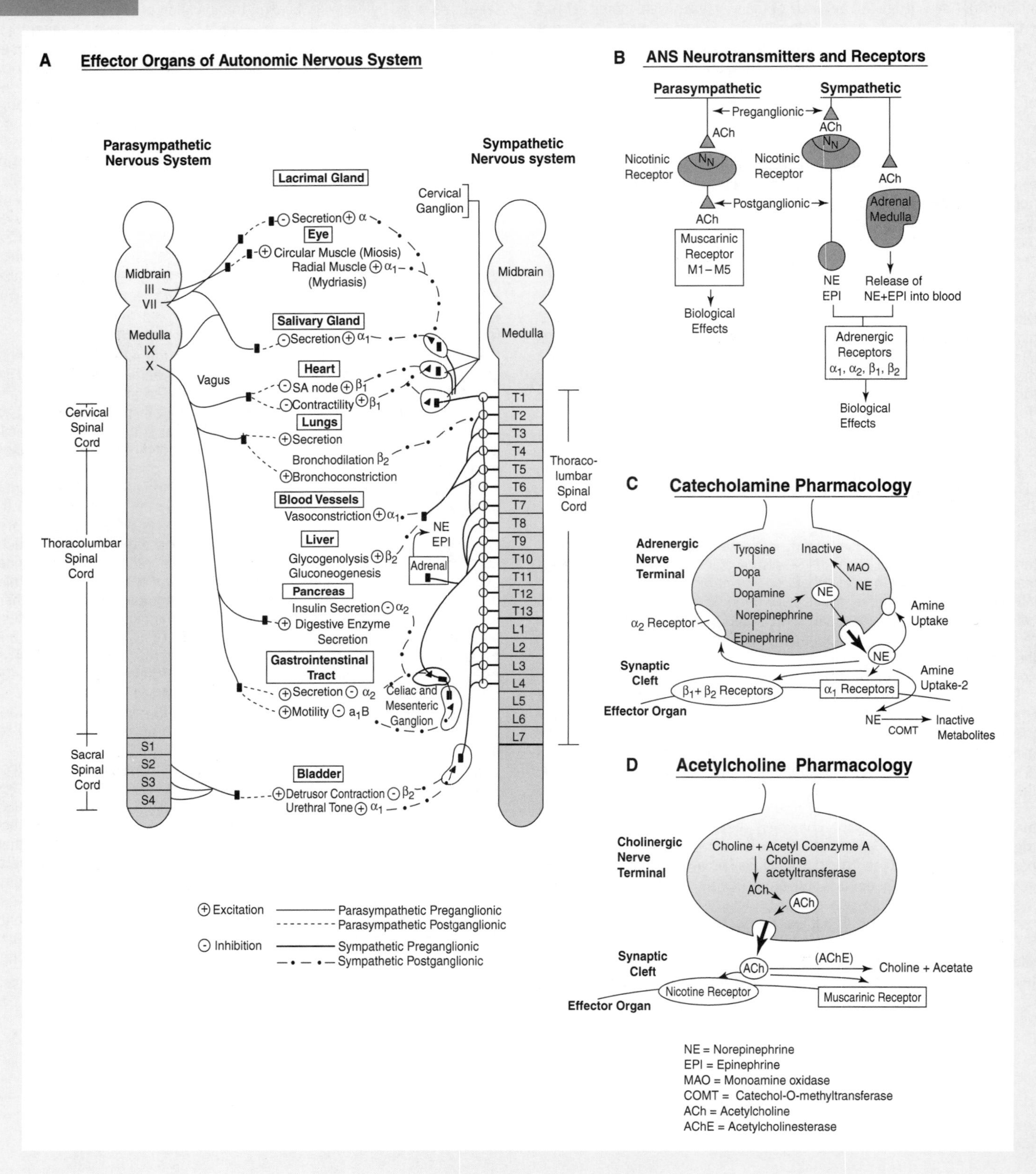

The autonomic nervous system (ANS) regulates the function of the heart, smooth muscle, and glands. Most visceral organs receive both sympathetic and parasympathetic input and this input often controls opposing effects. The efferent portion of the ANS is the most clinically significant and is composed of 2 neurons. The first neuron, the preganglionic neuron, originates in the CNS and synapses with the second neuron, the postganglionic neuron, in a ganglion located peripherally. The ANS is divided into 2 systems, the parasympathetic and sympathetic.

Parasympathetic ANS

The parasympathetic preganglionic neurons originate from the midbrain, medulla oblongata, or the sacral portion of the spinal cord (**Part A**). They terminate on postganglionic neurons, which are located close to the innervated organ. The vagus nerve (cranial nerve X) is the most important parasympathetic nerve trunk. It originates in the medulla and sends efferent fibers to the thoracic and abdominal viscera. The facial (cranial nerve VII), glossopharyngeal (cranial nerve IX), and oculomotor (cranial nerve III) nerves carry efferent output from the medulla to the eyes, face, and glands of the head. The sacral portion of the parasympathetic system sends efferent input through the pelvic nerve to the bladder, colon, and sex organs.

Sympathetic ANS

The preganglionic neurons of the sympathetic system originate in the thoracic or lumbar portion of the spinal cord (**Part A**). The nerves leave the spinal cord with the ventral nerve roots and then synapse with postganglionic neurons in the paravertebral ganglionic chain just outside of the spinal canal. Some preganglionic neurons synapse in prevertebral ganglia located more peripherally, such as the cervical, celiac, cranial mesenteric, and caudal mesenteric ganglia. The adrenal medulla is another component of the sympathetic nervous system. It is functionally homologous to a sympathetic ganglion although it has no postganglionic neuron; instead it secretes epinephrine and norepinephrine into the systemic circulation.

Neurotransmission

The ANS uses 2 neurotransmitters (NE) and several receptors. The 2 neurotransmitters are norepinephrine and acetylcholine (ACh). Sympathetic and parasympathetic preganglionic neurons release ACh, which binds to nicotinic receptors on postganglionic neurons (**Part B**). Sympathetic postganglionic neurons release NE, which acts on adrenergic receptors on the effector organ. Parasympathetic postganglionic neurons release ACh, which binds to muscarinic receptors on the effector organ. Nerves that release ACh are collectively called *cholinergic neurons* and those that release norepinephrine, *adrenergic neurons*. Epinephrine is the hormone preferably released upon stimulation of the adrenal medulla.

Receptors

There are 5 types of muscarinic (M1, M2, M3, M4, and M5) receptors, 2 types of nicotinic (N_M, N_N) receptors, and 4 types of adrenergic (α_1, α_2, β_1, β_2) receptors. The N_M receptors are present on the neuromuscular junction while the N_N receptors are located in autonomic ganglia, the adrenal medulla, and the CNS. M1 receptors are present on autonomic ganglia and in secretory glands, M2 receptors are present on the myocardium and smooth muscle, and M3 receptors are present on smooth muscle and secretory glands. The cellular localization of M4 and M5 receptors has not been well characterized. All of the muscarinic receptor subtypes are found in the CNS.

Catecholamine Pharmacology

Clinically relevant drugs that act on the ANS exert their action by interfering with synaptic neurotransmission by either binding at the receptor or interfering with neurotransmitter metabolism. An understanding of neurotransmitter biosynthesis and degradation simplifies an understanding of drug action. Catecholamines (dopamine, norepinephrine, NE) are synthesized from the aromatic amino acid tyrosine (*see* **Part C**). Epinephrine is an agonist at all 4 adrenergic receptors, while norepinephrine is an agonist at α_1, α_2, and β_1 receptors. Dopamine can stimulate the release of norepinephrine from adrenergic neurons and also binds to dopamine receptors in the renal, mesenteric, and coronary circulation (D1 receptors) and in the ganglia, adrenal cortex, and certain areas of the CNS (D2 receptors). The action of catecholamines is terminated by reuptake of the neurotransmitter in the synaptic cleft by the presynaptic neuron or by uptake by extraneuronal effector tissues. Catecholamines are taken up by an amine transport system referred to as the uptake-1 system. Once within the nerve terminal, catecholamines can be stored in granules or targeted for metabolic degradation. Monoamine oxidase (MAO) is a mitochondrial enzyme that catalyzes the oxidative deamination of catecholamines. There is an extraneuronal amine transporter known as the *uptake-2* system, which is present in the liver, myocardium, and other cells. Termination of the action of catecholamines may also be aided by diffusion of the neurotransmitter from the synaptic cleft to the extracellular fluid, where it is then methylated by the cytoplasmic enzyme catechol-*O*-methyltransferase (COMT). COMT is found in most peripheral tissues, with the highest concentration in the liver and kidneys. The breakdown products formed by the action of MAO and COMT are further metabolized and the inactive metabolites, primarily vanillylmandelic acid, are excreted in the urine.

The result of stimulation of the various receptors in the ANS is summarized in **Part A**. In general, β_1 adrenergic receptors are primarily present in the myocardium and renal juxtaglomerular apparatus and elicit excitatory responses. The β_2 receptors are primarily present in the smooth muscle (vascular, bronchial, gastrointestinal, genitourinary) and elicit inhibitory responses. The β_2 receptors are also present in the liver and skeletal muscle and promote glycogenolysis. The α_1 receptors are present primarily in vascular and genitourinary smooth muscle where they cause contraction. The α_2 receptors are present on vascular smooth muscle where they mediate vasodilation, on presynaptic adrenergic nerve terminals where they decrease the release of norepinephrine, and on pancreatic β cells where they decrease insulin release.

Acetylcholine Pharmacology

ACh is a quaternary compound synthesized by the enzyme choline acetyltransferase from choline and acetyl coenzyme A (**Part D**). ACh stimulates both muscarinic and nicotinic receptors. Termination of the action of ACh is brought about by rapid hydrolysis by an acetylcholinesterase (AChE) present in the synaptic cleft. A similar enzyme, pseudocholinesterase, is present in various body tissues. Binding of ACh to nicotinic receptors usually initiates an excitatory response, whereas binding to muscarinic receptors may elicit an excitatory or inhibitory response depending on the tissue (*see* **Part A**).

8 Autonomic Nervous System: Cholinergic Pharmacology

A Cholinergic Pharmacology

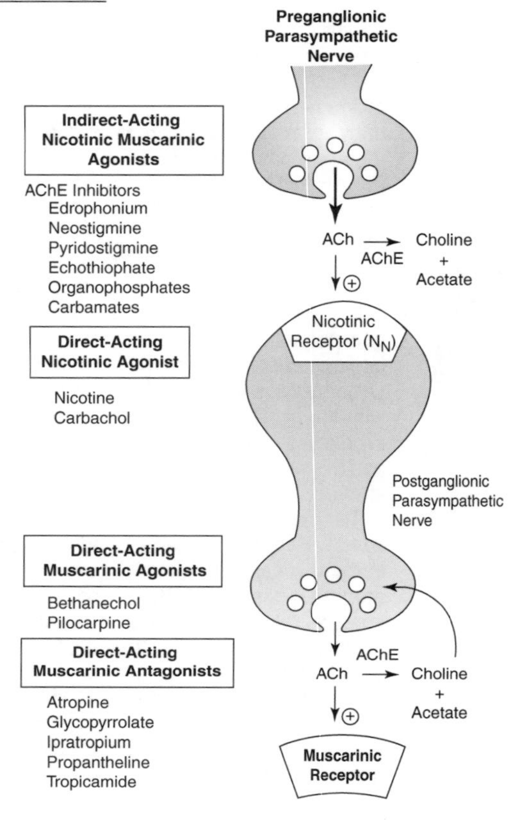

B Clinical Use of Cholinergic Drugs

Drug Group	Drug	Receptor	Clinical Use
Agonists			
Direct-Acting	Bethanechol	M	Bladder + gastrointestinal smooth muscle contraction
	Pilocarpine	M	Topical in eye for miosis and stimulation of lacrimal gland secretion
Indirect-Acting AChE Inhibitors	Edrophonium	M + N	Diagnosis of myasthenia gravis (MG)
	Neostigmine	M + N	Treatment of MG Stimulation of gastrointestinal motility
	Pyridostigmine	M + N	Treatment of MG
	Echothiophate	M + N	Topical in eye for miosis
	Organophosphates/ Carbamates	M + N	Topical antiparasitics

Drug Group	Drug	Receptor	Clinical Use
Antagonists	Atropine	M	Systemic: Control salivary and respiratory secretions during anesthesia; Antidote for organophosphate intoxication Treat bradyarrhythmias Topically in eye for long-acting mydriasis and cycloplegia
	Glycopyrrolate	M	Systemic: Control respiratory and salivary secretions during anesthesia
	Ipratropium	M	Inhalation, bronchodilation
	Propantheline	M	Systemic: Treat bradyarrhythmias and decrease gastrointestinal and bladder smooth muscle spasm
	Tropicamide	M	Topically in eye as short-acting mydriatic

Cholinergic drugs inhibit or excite effector cells that are innervated by the postganglionic parasympathetic neurons. Cholinergic drugs that mimic the action of acetylcholine are referred to as *parasympathomimetic agents*. Parasympathomimetic agents can exert their action by acting as direct cholinergic receptor agonists or as indirect agents that inhibit the breakdown of ACh (**Part A**).

Direct-Acting Cholinergic Agonists

Direct-acting cholinergic agonists include several synthetic choline esters and some naturally occurring alkaloids and their synthetic counterparts. ACh itself has no therapeutic applications because it is destroyed quickly by AChE and plasma cholinesterases. Several choline esters that withstand degradation by these enzymes have been synthesized. These choline esters include methacholine, a β methyl analogue of choline; carbachol, a carbamyl ester of choline; and bethanechol, the β methyl carbamyl ester of choline. Methacholine is degraded by AChE, but is resistant to plasma cholinesterases. Carbachol and bethanechol are resistant to hydrolysis with all cholinesterases. Bethanechol works preferentially at muscarinic receptors and has some selectivity for bladder and gastrointestinal tract smooth muscle. Bethanechol is used clinically to increase bladder detrusor muscle contraction and to augment gastrointestinal motility (**Part B**). Carbachol has both muscarinic and nicotinic receptor activity. There are 3 natural parasympathomimetic alkaloids: pilocarpine, muscarine, and arecoline. Muscarine works primarily on muscarinic receptors while arecoline acts on nicotinic receptors. Pilocarpine has predominantly muscarinic actions with a selectivity for the sweat glands. The use of these natural alkaloids clinically is limited by nicotinic side effects. Pilocarpine is used as a topical miotic agent or as an aid to augment lacrimal secretion in dry eye syndromes.

Indirect-Acting Cholinergic Agonists

Indirect-acting cholinergic agonists work by inhibiting AChE and thereby preventing the breakdown of ACh in the synaptic cleft (*see* **Part A**). This results in the accumulation of the ACh at cholinergic receptor sites. The AChE inhibitors are divided into 3 categories based on whether they: 1) reversibly inhibit AChE, 2) carbamylate AChE, or 3) phosphorylate AChE. They can have muscarinic or nicotinic actions, or both.

The quarternary amine edrophonium is the only clinically applicable reversible inhibitor of AChE. It works by binding directly to the active center of the enzyme. It has a short duration of action (10–15 min) due to the reversibility of its binding and rapid renal elimination. It is used in veterinary medicine as an aid in diagnosing myasthenia gravis (*see* **Part B**). Two other quarternary amines, physostigmine and neostigmine, have a carbamyl ester linkage that is hydrolyzed by AChE. As alternative substrates for the enyzme, their hydrolysis leads to the formation of a carbamylated form of the enzyme. This form of the enzyme is far more stable to hydrolysis than is the acetylated form ($t^{1}/_{2}$ = 30 min vs 1 min). Sequestration of the enzyme in this carbamylated form prevents further hydrolysis of ACh. Physostigmine is used clinically as a topical miotic agent and to reduce intraocular pressure in glaucoma. Neostigmine is used to treat myasthenia gravis in dogs and cats and as an aid for gastrointestinal or ruminal atony in horses and cattle, respectively. It can also be used to reverse the action of nondepolarizing neuromuscular blocking agents. Neostigmine is poorly absorbed orally and is best administered parenterally (SC or IM). It is metabolized in the liver and excreted in the urine. Pyridostigmine is structurally and functionally related to neostigmine. It is frequently used to treat myasthenia gravis. Although it is poorly absorbed from the gastrointestinal tract, this is the preferred route of administration. The onset of action after oral administration is within 1 hour and the drug is metabolized by the liver and excreted in the urine. Overdosage of neostigmine or pyridostigmine will cause a nicotinic/muscarinic cholinergic crisis marked by gastrointestinal signs (nausea, vomiting, diarrhea), respiratory effects (increased secretion, bronchospasm, respiratory paralysis), ophthalmic effects (miosis, blurred vision), cardiovascular effects (bradycardia or tachycardia, hypotension, cardiac arrest), and muscle weakness.

Carbaryl inhibits AChE in the same way as pyridostigmine and neostigmine. Carbaryl is used topically to control ectoparasites. There is little, if any, dermal absorption of this compound when applied at the appropriate dose.

Organophosphates comprise a large group of compounds that irreversibly inhibit AChE. They serve as substrates and result in the formation of a phosphorylated enzyme that is extremely stable. These compounds include the nerve gases sarin and soman and the insecticides parathion, fenthion, diazinon, and malathion. Echothiophate is a topical anticholinesterase agent used in the treatment of glaucoma.

Although phosphorylation of the AChE enzyme yields an extremely stable complex, compounds have been developed that can cause the dissociation of this complex. Pralidoxime (2-PAM) effectively removes the phosphate group from the enzyme and restores its activity. It is used clinically to treat organophosphate overdosages.

Direct-Acting Anticholinergics

Cholinergic agents that inhibit the action of ACh are called *parasympatholytics* (*see* **Part A**). Direct-acting parasympatholytics bind to muscarinic ACh receptors and act as direct antagonists. The natural anticholinergic drugs atropine and scopolamine are alkaloids derived from belladonna plants. Scopolamine has prominent actions in the CNS at low doses, making it useful for the treatment of motion sickness. Its action is likely mediated via inhibition of cholinergic neurotransmission in the vestibular system. Atropine is used clinically to decrease respiratory and salivary hypersecretion during anesthesia; topically in the eye to produce cycloplegia and mydriasis; as an antidote to anticholinesterase overdosage; as a large airway bronchodilator; and to treat conduction disturbances in the heart associated with increased vagal tone (sinus bradycardia, sinoatrial arrest, incomplete AV block) (*see* **Part B**). At high doses it will cross into the CNS and may cause drowsiness, stimulation, ataxia, or seizures. Atropine is metabolized in the liver and excreted in the urine. Propantheline is a synthetic quarternary ammonium anticholinergic agent whose pharmacological actions are similar to atropine except that it does not pass into the CNS. In small animals it is used to decrease smooth muscle spasm in the intestinal and urinary tracts and to treat bradycardias. In horses it is used to reduce colonic peristalsis to facilitate rectal examination. It is metabolized in the liver and gastrointestinal tract, and inactive metabolites are excreted in the urine. Glycopyrrolate is a synthetic anticholinergic agent that does not enter the CNS. It is approved for use in dogs and cats as a preanesthetic agent to control respiratory secretions. The majority of glycopyrrolate is eliminated unchanged in the urine and feces. Ipratropium is a quarternary ammonium compound that is administered by aerosol inhalation to horses for bronchodilation. When it is inhaled, its actions are almost completely limited to the respiratory tract. Tropicamide is a synthetic tertiary amine that is used as a topical mydriatic agent in ophthalmology.

9 Autonomic Nervous System: Adrenergic Pharmacology: Agonists

A Adrenergic Pharmacology

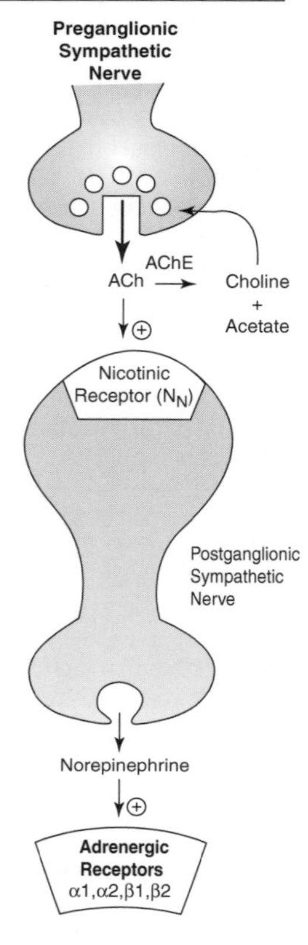

B Synthesis of Endogenous Catecholamines

Tyrosine —(*Tyrosine Hydroxylase*)→ Dopa —(*Dopa Carboxylase*)→ *Dopamine* —(*Dopamine - B - Hydroxylase*)→ Norepinephrine —(*Phenylethanolanine - - N - Methyltransferase*)→ Epinephrine

C Summary of Major Effects at Adrenergic Receptors

α_1 Vasoconstriction
Iris radial muscle contraction (mydriasis)
Lacrimal and sweat gland secretion
Urethral and gastrointestinal smooth muscle contraction
Splenic contraction

α_2 Presynaptic in CNS: Decrease NE release, causing analgesia, sedation, and muscle relaxation
Inhibit gastrointestinal secretion
Decrease insulin secretion

β_1 Increase heart rate and cardiac automaticity and contractility
Increase renin secretion

β_2 Bronchodilation
Relax ocular ciliary muscle
Vasodilation (coronary, skeletal muscle)
Relax detrusor muscle
Stimulate hepatic glycogenolysis/ gluconeogenesis

D Clinical Use of Adrenergic Drugs

Drug Group	Drug	Receptor	Clinical Use
Agonists			
Direct-Acting	Epinephrine	α_1, β_1, β_2	Systemic administration: Cardiopulmonary arrest Anaphylaxis Local administration: Hemostasis Topical administration: Vasoconstriction Long acting mydriatic
	Dopamine	D, α_1, β_1	Renal vasodilation Oliguric renal failure Circulatory shock Vasoconstriction to maintain blood pressure
	Dobutamine	β_1	Myocardial failure
	Terbutaline Albuterol Clenbuterol Salmeterol	β_2	Bronchodilation, given systemically or by inhalation
	Phenylephrine	α_1	Topical Nasal decongestant Ocular for vasoconstriction and mydriasis

Drug Group	Drug	Receptor	Clinical Use
Agonists			
Indirect-Acting	Phenylpropanolamine Ephedrine	α_1, β_1, β_2	Increase urethral smooth muscle tone Nasal decongestant
	Phenoxybenzamine	α_1, α_2	Decrease urethral smooth muscle tone Control hypertension
Antagonists **Direct-Acting**	Prazosin	α_1	Decrease urethral smooth muscle tone Control hypertension Afterload reduction in congestive heart failure
	Propranolol	β_1, β_2	Control hypertension Antiarrhythmic
	Atenolol Metoprolol Esmolol	β_1	Antiarrhythmic
	Timolol	β_2	Topical to decrease intraocular pressure
	Yohimbine Atipamezole Tolazoline	α_2	Antagonize α_2 agonists used for anesthesia/sedation

An important factor in the response of any cell or tissue to sympathetic modulation depends on the number and proportion of α and β adrenergic receptors that are present (**Part A**). The major effects of activation of adrenergic receptors are summarized in **Part C**.

Adrenergic Agonists

Endogenous Catecholamines

The endogenous catecholamines are epinephrine, norepinephrine (NE), and dopamine (**Part B**). Epinephrine is an agonist at both α and β receptors while norepinephrine is more active at α receptors (**Part D**). The major physiological effects of norepinephrine administration are vasoconstriction due to activation of α_1 receptors on vascular smooth muscle. Additional effects include mydriasis, stimulation of sweat gland secretion (horses) and increased urethral smooth muscle tone. Epinephrine administration is associated with bronchodilation (β_2) and increased myocardial contraction and heart rate (β_1). In vascular beds with predominantly α_1 receptors (abdominal viscera), epinephrine causes vasoconstriction, whereas at vascular beds with predominantly β_2 receptors (skeletal muscle), it causes relaxation. Large IV boluses of epinephrine have a predominant effect at α_1 receptors, resulting in vasoconstriction and increases in blood pressure. When epinephrine is given at lower doses, its predominant effect is a decrease in blood pressure mediated by excitation of β_2 receptors. Epinephrine also increases hepatic and muscle glycogenolysis and inhibits pancreatic β cell insulin secretion by activating β_2 receptors. Both norepinephrine and epinephrine increase myocardial irritability and predispose to the development of arrhythmias (especially if given concurrently with thyroid therapy, inhalation anesthetics, digitalis, or thiobarbiturates that sensitize the myocardium to epinephrine), cause splenic contraction (α_1), and when applied topically, cause pupillary dilation (α_1) and constriction of conjunctival and scleral vessels (α_1). Neither norepinephrine nor epinephrine are active orally or cross the blood-brain barrier. When administered parenterally, these drugs are inactivated by metabolism by monoamine oxidase (MAO) and carbachol-o-methyl transferase (COMT) in various tissues, and the resultant inactive metabolites are excreted in the urine (*see* Chapter 7 **Part C**). The clinical applications of epinephrine are cardiopulmonary resuscitation, treatment of anaphylaxis, prolongation of the action of local anesthetics, as a pressor agent, and as a topical hemostatic agent. Norepinephrine is used clinically as a pressor agent.

Dopamine is a precursor of norepinephrine. It has dose-dependent actions on α, β_1, and dopamine receptors. At low doses (~2.0 μg/kg/min), dopamine acts primarily on dopaminergic receptors in the renal, mesenteric, coronary, and intracerebral vascular beds, causing vasodilation. At intermediate doses (~5 μg/kg/min), the predominant effects are on β_1 receptors, with a resultant increase in heart rate and cardiac contractility. At higher doses (~10 μg/kg/min), dopamine activates α_1 receptors, causing vasoconstriction and increased total peripheral resistance and the net effect of dopamine on the kidney is vasoconstriction. Dopamine is not active orally and is given as a continuous-rate IV infusion. Its onset of action is immediate and its $t^{1/2}$ is short (5–10 min). The primary clinical indications for the use of dopamine are to promote renal blood flow in oliguric or anuric renal failure, to correct hypotension during shock that persists despite adequate volume replacement, and to augment myocardial contractility in heart failure (*see* Chapter 13). It is metabolized by MAO and COMT. A significant portion of the administered dose of dopamine is metabolized in nerve terminals to norepinephrine. This release of norepinephrine accounts for some of dopamine's effects at α adrenergic receptors. Side effects of dopamine include cardiac arrhythmias, arterial hypertension, and severe perivasculitis with inadvertent extravasation.

β Adrenergic Agonists

Dobutamine is a synthetic agent related to dopamine that preferentially activates β_1 receptors (*see* **Part D**). It has only mild β_2 and α_1 receptor-activating activity. These effects tend to balance one another and the net effect on the vasculature is minimal. Dobutamine does not cause the release of norepinephrine as does dopamine. Its main action is to increase myocardial contractility and it is used for short-term inotropic support in myocardial failure (*see* Chapter 10). It is less arrhythmogenic than dopamine. Dobutamine is given by continuous-rate IV infusion. It is metabolized in the liver. Side effects include gastrointestinal upset and tachyarrhythmias.

Several selective β_2 adrenergic agonists have been developed as bronchodilators (*see* **Part D**). Although both epinephrine and isoproterenol result in bronchodilation and are useful in the treatment of acute bronchoconstrictive disorders (anaphylaxis), their effect at β_1 and α receptors can cause cardiac and vascular side effects. Selective β_2 agonists include terbutaline, albuterol, clenbuterol, and salmeterol (*see* Chapter 21). These drugs relax bronchial, uterine, and vascular smooth muscle. Side effects may be related to mild stimulation of β_1 receptors (increased heart rate and development of cardiac conduction disturbances). When administered systemically, these agents may also cause CNS excitement. They should be used cautiously in animals with preexisting cardiac disease, hyperthyroidism, hypertension, seizure disorders, and diabetes mellitus. In horses that can be fitted with inhalers, these agents are preferentially given by inhalation. All extralabel use of clenbuterol in food animals is illegal, owing to the high incidence of toxic reactions in humans consuming meat with drug residues.

α Adrenergic Agonists

Phenylephrine is a synthetic α agonist that has predominantly postsynaptic α adrenergic effects (*see* **Part D**). At very high doses, β receptor effects may be seen. It is used topically in the eye for its vasoconstrictive properties and is applied intranasally to reduce nasal congestion in viral respiratory diseases or as an aid to control epistaxis (*see* **Part C**).

Ephedrine and phenylpropranolamine are synthetic sympathomimetic agents. They indirectly stimulate α and β receptors by enhancing the release of norepinephrine from sympathetic neurons. Prolonged use of either drug can deplete the nerve terminals of norepinephrine and lead to tachyphylaxis (decreased response) with chronic use. Their principal use in veterinary patients is to increase urethral sphincter tone. Both drugs may cause CNS stimulation. Other side effects include arterial hypertension, increased heart rate, bronchodilation, appetite suppression, increased intraocular pressure, and relief of nasal decongestion (**Part D**).

CNS α_2 receptors are important in modulating pain sensation and the level of arousal. In veterinary medicine, selective α_2 adrenergic agonists, such as xylazine and medetomidine, are used for sedation and analgesia (*see* Chapter 29).

10 Autonomic Nervous System: Adrenergic Pharmacology: Antagonists

A Adrenergic Pharmacology

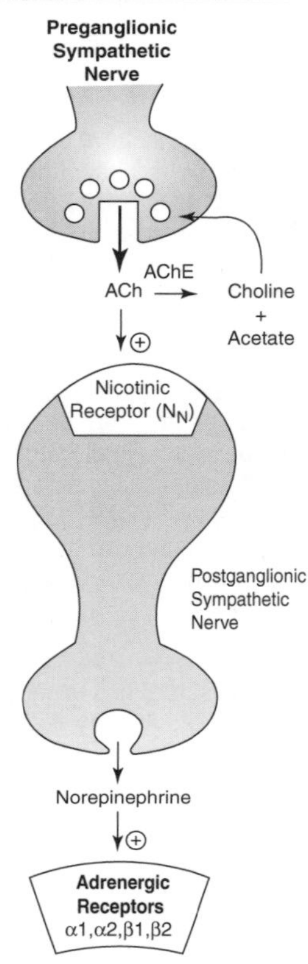

B Synthesis of Endogenous Catecholamines

Tyrosine —(*Tyrosine Hydroxylase*)→ Dopa —(*Dopa Carboxylase*)→ *Dopamine* —(*Dopamine - B - Hydroxylase*)→ Norepinephrine —(*Phenylethanolanine - - N - Methyltransferase*)→ Epinephrine

C Summary of Major Effects at Adrenergic Receptors

α_1 Vasoconstriction
- Iris radial muscle contraction (mydriasis)
- Lacrimal and sweat gland secretion
- Urethral and gastrointestinal smooth muscle contraction
- Splenic contraction

α_2 Presynaptic in CNS: Decrease NE release, causing analgesia, sedation, and muscle relaxation
- Inhibit gastrointestinal secretion
- Decrease insulin secretion

β_1 Increase heart rate and cardiac automaticity and contractility
- Increase renin secretion

β_2 Bronchodilation
- Relax ocular ciliary muscle
- Vasodilation (coronary, skeletal muscle)
- Relax detrusor muscle
- Stimulate hepatic glycogenolysis/ gluconeogenesis

D Clinical Use of Adrenergic Drugs

Drug Group	Drug	Receptor	Clinical Use
Agonists			
Direct-Acting	Epinephrine	α_1, β_1, β_2	Systemic administration: Cardiopulmonary arrest; Anaphylaxis. Local administration: Hemostasis. Topical administration: Vasoconstriction; Long acting mydriatic
	Dopamine	D, α_1, β_1	Renal vasodilation: Oliguric renal failure. Circulatory shock: Vasoconstriction to maintain blood pressure
	Dobutamine	β_1	Myocardial failure
	Terbutaline Albuterol Clenbuterol Salmeterol	β_2	Bronchodilation, given systemically or by inhalation
	Phenylephrine	α_1	Topical: Nasal decongestant; Ocular for vasoconstriction and mydriasis

Drug Group	Drug	Receptor	Clinical Use
Agonists			
Indirect-Acting	Phenylpropanolamine Ephedrine	α_1, β_1, β_2	Increase urethral smooth muscle tone Nasal decongestant
	Phenoxybenzamine	α_1, α_2	Decrease urethral smooth muscle tone Control hypertension
Antagonists **Direct-Acting**	Prazosin	α_1	Decrease urethral smooth muscle tone Control hypertension Afterload reduction in congestive heart failure
	Propranolol	β_1, β_2	Control hypertension Antiarrhythmic
	Atenolol Metoprolol Esmolol	β_1	Antiarrhythmic
	Timolol	β_2	Topical to decrease intraocular pressure
	Yohimbine Atipamezole Tolazoline	α_2	Antagonize α_2 agonists used for anesthesia/sedation

Adrenergic Antagonists

α Adrenergic Antagonists

The α adrenergic blocking agents are a heterogeneous group of agents that include haloamines (phenoxybenzamine), imidazole derivatives (phentolamine), piperazinyl quinazolines (prazosin), and indoles (yohimbine) (**Part D**). These agents differ in their relative specificity for α_1 and α_2 receptors. Phentolamine and phenoxybenzamine irreversibly block both α_1 and α_2 receptors. Their major effect is blockage of α_1 receptors, resulting in vasoconstriction and leading to a decrease in arterial blood pressure and a compensatory increase in heart rate. They have been used to treat hypertension associated with pheochromocytomas. Phentolamine is only available for parenteral administration. Phenoxybenzamine, which is available orally, is also used in veterinary medicine to relax urethral smooth muscle and to prevent or treat laminitis in horses. It has the added action of inhibiting the uptake of catecholamines into adrenergic nerve terminals. At high doses, greater than those necessary for α blockade, it has antiserotonergic, anticholinergic, and antihistaminergic actions. Although the $t^{1/2}$ of phenoxybenzamine is <24 hours, its duration of action may depend more on the rate of synthesis of new receptors. After discontinuation of the drug, phenoxybenzamine's effect may persist for a few days. The major side effects are hypertension, miosis, increased intraocular pressure, tachycardia, and constipation (horses).

Prazosin is a selective α_1 antagonist that is employed in the treatment of hypertension and as a balanced vasodilator to decrease afterload in cardiac failure. It is less likely to cause reflex tachycardia when used as an afterload reducer, as it does not block α_2 receptors and thus preserves their inhibitory effect on norepinephrine release. Prazosin may cause a first-dose phenomenon marked by profound hypotension and syncope. The exact mechanism accounting for this is unknown, but may be related to its inability to elicit reflex tachycardia.

Selective α_2 blockers, yohimbine, tolazoline, and atipamezole, are primarily used to antagonize the sedative effects of the α_2 agonists (*see* Chapter 29).

β Adrenergic Antagonists

Propranolol is a nonselective β adrenergic antagonist. Atenolol and metoprolol are β_1-selective agents (*see* **Part D**). Timolol is a selective β_2 antagonist. In the normal heart, blockage of β_1 receptors causes a minimal decrease in heart rate and contractility, since the heart at rest is not strongly influenced by sympathetic tone. When the heart is subject to increased sympathetic control such as during exercise, stress, or compensated cardiac failure, β_1 receptor antagonists cause a greater decrease in heart rate and myocardial contractility. Since vascular β_2 receptors only participate in a limited fashion in the control of vascular tone, β_1 receptor antagonists tends to have little effect on blood pressure. However, the β_1 receptor antagonists do lower blood pressure in hypertensive patients. This effect may be related to blockage of β_2-stimulated release of renin or inhibition of presynaptic β_2 adrenergic receptors that normally potentiate the release of norepinephrine from sympathetic neurons. Blockage of β_2 receptors in the bronchial smooth muscle results in bronchoconstriction, and thus β_2 blockers are contraindicated in patients with reactive airway disease. Agents that block β_2 receptors disrupt hepatic glucose metabolism and should not be used in patients with diabetes mellitus because they may delay recovery from a hypoglycemic episode. Topically administered β_2 blockers (timolol) are used to decrease intraocular pressure in patients with glaucoma.

The β_1 adrenergic receptor antagonists have several applications in the treatment of cardiac disease. Since they reduce sinus rate, slow conduction in the atria and atrioventricular (AV) node, and increase the functional refractory period of the AV node, they are useful in the treatment of tachyarrhythmias. Because of their negative inotropic effects, however, they should not be used in patients in congestive heart failure. In human patients with heart disease, β_1 receptor antagonists have been shown to improve survival by blunting the negative effects of chronic sympathetic activation on the myocardium. In hypertrophic cardiomyopathy, β_1 receptor β blockers may provide partial relief of the pressure gradient along the outflow tract. The β_1 receptor antagonists are also used to control cardiac side effects associated with hyperthyroidism.

11 Local Anesthetics

A Changes in Membrane Potential and Na^+ and K^+ Conductance During an Action Potential

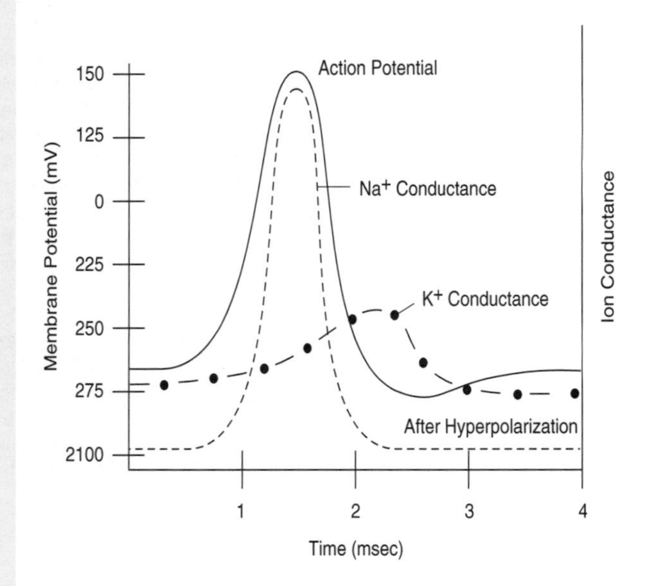

B Local Anesthetics

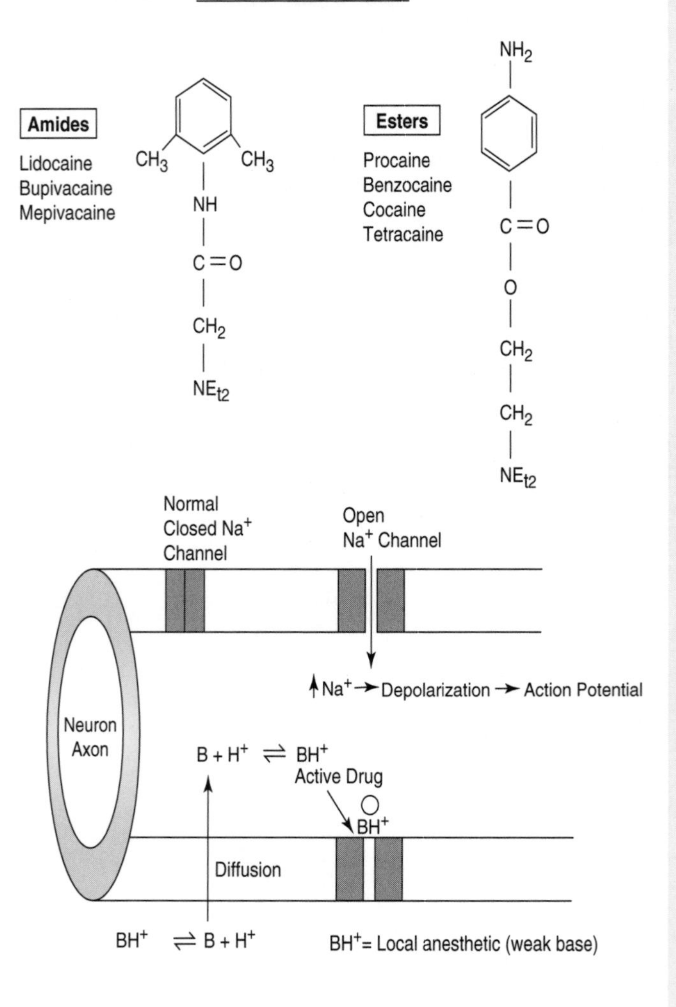

C Characteristics of Some Local Anesthetic Drugs

Agent	Lipid Solubility	pK_a	Onset of Action	Duration (Min)
Procaine	1	8.9	Slow	60–90
Lidocaine	3.6	7.7	Fast	120–240
Mepivacaine	2	7.6	Fast	90–200
Tetracaine	80	8.6	Slow	180–600
Bupivacaine	30	8.1	Intermediate	180–600

Action Potentials

Nerve impulses are conducted by action potentials, which result from changes in the permeability of the resting cell membrane to Na^+ and K^+. The normal neuronal resting membrane potential is maintained by the plasma membrane Na^+/K^+ ATPase pump, which pumps 3 Na^+ out of the cell for every 2 K^+ pumped into the cell. This results in a resting potential (voltage between the inside and outside of the cell) that is −70 mV. An action potential is generated when a change in membrane permeability permits the entry of large amounts of Na^+ through Na^+-gated channels in the cell membrane (**Part A**). This results in transient depolarization of the cell as the membrane potential becomes positive. Termination of the action potential is mediated by the closure of Na^+ channels and opening of slow K^+ channels that allow K^+ to escape from the cell, thereby repolarizing the membrane.

Local Anesthetics

Local anesthetics are drugs used to prevent pain by blocking the conduction of action potentials in nerve fibers. They are applied topically, infiltrated into an area, or injected directly around nerves, in the subarachnoid or epidural space or in joints to reversibly paralyze motor and sensory nerves. The local anesthetics act by blocking the opening of voltage-sensitive Na^+ channels in the nerve cell membrane and by raising the threshold at which an action potential can be generated.

Local anesthetics share 3 basic structural characteristics: a hydrophobic aromatic substituted benzene ring linked by an intermediate bond to a hydrophilic tertiary amine group (**Part B**). The two clinically important groups of local anesthetics are differentiated by the nature of the intermediate bond, which is either an ester (procaine, benzocaine, tetracaine) or an amide (lidocaine, bupivacaine, mepivacaine). Ester-linked local anesthetics undergo rapid hydrolysis by plasma pseudocholinesterases, while amide-linked agents are metabolized primarily by the hepatic CP450 enzyme system. Ester-type agents have a longer duration of action when given into the epidural space, since they are protected from degradation by esterases.

Anesthetic potency and the speed of onset of action are determined by the agent's hydrophobicity or lipid solubility. Since local anesthetics are weak bases ($pK_a = 7–9$), they exist in both ionized and nonionized forms, depending on their pK_a and surrounding pH. The lower the pK_a of the agent, the greater amount of drug present in the nonionized lipid soluble form at physiological pH and the more rapid the penetration of the nerve cell membrane and the faster the rate of onset (*see* **Part B** and **C**). The degree of protein binding of the anesthetic agent determines its duration of action. Once within the nerve cell, the cationic species of the local anesthetic binds to an intracellular protein that then interacts with and inactivates the Na^+ channel.

As a general rule, small unmyelinated nerve fibers are more susceptible to the action of local anesthetics than are large myelinated fibers. As a consequence, small, unmyelinated pain fibers are blocked first, followed by the fibers that control sensations such as cold, touch, and pressure. Last to be affected are large myelinated motor nerves.

Although local anesthetics are not administered systemically, a degree of systemic absorption occurs. The extent of systemic absorption is related to the vascularity at the injection site. Addition of a vasoconstricting agent, such as epinephrine, to the local anesthetics prolongs their duration of action by delaying vascular absorption.

Local anesthetics are generally associated with minimal toxic effects. The most frequent side effects are due to direct effects of systemically absorbed drugs on the cardiovascular system and the CNS. Signs of CNS toxicity include excitement, skeletal muscle twitching, and seizures. With high doses, CNS depression predominates. High plasma concentrations can be associated with cardiovascular changes, including a decrease in electrical excitability, conduction rate, and myocardial contractility. Hypersensitivity reactions, particularly to the ester compounds, may occur. The most serious systemic toxicity occurs with accidental IV injection.

Individual Local Anesthetic Agents

Lidocaine is the most widely used local anesthetic agent. It has a rapid onset (1–2 min) and a duration of action of 1–2 hours (**Part C**). Lidocaine is also used intravenously as an antiarrhythmic agent (*see* Chapter 16). It is metabolized by the hepatic CP450 system by *N*-demethylation and excreted in the urine. Bupivacaine is an amide-type local anesthetic that has a slow onset, but a long duration of action (up to 8 hours). It is most frequently used for regional and epidural nerve blocks and for long surgical procedures. Mepivacaine is another amide-type local anesthetic whose properties are similar to lidocaine. It is equal in potency with a slightly longer duration of action. Proparacaine is an ester-type agent that is used for topical application as a corneal anesthetic agent. It is less irritating to tissues than most local anesthetic agents. Benzocaine is a water-soluble ester type agent used for surface anesthesia of non-inflamed tissues.

12 Neuromuscular Blocking Agents

A Action of Neuromuscular Blocking Agents

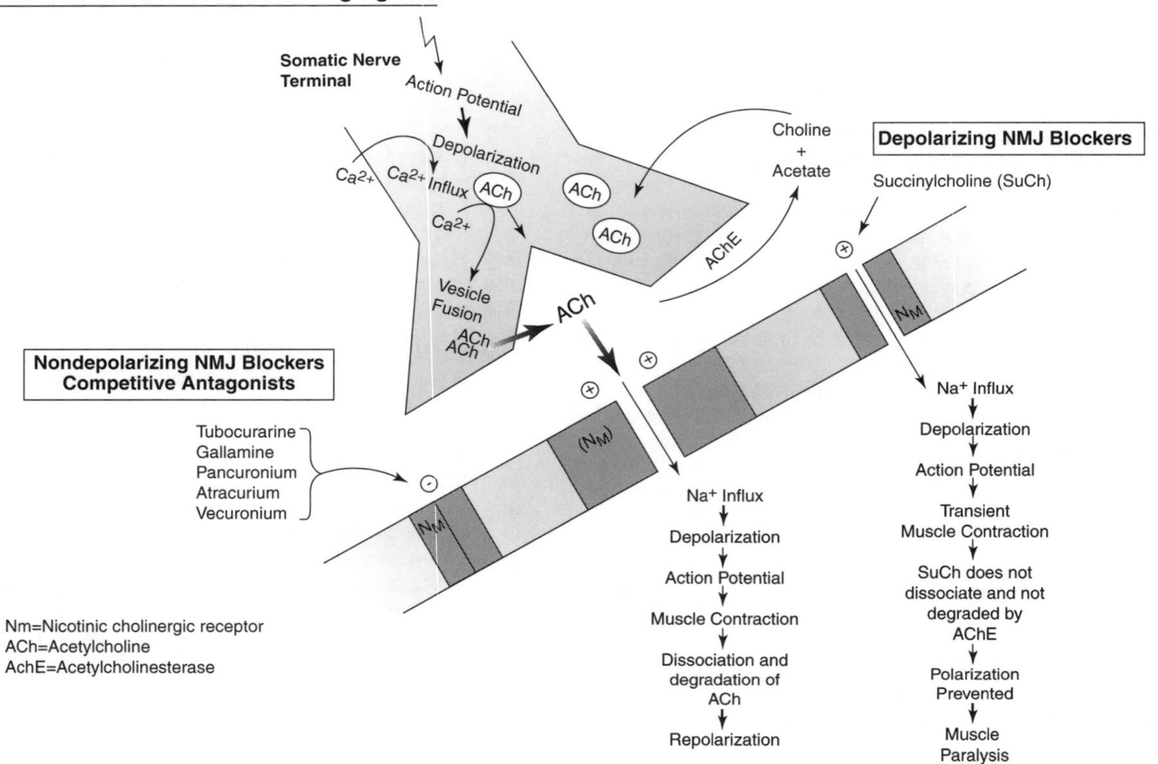

B Characteristics of Neuromuscular Blocking Agents

Agent	Duration*	Excretion	Ganglionic Block	Sympatho-mimetic action	Histamine Release	Muscarinic Antagonist	Muscarinic Agonist
Succinylcholine	2-8 (E) 4-6 (F) 22-29 (C)	Hydrolysis by plasma pseudocholinesteriases	-	-	-	-	+
Tubocurarine	100 (C) 20 (F)	Unchanged in urine	+	-	+++	-	-
Atracurium	8-12 (E) 17-29 (F + C)	Hydrolysis by plasma esterases Spontaneous degradation	-	-	+	-	-
Gallamine	14-40 (E) 24-29 (F + C)	Unchanged in urine	-	+	++	++	-
Pancuronium	20-35 (E) 30-100 (F + C)	50% unchanged in urine 50% in bile with some hepatic metabolism	+	+	Rare	+	-
Vecuronium	20-40 (E) 15-40 (C) 5-9 (F)	Hepatic metabolism with biliary excretion	-	-	-	-	-

* E = equine; F = feline; C = canine

Nerve impulses are coupled to skeletal muscle contraction at the neuromuscular junction (NMJ) (**Part A**). The arrival of the action potential at the end of a nerve fiber stimulates the opening of Ca^{2+} channels in the synaptic membrane. Calcium influx stimulates the fusion of vesicles containing the neurotransmitter acetylcholine (ACh) with the plasma membrane, resulting in the release of ACh into the synaptic cleft. ACh binds to nicotinic cholinergic receptors on the muscle end plate. Receptor engagement opens a membrane channel that is permeable to both Na^+ and K^+. The movement of Na^+ in and K^+ out of the cell generates the muscle action potential. After receptor binding, ACh is released into the synaptic cleft and degraded by acetylcholinesterase (AChE).

Neuromuscular Blocking Agents

Neuromuscular blocking agents interfere with ACh activation of nicotinic cholinergic receptors (N_m) on skeletal muscle cells (*see* **Part A**). The result is motor paralysis and increased muscle relaxation. These agents are given IV and used to facilitate mechanical ventilation and tracheal intubation, and to increase muscle relaxation during surgery without the need for higher levels of general anesthetic. All of the neuromuscular agents have the capacity to cause apnea, which may necessitate artificial respiration. Since they do not cross the blood-brain barrier, they have no effects in the CNS. Neuromuscular blocking agents have no analgesic activity. Side effects of these agents are due to either their tendency to induce histamine release or adverse action at autonomic nicotinic receptors (N_N) or cholinergic muscarinic receptors.

Neuromuscular blocking agents are divided into two categories: depolarizing agents (succinylcholine, decamethonium) and nondepolarizing agents (tubocurarine, gallamine, pancuronium, and atracurium) (**Part B**).

Depolarizing Agents

Depolarizing agents mimic the action of ACh at the nicotinic receptor, but are hydrolyzed by AChE at a much slower rate. The initial binding of these agents to the nicotinic ACh receptor causes depolarization of muscle and transient muscle contraction (evidenced by the presence of muscle fasciculations). Repolarization of the postsynaptic membrane is blocked by these agents, which prevents further impulse generation and causes flaccid paralysis (*see* **Part A**).

Succinylcholine is composed of two ACh molecules joined back to back. When administered IV succinylcholine results in a rapid onset of muscle paralysis (within 1 min) that lasts 2–8 min in pigs, horses, cats, and ruminants and up to 25 min in dogs. It is metabolized by plasma pseudocholinesterases. There is considerable species variation in the amount of plasma pseudocholinesterase activity, which may account for the different $t^1/_2$ in various species. The adverse effects of succinylcholine administration include the induction of painful contractions associated with the initial period of muscle depolarization, increases in bronchiolar and salivary secretion due to stimulation of cholinergic muscarinic receptors, and hyperkalemia. Malignant hyperthermia can occur in susceptible individuals. In horses, severe hypertension, bradycardia followed by tachycardia, and AV conduction disturbances have been reported. Succinylcholine also increases intraocular and intracranial pressure.

Nondepolarizing Agents

The nondepolarizing agents are competitive antagonists at the nicotinic cholinergic receptor on the motor end plate (N_m). Unlike the depolarizing agents, the action of nondepolarizing agents can be antagonized by the use of anticholinesterases such as neostigmine or edrophonium. These AChE inhibitors prevent the breakdown of ACh, permitting the neurotransmitter to accumulate in the synaptic cleft and thus compete with the nondepolarizing agents for binding to the nicotinic receptor.

The nondepolarizing agents include the benzylisoquinoliniums (tubocurarine, atracurium), the quarternary amine (gallamine), and the aminosteroids (pancuronium, vecuronium) (*see* **Part A**). Their onset of action is typically within 2–3 min and they last for 20–40 min with some species variation (*see* **Part B**). Their duration of action is dependent on redistribution away from the NMJ. Repeat injections may saturate redistribution sites and prolong the action of nondepolarizing agents. This cumulative effect may necessitate giving smaller successive doses to maintain the same amount of muscle relaxation.

The choice of an individual agent is determined by the agent's duration of action, site of metabolism and side effect profile. Side effects of neuromuscular blocking agents include blockage of muscarinic cholinergic receptors (tachycardia), indirect sympathomimetic effects (hypertension, tachycardia), and induction of histamine release (skin erythema and pruritus, hypotension, tachycardia, bronchoconstriction).

Tubocurarine is a naturally occurring substance with a relatively long duration of action (20–100 min). It may block nicotinic neurotransmission at autonomic ganglia (N_N receptors) resulting in hypotension and bronchospasm. It can also induce the release of histamine, especially in small animals. Tubocurarine is excreted unchanged in urine and feces. Atracurium is a synthetic isoquinolone that is rapidly inactivated by plasma esterases and by spontaneous degradation in the plasma (Hoffman reaction). Atracurium is therefore noncumulative and its duration of action on repeated injection does not change. It is the neuromuscular blocking agent of choice in patients with renal or hepatic disease. Atracurium causes less histamine release than tubocurarine and has minimal cardiovascular effects.

Gallamine increases heart rate by blocking muscarinic receptors in the heart and may cause histamine release. It is excreted unchanged in the urine.

Pancuronium is a synthetic steroid molecule. It is infrequently associated with histamine release, but may cause mild tachycardia, hypertension, and hypersalivation. About 50% of the drug is excreted unchanged in the urine and the rest is eliminated unchanged or as an active metabolite in feces. Pancuronium is highly protein bound. Vecuronium is an analogue of pancuronium that was developed to overcome the vagolytic and indirect sympathomimetic sides effects of the parent drug. It undergoes hepatic transformation and is excreted primarily in the bile.

A number of factors can influence the degree and duration of muscle relaxation produced by neuromuscular blocking agents. Volatile anesthetics (enflurane $>$ isoflurane $>$ halothane) cause a dose-dependent potentiation of neuromuscular block. Hypomagnesemia, hypocalcemia, and hypokalemia potentiate neuromuscular block as well. Hypothermia can delay the onset of action and prolong the duration of neuromuscular block. Older individuals are more sensitive to the effect of neuromuscular block of nondepolarizing agents. A number of antibiotics (among them the aminoglycosides, clindamycin, lincomycin, and tetracyclines) potentiate the action of neuromuscular blocking agents. The aminoglycosides reduce presynaptic ACh release and stabilize postsynaptic membranes. The other antibiotics have a prejunctional action and also decrease muscle cell contractility. A number of other drugs that decrease ACh release or stabilize postsynaptic motor-end plate membranes such as local anesthetics, phenytoin, quinidine, procainamide, propranolol, and barbiturates also potentiate the action of neuromuscular blocking agents.

13 Cardiovascular Drugs: Inotropic Agents

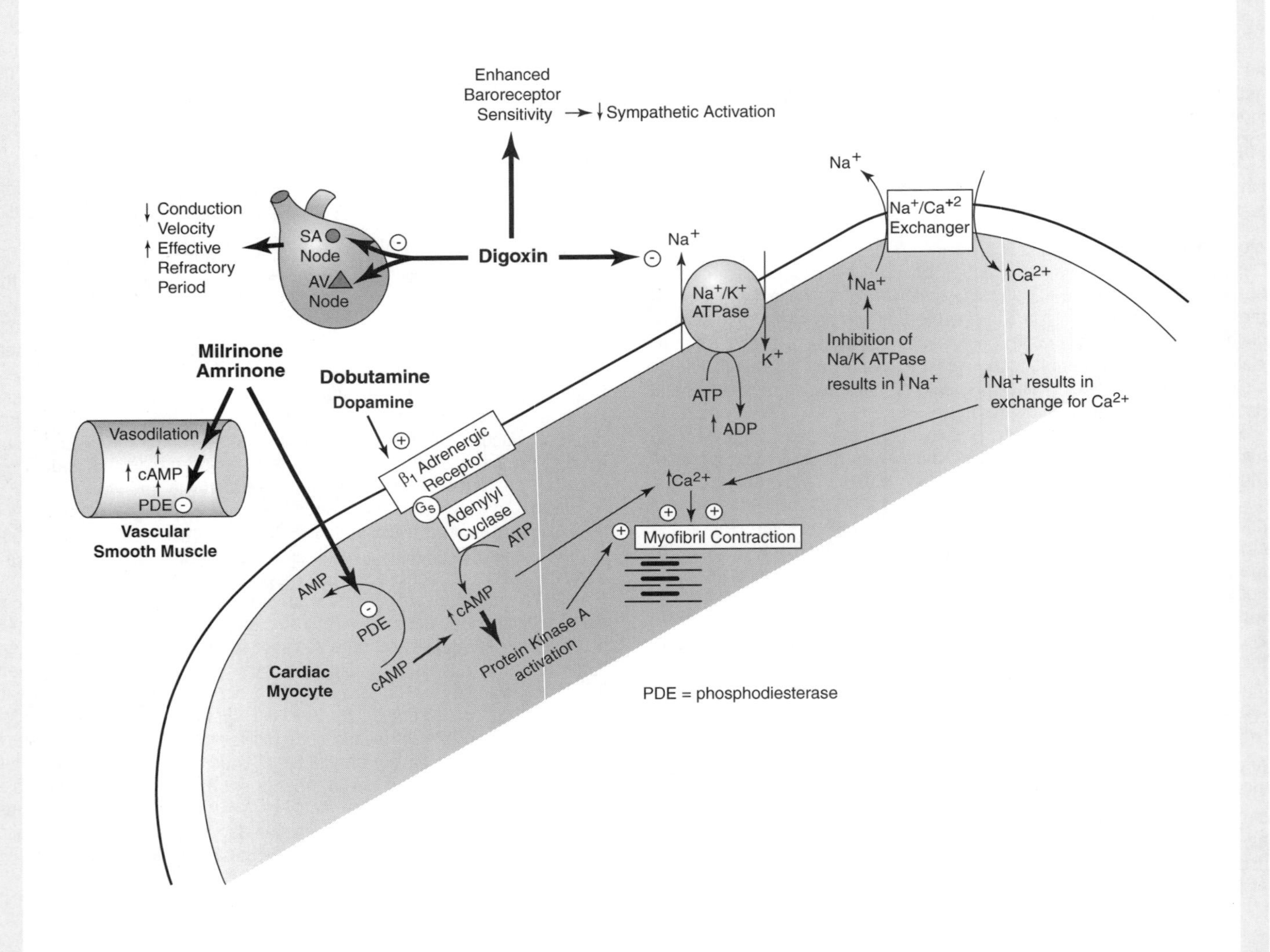

Overview of Cardiac Pharmacology

Cardiac drugs are used in the clinical management of congestive heart failure (CHF) and the management of arrhythmias. Cardiac failure is characterized by inadequate cardiac output resulting in insufficient delivery of oxygen to meet body tissue requirements. This may occur because of cardiac muscle damage, valvular weakness, coronary insufficiency, pericardial disease, or rhythm abnormalities. Clinically, CHF is recognized as fluid accumulation in body cavities, exercise intolerance, and coughing and dyspnea secondary to pulmonary edema. During cardiac failure the body attempts to compensate for fading cardiac output with 3 intrinsic homeostatic mechanisms. The first is to activate the sympathoadrenal system, which results in an increase in heart rate and cardiac contractility. The second mechanism is activation of the renin-angiotensin system (RAS), which causes systemic vasoconstriction to maintain arterial blood pressure. The RAS also stimulates the release of aldosterone, which promotes sodium and water retention to increase cardiac filling preload and thus cardiac stroke volume. The third mechanism is the development of myocardial hypertrophy. In the short term, these compensatory mechanisms are beneficial, but in the long term they result in increased cardiac afterload, leading to increased myocardial demand and decompensated heart failure. Medical therapy for heart failure involves: 1) diuretics coupled with a reduction in sodium intake to mobilize excess fluid accumulation, 2) vasodilators to decrease cardiac preload and afterload, 3) β-adrenergic receptor blockade to interrupt the toxic effects of chronic sympathetic stimulation, and 4) inotropic agents to increase cardiac contractility.

Digoxin

Digoxin has several beneficial actions in heart failure (**Figure**). First, it increases the strength of myocardial contraction. This inotropic effect is due to inhibition of the plasma membrane Na^+/K^+ ATPase, which results in increased intracellular concentrations of Na^+, which in turn increase the transmembrane exchange of intracellular Na^+ for extracellular Ca^{2+}. The resultant increase in intracellular Ca^{2+} increases the amount of Ca^{2+} available for cardiac myocyte contraction. Digitalis also decreases heart rate by decreasing the rate of discharge from the sinoatrial (SA) node. This effect is due to stimulation of vagal afferents and direct depression of SA node conduction. Digoxin also slows atrioventricular (AV) conduction and prolongs the refractory period for AV conduction. Digoxin sensitizes baroreceptors and induces a sustained inhibition of sympathetic activation.

Absorption of digoxin after oral administration varies with formulation, with the capsule and elixir having better bioavailability (60%–80%) than the tablets (50%–60%). The drug is widely distributed, with the highest levels in the kidneys, heart, intestines, liver, and skeletal muscle and the lowest concentrations in plasma. Since digoxin is not well distributed to fat, it is dosed based on lean body weight to avoid overdose in obese patients. Digoxin is eliminated in the urine by both glomerular filtration and tubular secretion. The plasma $t^1/_2$ shows high species variability, with approximate values of 30, 33, and 23 hours for dogs, cats, and horses, respectively. It is because of this variability and the low therapeutic index for digoxin that steady-state serum levels (which are obtained after 5–7 days) should be monitored. Trough concentrations obtained 8 hours post pill should be kept within the therapeutic range (1–2 ng/mL).

Several drug interactions have been reported with digoxin. Diazepam, quinidine, anticholinergics, succinylcholine, verapamil, tetracycline, and erythromycin either increase the serum level or enhance the toxic effects of digoxin. Antacids, cimetidine, metoclopramide, and neomycin may decrease the amount of digoxin absorbed from the gastrointestinal tract. Drugs that deplete body potassium (diuretics, glucocorticoids, amphotericin B) enhance the action of digoxin, while those that predispose to hyperkalemia (spirolactone) may decrease the effect of digoxin. Hypothyroid patients on thyroid supplementation need lower doses of digoxin. Cats on concurrent therapy with diuretics, aspirin, and low-sodium diet require lower doses of digoxin.

Adverse effects of digoxin are classified as cardiac and extracardiac. The cardiac side effects include partial or complete heart block due to digoxin's anticholinergic effects. Digitalis also increases the excitability of cardiac tissue and may cause the development of ventricular arrhythmias. Inhibition of the Na^+/K^+ ATPase results in reduced intracellular concentrations of K^+ and increased intracellular concentrations of Na^+, which results in a partially depolarized cell. This brings the cardiac myocytes closer to the threshold potential. Extracardiac side effects include gastrointestinal upset (anorexia, vomiting, diarrhea), muscle weakness, disorientation, and depression. Digoxin toxicity is treated with temporary discontinuation of the drug, institution of appropriate antiarrhythmic therapy (phenytoin is the drug of choice), correction of AV block (parasympatholytic drugs, pacemaker), and correction of electrolyte abnormalities. In humans, specific antidigoxin antibodies are used to treat digoxin toxicity.

β Adrenergic Agonists

Dopamine is an endogenous catecholamine that acts on α- and β-adrenergic receptors and dopamine receptors. Dopamine's positive inotropic effects are principally due to stimulation of cardiac β_1 receptors. Dopamine must be given as a continuous-rate IV infusion because its $t^1/_2$ is only 2 min. It is metabolized in the liver, kidneys, and plasma (*see* Chapter 7 **Part C**). Dopamine's pharmacological effects are dose dependent. At low doses (2 μg/kg/min), dopamine stimulates dopaminergic receptors to increase mesenteric, coronary, and renal blood flow. Intermediate doses of dopamine (5 μg/kg/min) cause primarily cardiac β adrenergic stimulation, resulting in increased cardiac output without major effects on the systemic vasculature. Higher doses of dopamine (>10 μg/kg/min) result in stimulation of α adrenergic receptors, which results in peripheral vasoconstriction and increased blood pressure. Potentially deleterious effects of using intermediate doses of dopamine in CHF are excessive tachycardia, induction of arrhythmias, and an increase in myocardial oxygen demand.

Dobutamine, a synthetic catecholamine, is a direct β_1 agonist with only mild β_2 and α_1 adrenergic effects (*see* **Figure**). Dobutamine does not stimulate renal dopamine receptors. The net effect of dobutamine administration is to increase myocardial contractility and cardiac output without significantly affecting systemic blood pressure and heart rate. It is less arrhythmogenic than dopamine. Dobutamine must be administered as a continuous-rate IV infusion because it has a very short $t^1/_2$ (2 min). It is metabolized in the liver. It is given for only a short time period (48–72 hours) as tolerance (β_1 receptor desensitization) develops with more prolonged use. The beneficial effects of one short-term infusion, however, may extend over several weeks.

Phosphodiesterase Inhibitors

Amrinone and milrinone are referred to as nonglycoside, nonsympathomimetic positive inotropic agents that also have potent balanced vasodilatory action. These agents inhibit phosphodiesterase type III, the cardiac-specific phosphodiesterase (*see* **Figure**). Inhibition of phosphodiesterase leads to an increase in intracellular cAMP levels, which results in phosphorylation events that increase myocardial contractility and cause vasodilation. Since increased cAMP levels may also increase the automaticity of SA pacemaker cells, one of the limiting side effects of therapy with these drugs is the development of tachyarrhythmias. Other adverse side effects include gastrointestinal disturbances, thrombocytopenia, hepatotoxicity, and fever. Milrinone, which is 10–20 times more potent than amrinone, has been shown to have beneficial effects in dogs with CHF that were evaluated over a 4-week period. Additional long-term studies will be necessary to determine if arrhythmic potential of these drugs will limit their use in CHF.

14 Cardiovascular Drugs: Vasodilators I

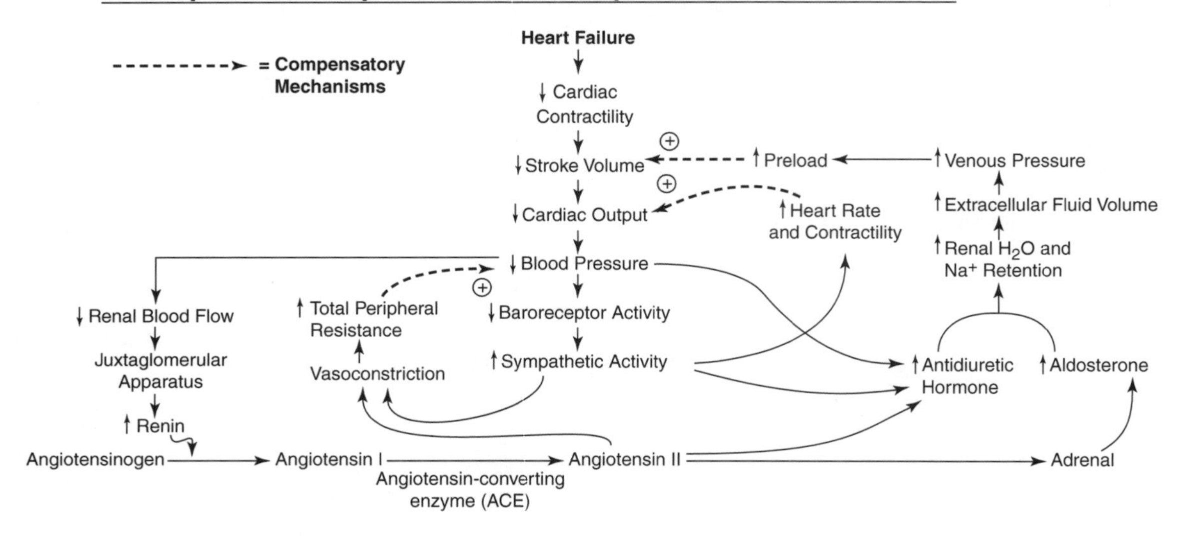

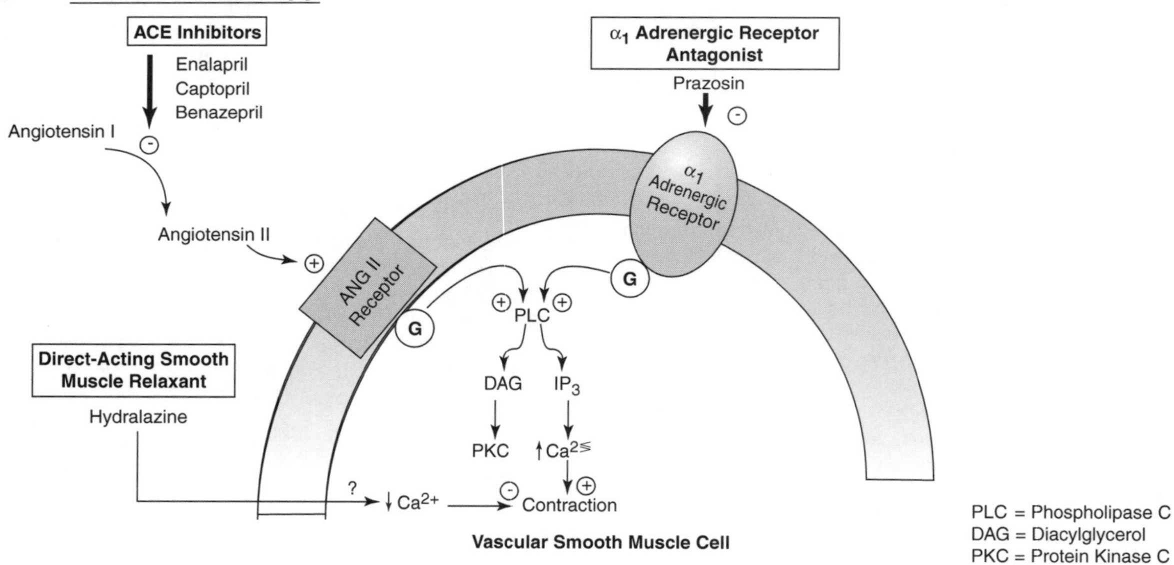

Drugs that result in peripheral vasodilation are used in the treatment of congestive heart failure (CHF) and in the control of systemic hypertension. Vasodilators can selectively dilate arteries or veins or have activity on both (balanced vasodilators). Arteriolar dilators decrease systemic vascular resistance, which decreases afterload. Afterload is the force that the ventricular muscle must overcome to shorten and move blood into the aorta. A reduction in afterload facilitates myocardial shortening, decreases myocardial oxygen demand, and increases cardiac output. Venodilators decrease hydrostatic pressure in the venous system, leading to a reduction in both systemic and pulmonary vascular resistance. This decreases preload, the filling pressure of the heart, which is also an important determinant of ventricular systolic function. The primary beneficial effect of preload reduction is decreased edema formation. Clinically the optimal benefit to patients in CHF is obtained with balanced vasodilation.

Angiotensin-Converting Enzymes

Activation of the renin-angiotensin system (RAS) occurs in heart failure (**Part A**). Decreases in renal blood flow to the proximal convoluted tubule result in the release of renin from the renal juxtaglomerular apparatus. Renin released from the kidney converts angiotensinogen to angiotensin I. Angiotensin I is enzymatically converted to angiotensin II by the action of angiotensin-converting enzymes (ACE), which are found in the lungs and vascular endothelium. Activation of the RAS results in systemic vasoconstriction, aldosterone release, and activation of the sympathetic nervous system. The initial compensatory advantages of RAS activation include maintenance of systemic blood pressure and increased cardiac contractility. In addition, aldosterone release results in sodium and water retention, which increases the filling pressure of the heart and thus cardiac output via Frank-Starling mechanisms. Over the long term, however, RAS activation is detrimental in heart failure because further increases in afterload result in a decrease in stroke volume and an increase in myocardial oxygen demand. In addition, β adrenergic receptor downregulation occurs owing to sustained sympathetic activation.

Inhibition of RAS activation with ACE inhibitors (captopril, enalapril, benazepril) has emerged as an important therapeutic intervention in heart failure (**Part B**). Clinical trials in humans and veterinary patients with congestive heart failure have demonstrated increased exercise capacity and prolonged survival with ACE inhibitor therapy. ACE inhibitors are balanced vasodilators. Captopril was the first-available ACE inhibitor. It was not well tolerated by veterinary patients due to a high incidence of gastrointestinal disturbances. Enalapril is the preferred ACE inhibitor in veterinary patients since it is approved by the FDA for use in dogs and cats. The beneficial effects of enalapril in cardiovascular disease, which include decreases in total peripheral and pulmonary vascular resistance, pulmonary wedge pressure, and systemic blood pressure, occur within the first 24 hours after initiation of therapy and are maintained in the long term. After 3–4 weeks of therapy, improvement is seen in clinical markers of hemodynamic function such as increased exercise tolerance, general well-being, and a reduction in signs associated with pulmonary congestion. Enalapril is well absorbed after oral administration, although its bioavailability in the dog is only 64%. Enalapril is metabolized in the liver to the active drug enalaprilat, which is excreted in urine. The most common side effect of therapy with enalapril is gastrointestinal upset. Unlike enalapril, benazepril is highly protein bound (95%). Although primarily eliminated in the urine, in the face of renal dysfunction, biliary clearance becomes more important. Since angiotensin is important in maintaining renal perfusion in heart failure (via preferential efferent arteriolar vasoconstriction), renal function should be monitored carefully during therapy with ACE inhibitors. Other side effects include hypotension and hyperkalemia due to inhibition of aldosterone secretion.

ACE inhibitors are considered one of the first-line agents for treatment of systemic arterial hypertension in small animals. ACE inhibitors also have renoprotective effects independent of their ability to control glomerular hypertension. They have been shown to reduce glomerular protein loss and decrease mesangial cell proliferation and glomerular fibrosis in animal models of glomerulonephritis. Nonsteroidal anti-inflammatory agents reduce the clinical efficacy of ACE inhibitors as antihypertensive agents.

Hydralazine

Hydralazine is a potent arteriolar dilator that has a direct action on vascular smooth muscle (*see* **Part B**). Its mechanism of action is not fully understood, but may involve alterations in calcium metabolism and thereby interference with smooth muscle contraction. Hydralazine decreases systemic vascular resistance and increases cardiac output in patients with CHF. When hydralazine is given to patients with normal heart function, the drug-induced initial drop in systemic blood pressure may be accompanied by sympathetic activation. Sympathetic activation results in increased heart rate and increased myocardial contractility, both of which restore blood pressure to normal. Addition of a β adrenergic antagonist blocks sympathetic activation and preserves the decrease in blood pressure. In patients with CHF, the sympathetic nervous system is already fully activated and β adrenergic receptor desensitization is often present. Thus, the vasodilatory effects of hydralazine are preserved. Hydralazine is rapidly absorbed from the gastrointestinal tract. It has a high hepatic first-pass effect that is exacerbated by the presence of food. The drug is metabolized in the liver and the metabolites undergo renal excretion. The primary use of hydralazine in small animals is for severe CHF associated with mitral valve disease, for which it is often used in combination with diuretic therapy. The main side effects of hydralazine are hypotension, reflex tachycardia, and gastrointestinal upset.

Prazosin

Prazosin is an α_1 adrenergic antagonist that acts as a balanced vasodilator (*see* **Part B**). In patients with CHF, prazosin reduces filling pressure and increases cardiac output. Prazosin is also used as an antihypertensive agent. The drug is extensively metabolized by the liver and excreted in the feces. Prazosin is highly protein bound and may displace or be displaced by other highly protein-bound drugs. Side effects include gastrointestinal upset and syncopy due to orthostatic hypotension after the first dose (first-dose effect). The development of drug tolerance has been reported and limits the use of prazosin for the long-term treatment of CHF and hypertension.

15 Cardiovascular Drugs: Vasodilators II

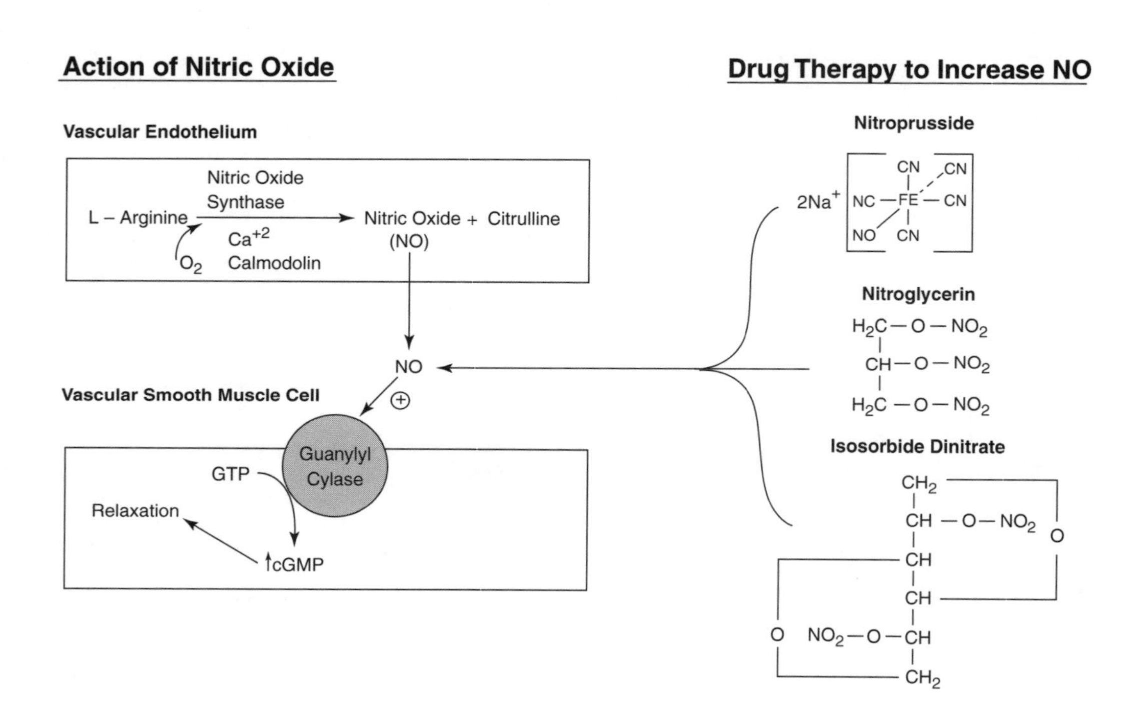

Nitric oxide (NO) is formed endogenously by the action of nitric oxide synthetase on L-arginine (**Figure**). This reaction, which occurs in vascular endothelial cells, requires molecular oxygen, calcium, and calmodulin. Nitric oxide is a nonpolar gas that does not require interaction with a surface receptor. After diffusion into vascular smooth muscle cells, NO activates guanylyl cyclase through nitrosylation. Guanylyl cyclase catalyzes the conversion of GTP to cGMP, which in turn mediates smooth muscle relaxation and thus vasodilation.

Organic compounds containing nitrate or nitroso moieties such as nitroglycerin, nitroprusside, and isosorbide dinitrate, undergo tissue catabolism or spontaneously decompose to yield nitric oxide. Unlike other vasodilators the effect of these compounds is independent of the state of autonomic innervation in the vascular system.

Nitroprusside is a potent balanced vasodilator. It is administered as a continuous rate IV infusion and results in an almost immediate decrease in systemic vascular resistance and systemic arterial blood pressure. The result is a decrease in ventricular filling pressure and a reduction in pulmonary capillary wedge pressure. Nitroprusside is used in the short-term management of acute, severe, life-threatening congestive heart failure (CHF), usually in combination with diuretics and dobutamine. It may also be used to treat an acute hypertensive crisis.

Nitroprusside is metabolized in the blood and tissue to cyanogen. Cyanogen is converted in the liver to thiocyanate, which is eliminated in feces, urine, and exhaled air. Cyanide toxicity can occur with prolonged administration of nitroprusside. An early indicator of cyanide toxicosis is metabolic acidosis. Plasma levels of thiocyanate should be monitored in dogs with renal insufficiency or when the drug is given for longer than 2 days. A plasma concentration of >10 mg/dl is considered toxic. Since nitroprusside is a potent vasodilator, blood pressure should be monitored intensively to prevent hypotension. Other side effects of therapy include nausea, muscle twitching, and dizziness. Nitroprusside is degraded by light so bottles and IV lines should be wrapped in light protective covers.

Nitroglycerin is primarily a venodilator when it is administered transcutaneously. Its beneficial actions in CHF include a decrease in ventricular filling pressure and resolution of signs of pulmonary edema. Topical nitroglycerin is available as an ointment or patch for cutaneous application. The most common areas for application in veterinary patients are the groin, axilla, and pinna of the ear. Its onset of action is 1 hour, with a duration of action ranging from 2–12 hours. Nitroglycerin is metabolized in the liver by glutathione-organic nitrate reductase into more water soluble denitrated metabolites and inorganic nitrite. The liver has an enormous capacity to catalyze the reduction of organic nitrates. This transformation has important implications for oral bioavailability and duration of action. After topical administration, most of the drug bypasses the hepatic circulation because the liver receives only 20% of the cardiac output. Oral bioavailability is very low due to the large hepatic first pass effect. Controlled studies on the efficacy of nitroglycerin in canine and feline heart failure are lacking. Side effects of therapy include rash at the site of application and hypotension. Frequent repeated exposure to nitroglycerin leads to a decrease in the magnitude of its pharmacologic effect. Brief periods (overnight) of no therapy may be sufficient to avoid the development of tolerance.

Isosorbide Nitrate

Isosorbide dinitrate is metabolized in the liver by enzymatic denitration. Unlike nitroglycerin two of the major metabolites, isosorbide-2-mononitrate and isosorbide-5-mononitrate, have longer half lives than the parent drug. These active metabolites make it possible to administer isosorbide dinitrate orally to obtain therapeutic vasodilation.

16 Cardiovascular Drugs: Antiarrhythmics I

A Cardiac Conduction System

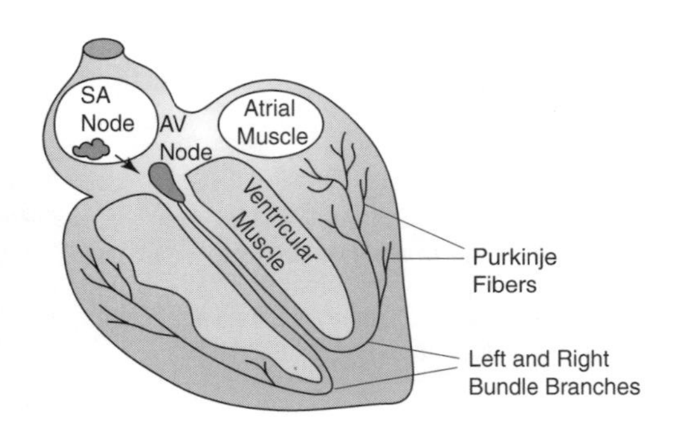

B1 Action Potential in Pacemaker Tissue

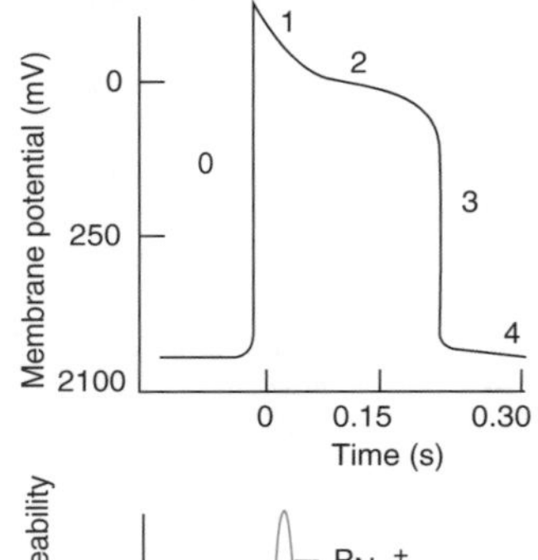

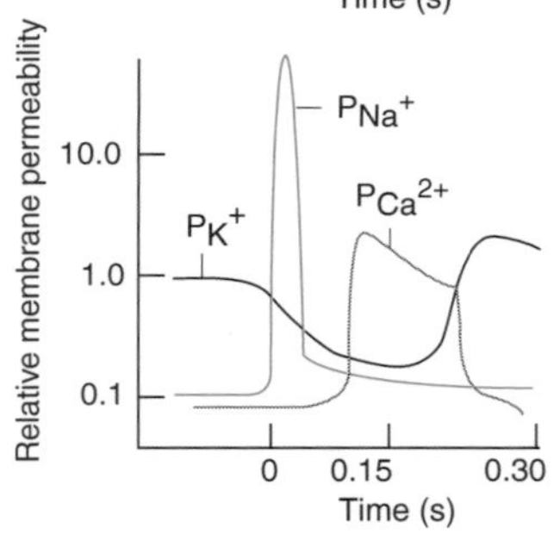

B2 Action Potential in Atrial and Ventricular Myocyte

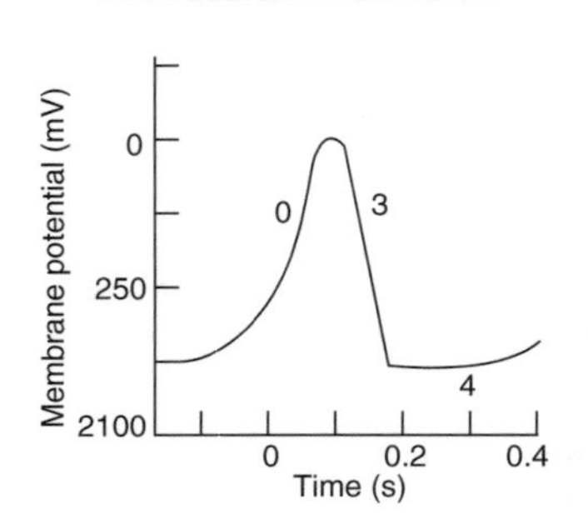

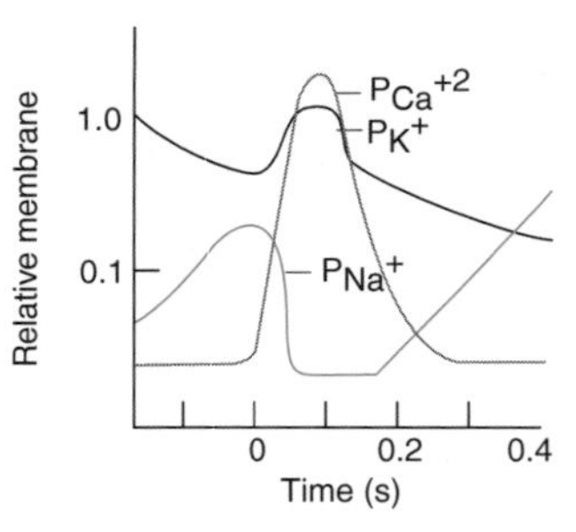

C Class 1 Antiarrhythmic Agents

Class	Na^+ Channel Blocker	K^+ Channel Blocker	Autonomic Effects	Hemodynamic Effects	Clinical Use	
Class IA						
Quinidine	⊕	⊕	Vagolytic α_1 adrenergic antagonist	Hypotension Tachycardia Negative inotrope	Oral IV IM	Supraventricular and ventricular arrhythmias
Procainamide	⊕	⊕ †	Ganglionic and NMJ blockade	Hypotension Mild negative inotrope	Oral IV IM	Supraventricular and ventricular arrhythmias
Class IB						
Lidocaine	⊕	NE	NE	NE	IV	Ventricular arrhythmias
Tocainide and Mexiletine	⊕	NE	NE	NE	Oral	Ventricular arrhythmias
Phenytoin	⊕	NE	NE	Hypotension (IV)	IV Oral	Digitalis-associated arrhythmias
Class IC						
Fecainide	⊕	⊕	NE	NE	Oral	Supraventricular arrhythmias (Proarrhythmic in ischemic heart disease)

* Results in prolongation of action potential.

† Due to hepatic metabolite in humans, N-acetyl procainamide, which may not be formed in canines due to deficiency in the conjugating enzyme acetyltransferase.

NE=No Effect

Normal cardiac rhythm is maintained by: 1) dominance of the sinoatrial (SA) node as the pacemaker; 2) rapid and uniform conduction of the electrical signal from the SA node sequentially through the atria, atrioventricular (AV) junction, the bundle branches, Purkinje fibers, and ventricular muscle; and 3) a long and uniform refractory period in the Purkinje fibers and cardiac myofibers (**Part A**). Arrhythmias arise from disturbances in automaticity (increase in phase 4 of the action potential) or disturbances in impulse conduction (re-entry arrhythmias). The action potential in working cardiac muscle and in pacemaker tissue is depicted in **Parts B1** and **B2**. In working muscle (*see* **Part B1**), phase 0 reflects rapid membrane depolarization due to the inward current of Na^+ through fast sodium channels. Phase 1 is the initial early repolarization phase. Phase 2 is the plateau of the action potential and is mediated in part by the slow inward current carried by Ca^{2+} through slow calcium channels. Phase 3 is the rapid repolarization phase of the action potential that returns the membrane potential to the diastolic level. Normally myocytes are not excitable between phase 1 and phase 3 of the cycle, and this is referred to as the *refractory period*. This period of refractoriness provides protection from re-excitation of this tissue by the initial stimuli. Phase 4 is the membrane potential during diastole. In pacemaker cells, the same phases of the action potential exist, but phase 4 is characterized by a slow spontaneous depolarization to threshold potential, leading to automatic discharge (*see* **Part B2**). The slope of phase 0 depolarization is not as steep in pacemaker tissue. This is due to the fact that this phase in pacemaker tissue is dependent on the entry of Ca^{2+} through slow calcium channels.

Antiarrhythmic drugs are pharmacodynamically organized into 4 classes: class I—membrane stabilizers, class II—β adrenergic antagonists, class III—agents that prolong the refractory period, and class IV—calcium channel blockers.

Class I Antiarrhythmics

Class I antiarrhythmics are local anesthetics that depress the rate of influx through fast Na^+ channels and thus decrease the maximal rate of depolarization in cardiac fibers (phase 0). They also inhibit the rate of the spontaneous depolarization (phase 4) in automatic cells and some agents prolong the refractory period. Class I antiarrhythmics are subdivided into class IA, IB, and IC agents (*see* **Part C**).

Quinidine is the prototypical class IA agent. It decreases the rate of phase 0 depolarization in both abnormal and normal cardiac cells and also prolongs the action potential duration and cardiac refractory period in both atrial and ventricular muscle. This latter effect may be mediated by blockage of K^+ channels. The capability of quinidine to prolong the refractory period in atrial muscle likely accounts for its efficacy in converting atrial fibrillation. Quinidine also has an atropine-like parasympatholytic effect. Since acetylcholine normally acts to shorten the atrial refractory period, this parasympatholytic action helps to further increase the atrial refractory period. This action combined with quinidine's direct effects on atrial muscle makes the drug particularly useful in controlling atrial arrhythmias. In horses, quinidine is used to convert atrial fibrillation to normal sinus rhythm. It seldom converts atrial fibrillation in small animals, but is useful to control ventricular arrhythmias.

Quinidine's vagolytic action tends to improve conduction through the AV node and may lead to an increased ventricular rate. This occurs most often after IV administration and can be prevented by pretreatment with a drug that slows AV conduction such as digoxin. Care should be used in administering these 2 drugs together because quinidine increases plasma digoxin levels. Other adverse hemodynamic effects of quinidine include a modest negative inotropic effect (due to interference with transmembrane Ca^{2+} flux), peripheral vasodilation (due to α_1 adrenergic blockade), and induction of AV block. These effects may lead to development of hypotension or worsening of congestive heart failure. Noncardiac side effects in small animals are manifested primarily as gastrointestinal disturbances, while horses may experience urticaria, gastrointestinal disturbances, inflammation of the nasal mucosa, and laminitis. Quinidine is rapidly and completely absorbed after oral or IM administration. It is highly protein bound and extensively metabolized in the liver, with $<20\%$ excreted unchanged in the urine. Drugs that induce hepatic microsomal enzymes enhance quinidine metabolism while those that inhibit the microsomal enzymes may increase the plasma concentrations. Quinidine itself inhibits hepatic microsomal enzymes.

Procainamide has similar pharmacological properties as quinidine, but lacks its vagolytic and α_1 adrenergic blocking actions. Procainamide is more effective against ventricular than atrial arrhythmias. After IV or IM administration, its onset of action is almost immediate, making it useful in the emergency treatment of ventricular arrhythmias. Procainamide's oral bioavailability and $t^1/_2$ is variable in humans and may be similar in veterinary patients. Food, delayed gastric emptying, or a decrease in stomach pH may delay gastrointestinal absorption. Procainamide is excreted primarily unchanged in the urine. Cimetidine will compete for urinary excretion and raise plasma levels. Procainamide has mild negative inotropic effects and may cause peripheral vasodilation and hypotension when given IV. Other side effects include gastrointestinal disturbances, AV block, and a proarrhythmic potential. A lupus-like syndrome has been reported with chronic administration in humans. Procainamide should be used with caution in patients with myasthenia gravis because it will antagonize the effects of acetylcholinesterase inhibitor therapy. It also potentiates the action of neuromuscular blocking agents.

Lidocaine is a class IB antiarrhythmic agent. The primary electrophysiological effect of class IB agents is to decrease conduction velocity in ventricular tissue with little to no effect on the SA or AV node or atrial muscle. Their effects are more pronounced in injured cardiac tissue than in normal tissue. Lidocaine has no anticholinergic activity and no effect on cardiac contractility. It is the drug of choice for the treatment of life-threatening ventricular arrhythmias.

Lidocaine is not effective orally, owing to high first-pass hepatic metabolism. When given IV, the onset of action is rapid (2 min), with a duration of action of 10–20 min. An initial IV loading dose should be given because it can take up to several hours to achieve therapeutic serum levels with a continuous-rate IV infusion. Lidocaine is extensively metabolized in the liver to active metabolites that are excreted in the urine. Lidocaine clearance is decreased in hepatic disease or with concurrent treatment with agents that decrease hepatic microsomal enzyme activity or hepatic blood flow. Side effects include vomiting and CNS signs (ataxia, depression, nystagmus, and seizures). Hypotension and adverse cardiac effects are rare unless an overdose is given.

Tocainide and mexiletine are orally available structural analogues of lidocaine that have been modified to decrease hepatic first-pass metabolism. These drugs have received limited use in veterinary patients. Their pharmacological properties are similar to those of lidocaine. Side effects of chronic use in dogs include gastrointestinal upset, development of progressive corneal dystrophy, and renal azotemia. Tocainide has been associated with fatal pulmonary and bone marrow toxicity in humans.

Phenytoin, another class IB agent, is better known as an anticonvulsant agent. It has limited use as an antiarrhythmic agent, primarily in the treatment of digitalis-induced arrhythmias.

Fecainide is the only class IC antiarrhythmic agent. It is available for oral administration. Fecainide's pharmacological actions are similar to the class IB antiarrhythmic agents but it also profoundly decreases conduction velocity. It has received limited use in veterinary patients. Recent studies in human patients have shown that chronic administration of fecainide after myocardial infarction is proarrhythmic.

17 Cardiovascular Drugs: Antiarrhythmics II

A Action of Antiarrhythmic Drugs

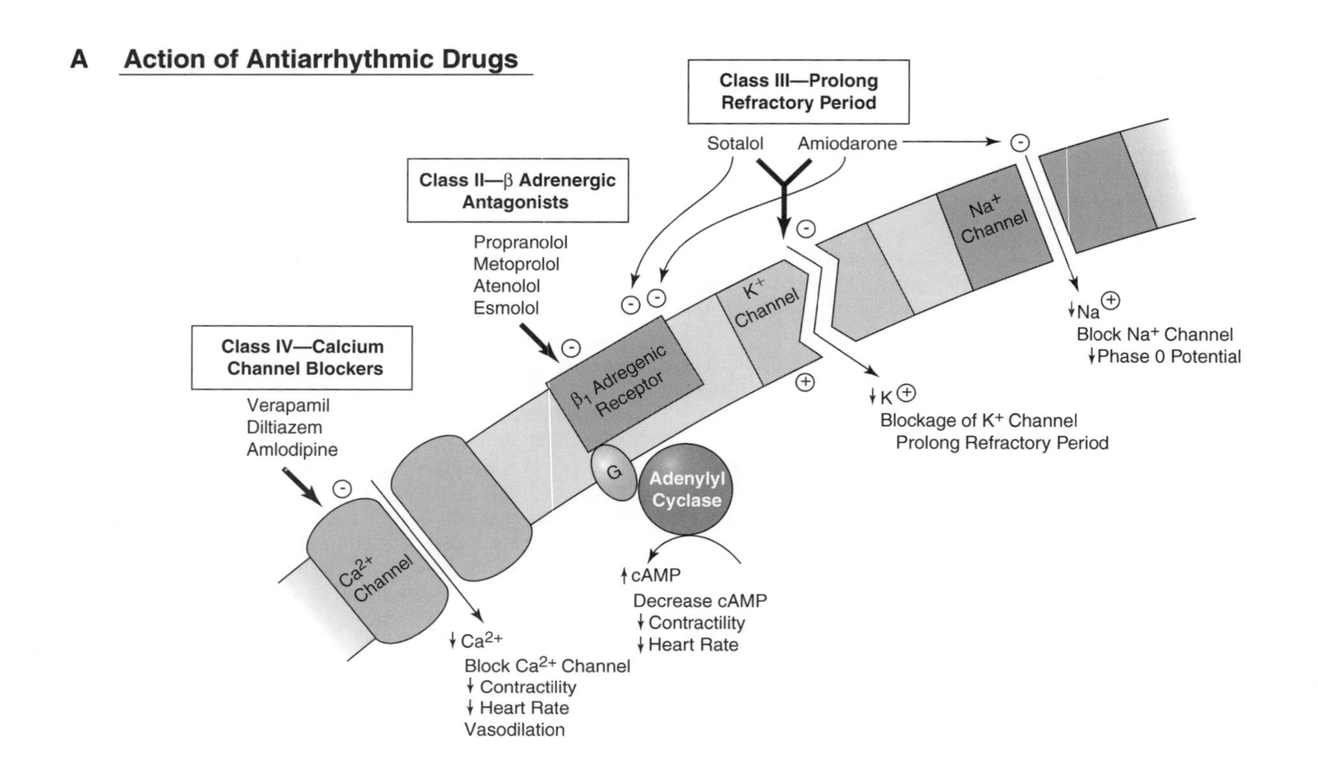

B Class II, III, and IV Antiarrhythmic Agents

Class	Drug	β Adrenergic Antagonist	K^+ Channel Blocker	Na^+ Channel Blocker	Ca^{+2} Channel Blocker	Clinical Use Against Arrhytmias
Class II	Propranolol	β_1, β_2	-	⊕*	NE	Oral IV Supraventricular and ventricular
	Metoprolol	β_1	-	-	-	Oral Supraventricular
	Atenolol	β_1	-	-	-	Oral Supraventricular
	Esmolol	β_1	-	-	-	IV Supraventricular ($t_{1/2}$= 2 min)
Class III	Amiodarone	β_1, β_2 Noncompetitive	⊕	⊕	⊕*	Oral Ventricular
	Bretylium		⊕	-	-	IV Ventricular
	Sotalol	β_1, β_2	⊕	-	-	Oral Ventricular
Class IV	Verapamil	-	-	-	⊕	Oral IV Supraventricular
	Diltiazem	-	-	-	⊕	Oral IV Supraventricular
	Amlodipine	-	-	-	⊕	Oral Hypertension

* Effects demonstrated "in vitro" at high doses.

Class II Antiarrhythmics

Class II antiarrhythmics are the β adrenergic receptor antagonists. These agents include propranolol, atenolol, metoprolol, and esmolol (**Parts A** and **B**). The β_1 adrenergic receptors are present in pacemaker tissues, conduction tissue, and myocardial cells. The net effect of stimulation of these receptors is to increase the slope of phase 0 and phase 4 depolarization (*see* Chapter 16 **Part B1**), increase the amplitude of the action potential, and decrease the refractory period. These actions result in an increase in conduction velocity and myocardial contractility. By blocking β_1 receptors, class II antiarrhythmics decrease the strength of contraction, decrease sinus heart rate, and depress AV conduction by increasing the refractory period of the AV node. These agents also decrease myocardial oxygen demand and systemic vascular resistance.

The β adrenergic receptor antagonists are used to control supraventricular tachyarrhythmia, but in cats have been used to treat both ventricular and supraventricular arrhythmias. They also slow the ventricular rate in atrial fibrillation. They decrease heart rate and myocardial oxygen demand in feline hypertrophic cardiomyopathy. They are also occasionally used to control systemic hypertension.

Propranolol is a nonselective β adrenergic receptor antagonist. After absorption from the gastrointestinal tract, extensive first-pass metabolism in the liver limits the drug's bioavailability. With chronic administration, hepatic enzyme saturation occurs and oral bioavailability increases. Drug metabolites are excreted in the urine. Propranolol decreases hepatic blood flow and therefore prolongs the elimination of drugs dependent on liver blood flow for metabolism such as lidocaine. Metoprolol and atenolol are orally available, relatively β_1-selective agents. Metoprolol has high hepatic first-pass metabolism and is excreted in the urine. Atenolol also undergoes hepatic metabolism, but a significant portion of the drug is eliminated in the feces. Metoprolol and propranolol are highly lipid soluble and enter the CNS; atenolol does not. Esmolol is a highly hydrophilic selective β_1 blocker with a short $t^1/_2$ that is only used IV. Side effects of β_1 blockage include bradycardia and hypotension, whereas nonselective agents that also block β_2 receptors may cause hypoglycemia and bronchoconstriction due to interference with hepatic and bronchial receptors, respectively. The β blockers should be used cautiously in patients with decreased cardiac reserve as the negative inotropic action of these drugs can precipitate CHF. Side effects of the β blockers that enter the CNS include depression and lethargy.

Class III Antiarrhythmics

The class III drugs (amiodarone, bretylium, and sotalol) increase the duration of the action potential as well as the refractory period (*see* **Parts A** and **B**). Their major pharmacological effect is to block K^+ channels. Their effects are greatest in Purkinje fibers and ventricular muscle, with less action in the atria and AV node. Amiodarone, a structural analogue of thyroid hormone, is used in humans to treat life-threatening ventricular arrhythmias and atrial fibrillation. Amiodarone is highly lipophilic and has the ability to disrupt membranes in which ion channels are embedded. For this reason, the drug also inhibits interaction at β receptors and Na^+ and Ca^{2+} channels on cardiac myocytes. The bioavailability of amiodarone is highly variable. The drug undergoes hepatic metabolism to an active metabolite, desethyl-amiodarone. The drug is slowly concentrated in tissue. It takes several weeks for the full therapeutic effect of the drug to become evident. The high incidence of serious side effects with chronic amiodarone therapy (pulmonary fibrosis, hepatotoxicity, gastrointestinal disturbances, cutaneous photosensitization, neuropathy, interference with thyroid metabolism) and the need for intensive monitoring limit the use of this drug in veterinary patients. Bretylium has been used with some success to bring about chemical defibrillation in patients with ventricular fibrillation, to stabilize ventricular irritability, and to facilitate resuscitation by electrical defibrillation. Sotalol is a nonselective β adrenergic receptor antagonist as well as a K^+ channel blocker that has similar indications as amiodarone.

Class IV Antiarrhythmics

Class IV antiarrhythmics are calcium channel blockers. These agents block influx of Ca^{2+} through slow calcium channels and thus decrease the transsarcolemmal flux of Ca^{2+} during phase 2 of the cardiac cycle. Tissues dependent on the slow calcium influx such as the SA and AV nodal tissues are most affected. The major electrophysiological effect of the calcium channel blockers on the heart is to slow SA and AV conduction. Thus, they are effective in treating supraventricular arrhythmias, especially those that involve AV nodal re-entry pathways. Calcium channel blockers decrease vascular smooth muscle contractility, causing systemic and coronary vasodilation. They also decrease myocardial contractility and thus may have clinically significant negative inotropism. These drugs should be used cautiously in patients with underlying cardiac dysfunction as they may lead to CHF. They may also precipitate fatal arrhythmias in patients with atrial fibrillation or Wolfe-Parkinson-White syndrome. Extracardiac side effects include constipation and nausea. Calcium channel blockers may potentiate the negative inotropic and chronotropic effects of β adrenergic antagonists.

There are 3 classes of calcium channel blockers: the phenylalkylamines (verapamil), the benzothiazepines (diltiazem), and the dihydropyridines (nifedipine and amlodipine). These agents differ in their selectivity for cardiac and vasculature calcium channels. Verapamil has pronounced effects on SA and AV conduction, with clinically significant negative inotropic effects. Verapamil increases the blood levels of digoxin and theophylline. It is metabolized in the liver and inactive metabolites are excreted in the urine. When used IV to treat life-threatening supraventricular arrhythmias, it may cause hypotension (due to effects on vascular smooth muscle) and heart block. For this reason, diltiazem has become the preferred drug in this situation. Diltiazem slows conduction velocity with a negligible effect on cardiac contractility and results in only mild peripheral vasodilation. The latter effect may actually be beneficial in patients with CHF because it decreases afterload. After oral administration, diltiazem undergoes extensive first-pass hepatic metabolism that limits bioavailability (30%–40%). The time-release formulation that is popular for use in cats may have even lower bioavailability. The drug is transformed in the liver to active metabolites that are excreted in the urine and feces. The drug undergoes some enterohepatic circulation. Nifedipine (first generation) and amlodipine (second generation) have potent vasodilatory action with little to no effect on AV or SA conduction or cardiac contractility. They are primarily used as agents to treat systemic hypertension. Amlodipine binds to and dissociates from receptors slowly so that it has a slow onset of action and prolonged duration of action. It is metabolized in the liver to inactive metabolites that are excreted in the urine. Oral bioavailability of amlodipine is high owing to a negligible hepatic first-pass effect.

18 Ocular Pharmacology I

A Ocular Pharmacology

Ciliary Body

Decrease Aqueous Humor Production
- Carbonic Anhydrase Inhibitors
  - Systemic: Daranide
  - Topical: Dorzolamide
- Osmotic Agents
  - Mannitol
- β_2 Adrenergic Antagonists
  - Timolol
- α_1 Adrenergic Agonists
  - (Secondary Vasoconstriction)
  - Epinephrine
  - Phenylephrine

Lacrimal Gland

Increase Secretion
- Parasympathomimetics
  - Pilocarphine
- Cyclosporine

Decrease Secretion
- Parasympatholytics
  - Atropine

Iris

Mydriasis
- Parasympatholytic
  - Atropine
  - Tropicamide
- α_1 Adrenergic Agonists
  - Epinephrine
  - Phenylephrine

Miosis
- Parasympathomimetics
  - Pilocarpine
  - Echothiophate
  - Demecarium
  - Physostigmine

Ocular Vasculature

Vasoconstriction
- α_1 Adrenergic Agonists
  - Epinephrine
  - Phenylephrine

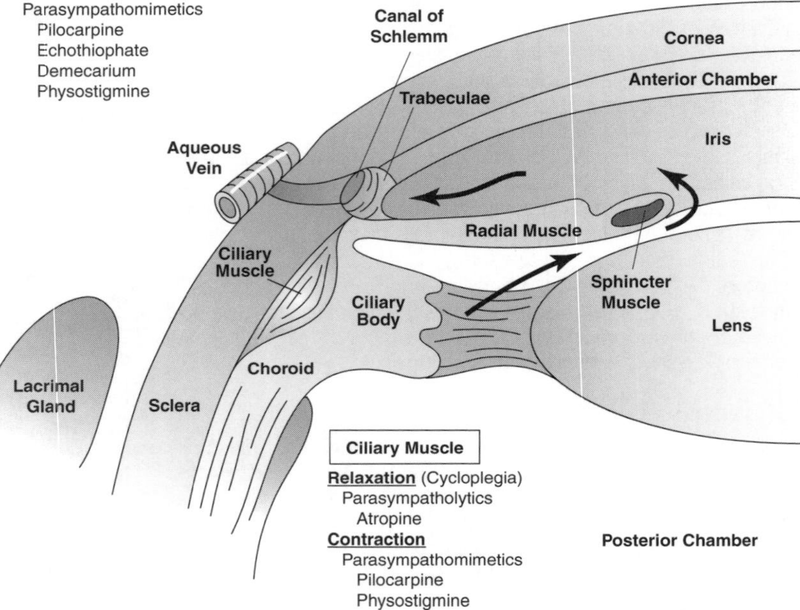

→ Flow of Aqueous Humor

B Autonomic Innervation of Eye

Function	Parasympathetic	Sympathetic
Iris		
Radial muscle	–	α_1: Contraction (Mydriasis)
Sphincter muscle	M3: Contraction (Miosis)	–
Ciliary Muscle	M3: Contraction (Accommodation)	β_2: Relaxation
Aqueous Humor Production	–	β_2: Increase production
Lacrimal Gland	M3: Secretion	α_1: Mild increase secretion

Many drugs used in ophthalmology are applied topically. Since topically applied drugs generally do not pass through the corneal epithelium, they are useful only in the treatment of disorders in the anterior segment of the eye. Topically applied ocular drugs are eliminated by dilution and washout from lacrimal secretions, evaporation, and absorption by the conjunctival circulation.

Autonomic Control of the Eye

Parasympathetic stimulation of the eye results in lacrimal gland secretion; contraction of the iris sphincter muscle, causing miosis; and contraction of the ciliary body (**Parts A** and **B**). The latter effect decreases resistance to the outflow of aqueous humor and decreases intraocular pressure (IOP). Adrenergic drugs relax the iris sphincter muscle, causing mydriasis; retract the third eyelid; and relax orbital smooth muscles. Activation of β adrenergic receptors increases aqueous humor production while stimulation of α_2 receptors decreases production.

Drug Used to Treat Glaucoma

Glaucoma is an abnormal elevation in IOP that can irreversibly damage the optic nerve and retina, if left untreated. In veterinary patients, glaucoma is most often the result of obstruction to aqueous humor outflow. Aqueous humor is produced by the ciliary body and normally is resorbed at the iridocorneal angle (*see* **Part A**). Primary glaucoma is associated with anatomic defects in the iridocorneal angle, while secondary glaucoma is associated with obstruction of the angle due to neoplasia, hemorrhage, or inflammation. Treatment of glaucoma is aimed at either decreasing aqueous production or opening the iridocorneal drainage angle with miotic agents.

Direct-Acting Parasympathomimetics

Pilocarpine is a topical miotic agent. A single application lowers IOP for up to 4 hours. Since pilocarpine may cause conjunctival irritation and repeated application may result in uveal inflammation, it is not used to treat secondary glaucoma. Systemic side effects include salivation, bronchospasm, cardiac arrhythmias, vomiting, and diarrhea. Pilocarpine is also used to stimulate lacrimal gland secretion in ocular conditions associated with dry eye.

Indirect-Acting Parasympathomimetics

Echothiophate and demecarium are acetylcholinesterase (AChE) inhibitors. A single instillation of either drug can lower IOP for 12–24 hours in dogs. Both drugs may cause ocular irritation and should not be used to treat secondary glaucoma. Cats are particularly sensitive to topical AChE inhibitors and may exhibit signs of a systemic cholinergic crisis.

Adrenergic Agents

Timolol, a β_2 adrenergic antagonist, decreases IOP. Since it results in a small drop in IOP (3–10 mm Hg), timolol is often combined with other therapeutic modalities. It can be used in the treatment of either primary or secondary glaucoma and in the prevention of glaucoma in a predisposed eye. Topical side effects are rare.

Epinephrine, an α and β adrenergic receptor agonist, causes ocular vasoconstriction, transient mydriasis, and decreased IOP. The latter effect is most likely due to decreased production of aqueous humor secondary to vasoconstriction. It is typically used in combination with pilocarpine to prevent epinephrine-induced mydriasis. Epinephrine may cause ocular irritation.

Carbonic Anhydrase Inhibitors

Dorzolamide, a topical carbonic anhydrase inhibitor, decreases aqueous humor production by decreasing the formation of HCO_3^- and H^+ from H_2CO_3 in ciliary epithelial cells. Decreased availability of H^+ leads to decreased movement of Na^+ and thus water into the aqueous humor. Dorzolamide may cause ocular irritation, but is safe to use in the treatment of primary or secondary glaucoma. Systemically administered carbonic anhydrase inhibitors are indicated in the emergency treatment of glaucoma.

Vasoconstrictors

Phenylephrine is a potent α adrenergic receptor agonist that causes vasoconstriction and mydriasis that lasts for about 2 hours. It may be used to differentiate conjunctival from deep episcleral injection, since the former will blanche with phenylephrine while the latter will not. It is helpful in the diagnosis of Horner's syndrome (sympathetic paralysis of the eye). Application of dilute phenylephrine to the normal eye has no effect on pupil size, whereas an eye with Horner's syndrome will dilate due to α adrenergic receptor upregulation. Phenylephrine may cause ocular irritation and chronic use can result in intraocular inflammation.

Mydriatics

Anticholinergics and sympathomimetic agents are used as mydriatics. Atropine blocks cholinergic input into the iris sphincter muscle and ciliary body, leading to mydriasis and accommodation paralysis (cycloplegia). It is useful in the treatment of uveitis to control painful ciliary muscle spasms and to dilate the pupil to prevent synechia formation. Mydriasis from atropine is long lasting (120–144 hours). Atropine increases IOP and decreases tear production. Systemic absorption of atropine may cause gastrointestinal ileus (colic) and tachycardia. Salivation may be noted after application, owing to entry of the bitter-tasting alkaloid in the mouth from the nasolacrimal system.

Tropicamide is a parasympatholytic agent that causes a rapid onset of mydriasis (15–30 min) that persists for 6–12 hours. It has minor cycloplegic action. It is used to dilate pupils for funduscopic examination and should not be used in patients with glaucoma.

Topical Anesthetics

Proparacaine is a rapid-acting (onset of action about 1 min) local anesthetic whose action persists for approximately 10 min. The drug is used prior to tonometry and gonioscopy to desensitize the cornea. It should not be used for long-term control of ocular pain because chronic use can damage the corneal epithelium.

19 Ocular Pharmacology II

A Ocular Pharmacology

Ciliary Body

Decrease Aqueous Humor Production
- Carbonic Anhydrase Inhibitors
  - Systemic: Daranide
  - Topical: Dorzolamide
- Osmotic Agents
  - Mannitol
- β_2 Adrenergic Antagonists
  - Timolol
- α_1 Adrenergic Agonists
  - (Secondary Vasoconstriction)
  - Epinephrine
  - Phenylephrine

Iris

Mydriasis
- Parasympatholytic
  - Atropine
  - Tropicamide
- α_1 Adrenergic Agonists
  - Epinephrine
  - Phenylephrine

Miosis
- Parasympathomimetics
  - Pilocarpine
  - Echothiophate
  - Demecarium
  - Physostigmine

Ocular Vasculature

Vasoconstriction
- α_1 Adrenergic Agonists
  - Epinephrine
  - Phenylephrine

Lacrimal Gland

Increase Secretion
- Parasympathomimetics
  - Pilocarphine
- Cyclosporine

Decrease Secretion
- Parasympatholytics
  - Atropine

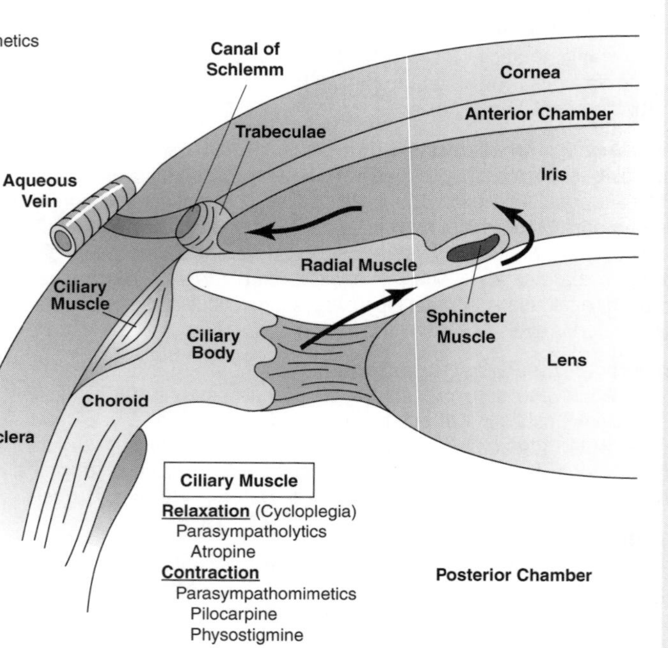

Ciliary Muscle

Relaxation (Cycloplegia)
- Parasympatholytics
  - Atropine

Contraction
- Parasympathomimetics
  - Pilocarpine
  - Physostigmine

B Autonomic Innervation of Eye

Function	Parasympathetic	Sympathetic
Iris		
Radial muscle	–	α_1: Contraction (Mydriasis)
Sphincter muscle	M3: Contraction (Miosis)	–
Ciliary Muscle	M3: Contraction (Accommodation)	β_2: Relaxation
Aqueous Humor Production	–	β_2: Increase production
Lacrimal Gland	M3: Secretion	α_1: Mild increase secretion

Anti-Inflammatory Agents

Corticosteroids

Corticosteroids are used to treat a variety of inflammatory ocular conditions including uveitis, episcleritis, allergic conjunctivitis, and nonulcerative keratitis. Since penetration of topical corticosteroids into the eyelids and posterior segment of the eye is poor, systemic corticosteroids should be used to treat inflammatory conditions in these areas. Prednisone acetate solution (1%) is the preferred topical preparation to treat uveitis, as its high lipid solubility permits entry into the uveal tract. Due to its relatively short duration of action, it must be applied frequently (every 4–6 hours). Although dexamethasone ointments do not penetrate the eye as well, they are used in large animals because they persist longer in the eye. The higher potency of dexamethasone (6X prednisone) compensates for the poorer penetration. Dexamethasone ointment is also used to treat episcleritis, scleritis, and nonulcerative keratitis. Topical ointments containing hydrocortisone acetate are effective, less expensive alternatives to treat allergic conjunctivitis. Side effects of topical corticosteroid use are impaired corneal healing, increased susceptibility to infection, and enhancement of collagenase activity. Long-term topical application of corticosteroids may suppress the pituitary-adrenal axis, resulting in iatrogenic hyperadrenocorticism.

Nonsteroidal Anti-inflammatory Agents

Intraocular prostaglandins have several effects including vasodilation, increased IOP, increased vascular permeability, disruption of the blood-aqueous humor barrier, and iris sphincter muscle contraction, leading to miosis. Topical nonsteroidal anti-inflammatory agents that inhibit prostaglandin production include flurbiprofen, ketorolac, suprofen, and diclofenac. These drugs can be used to control ocular inflammation in situations where corticosteroids are contraindicated such as in the presence of corneal ulcers or infections or in patients with diabetes mellitus who may be unable to tolerate systemic absorption or even small amounts of corticosteroids. NSAIDs are also used prior to cataract surgery to prevent miosis and post-surgically to control inflammation. Although these agents may cause some delay in corneal healing, unlike corticosteroids they do not predispose to infections or promote collagenase activity. Topical irritation may accompany application of these drugs. These drugs may be combined with topical corticosteroids in severe cases of uveitis.

Immunosuppressant Drugs

Cyclosporine is an immunosuppressive agent that inhibits T-lymphocyte production of interleukin-2 (*see* Chapter 61). Topical cyclosporine ointment (0.2%) is used in veterinary medicine to control keratoconjunctivitis sicca (KCS) (**Part A**). Cyclosporine restores normal tear production by decreasing lacrimal gland inflammation and directly stimulating lacrimal gland secretion. The topical $t^{1/2}$ of cyclosporine is about 8 hours. Most cases of canine KCS can be controlled with twice daily therapy. It may take 3–8 weeks of therapy to see the full therapeutic effect. Dogs with very low results on initial tear tests may fail to respond because of irreversible fibrosis of the lacrimal glands. Cyclosporine is also used to treat canine chronic superficial keratitis (pannus). No systemic toxicity has been associated with topical administration of cyclosporine and topical adverse effects are rare.

Anti-Viral Agents

Topical antiviral medications are commonly used in the treatment of feline ocular herpesvirus infections. Trifluridine, a virostatic pyrimidine nucleoside analogue, is the antiviral agent of choice. Since it penetrates the cornea poorly and is virostatic, it must be applied frequently (every 2 hours) during the first 2 days to allow establishment of adequate corneal levels. Treatment should continue 1 week after resolution of clinical signs.

Antifungal Agents

Fungal keratitis is most frequently diagnosed in the horse. Natamycin, clotrimazole, econazole, fluconazole, ketoconazole, and miconazole are available as topical ophthalmic suspensions. Since all of the antifungal drugs penetrate the cornea poorly, they must be applied frequently (every 1–2 hours) for 1–3 days, with gradual reduction in dosing as signs improve. Silver sulfadiazine is a topical antibacterial cream that has antifungal activity. It is used as an off-label topical antifungal agent in horses as an inexpensive alternative to other approved antifungal agents.

Antibacterial Agents

There are a number of topical antibacterial products used in the eye and the reader is referred to the chapters on antimicrobial therapy for a discussion of the action of these agents. These topical antibiotic agents are used to treat bacterial conjunctivitis and bacterial infections complicating corneal damage.

Commonly used topical ocular antibacterial agents include triple-antibiotic ointments (polymixin, bacitracin, and neomycin), aminoglycosides (gentamicin and tobramycin), chloramphenicol, erythromycin, and tetracyclines. Specific indications depend on the underlying etiology. Tetracycline, chloramphenicol, and erythromycin ointments are useful in treating *Mycoplasma* or *Chlamydia* infections in cats and the aminoglycosides (which obtain high corneal levels with frequent dosing) are useful in treating infections with *Pseudomona* sp.

Topical antibiotic preparations are frequently used in combination with topical hydrocortisone in cases of nonspecific, uncomplicated conjunctivitis in the canine and equine patient. In cats, however, these combination products should be avoided due to the high incidence of ocular herpes viral infection. Topical steroids can prolong herpes viral infections and predispose to corneal complications. Rare anaphylactic reactions to ophthalmic preparations containing bacitracin, polymixin, and neomycin have been reported in cats.

20 Control of Micturition

A Neuronal Control of Micturition

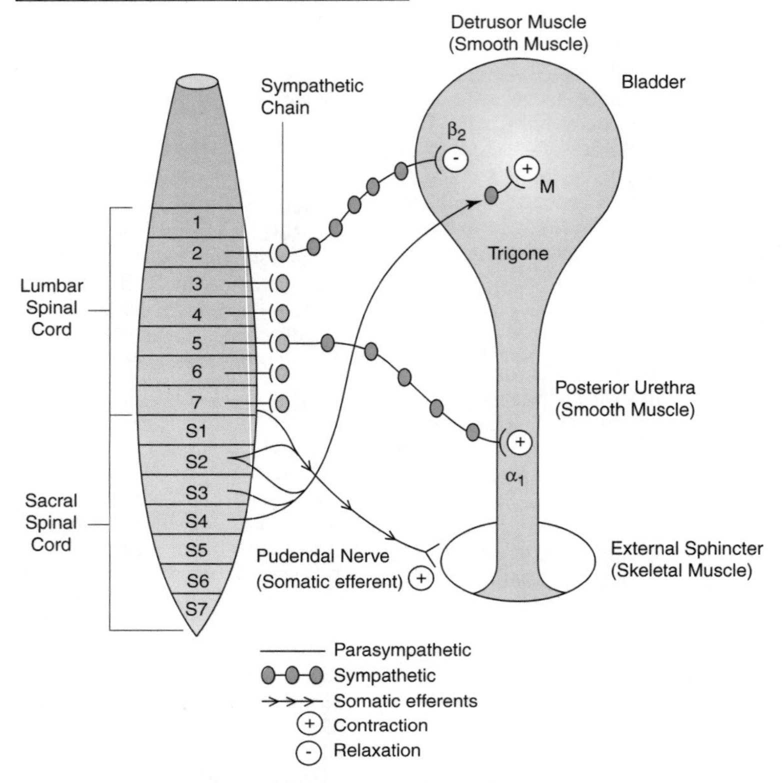

B Pharmacologic Manipulation of Micturition

Detrusor Atony

Parasympathomimetics
- Bethanechol

Detrusor Spasticity

Parasympatholytics
- Propantheline
- Flavoxate
- Oxybutynin

Urethral Insufficiency

α_1 Adrenergic Agonists
- Phenylpropranolamine
- Ephedrine
- Imipramine

Sex Hormones
- Estrogen
- Testosterone

Urethral Spasticity

α_1 Adrenergic Antagonists
- Phenoxybenzamine
- Prazosin

External Sphincter Spasticity

Skeletal Muscle Relaxants
- Dantrolene
- Benzodiazepines

Micturition is the process by which the urinary bladder empties. The urinary bladder is composed of 1) the body containing the detrusor muscle; 2) the trigone, a triangular area near the neck of the bladder through which both the ureters and the urethra pass; and 3) the bladder neck also called the posterior urethra (**Part A**). Beyond the bladder neck, the urethra passes through the urogenital diaphragm, which contains a layer of voluntary skeletal muscle called the external sphincter. The muscle of the bladder body and neck is smooth muscle. Parasympathetic stimulation contracts the detrusor muscle and relaxes the trigonal muscle. Sympathetic stimulation relaxes the detrusor muscle around the neck of the bladder (β_2 effect) and contracts (α_1 effect) the internal urethral sphincter (*see* **Part A**). The external sphincter is innervated by motor nerves supplied by the pudendal nerve and is under voluntary control.

Urethral Sphincter Incompetence

Estrogen and testosterone exert a permissive effect on α_1 adrenergic activity of the internal urethral sphincter to maintain adequate urethral sphincter tone. Diethylstilbestrol (DES) is a synthetic nonsteroidal estrogen agent used to treat incontinence in spayed dogs. It is well absorbed from the gastrointestinal tract after oral administration, metabolized in the liver to a glucuronide conjugate, and excreted in the urine and feces. Side effects of low-dose DES therapy are uncommon, but include induction of estrus and bone marrow suppression. Presently there are no commercially available formulations of DES, but it may be compounded. Testosterone is available as the cypionate, enanthate, and propionate ester for injection. After IM injection, the cypionate and enanthate ester are released slowly and have a duration of action of 2–4 weeks. The propionate ester has a much shorter duration of action and is the preferred preparation to use in treating urinary incontinence. Testosterone is metabolized in the liver and its metabolites are eliminated primarily in the urine. Side effects include prostatic hypertrophy, induction of perianal gland tumors, behavioral changes, polycythemia, and hepatic damage. Testosterone preparations are controlled substances.

Alpha adrenergic receptors, agonists that increase tone in the internal urethral sphincter, are the drugs of choice in treating urethral incompetence. Phenylpropranolamine and ephedrine are sympathomimetic amines that stimulate both α and β adrenergic receptors by causing the release of norepinephrine. A side effect of these drugs is systemic hypertension. Other side effects associated with drug entry into the CNS include restlessness, irritability, and inappetence (rarely). Stimulation of β adrenergic receptors may increase heart rate, raise intraocular pressure, and stimulate hepatic glycogenolysis. Due to their side effect profile, the use of these amines is contraindicated in patients with systemic hypertension, diabetes mellitus, or glaucoma. Prolonged use of the amines has been reported to deplete norepinephrine from storage sites, causing decreased responsiveness, although this does not appear to be a prominent problem in the treatment of canine urinary incontinence. Sympathomimetic agents may be combined with hormonal therapy for additive effects on urethral muscle tone. Phenylpropanolamine is partially metabolized in the liver to an active metabolite, but the majority is excreted unchanged in the urine. Recent findings indicate that phenylpropanolamine use in humans can increase the risk of stroke and have prompted withdrawal of the drug from the market. The drug will need to be compounded for future veterinary use.

Imipramine is a tricyclic antidepressant agent that increases sympathetic tone by inhibiting synaptic norepinephrine reuptake. Imipramine is metabolized in the liver to desipramine, which is pharmacologically active. Both of these agents block the amine reuptake pump leading to an increase in norepinephrine and serotonin, and have central and peripheral anticholinergic activity that promotes urinary retention. Side effects include sedation, behavioral changes (aggression, dysphoria), xerostomia, constipation, hypotension, and sinus tachycardia. Since tricyclic antidepressants depress cardiac conduction, their use is contraindicated in the presence of preexisting conduction disturbances. Jaundice, agranulocytosis, and dermatitis have been reported in human patients.

Detrusor Spasticity

Detrusor hyperspasticity may be associated with bladder infections or neurological disorders (**Part B**). Since the detrusor muscle is under parasympathetic control, hyperactivity is treated with anticholinergic drugs (propantheline, oxybutynin, dicyclomine, and flavoxate). Side effects of these agents include gastrointestinal stasis, xerostomia, and tachycardia. Oxybutynin and flavoxate, which enter the CNS, can cause excitement, sedation, ataxia, increased IOP, and mydriasis. Propantheline does not cross the blood-brain barrier and is not associated with CNS side effects. Oxybutynin and flavoxate are tertiary amines with weak antimuscarinic effects and spasmolytic effects on bladder and gastrointestinal smooth muscle. All of these anticholinergics are metabolized in the liver and excreted in the urine.

Detrusor Atony

If a detrusor reflex is present but insufficient to induce adequate voiding, cholinergic agents may stimulate bladder contraction. Bethanechol, a synthetic derivative of acetylcholine, is resistant to hydrolysis by acetylcholinesterase and has a prolonged duration of action. Its pharmacological effects are most profound on muscarinic cholinergic receptors. When it is administered orally, its effects are primarily limited to stimulation of smooth muscle contraction in the urinary and gastrointestinal tracts. Stimulation of gastrointestinal smooth muscle may result in diarrhea, vomiting, and anorexia. Hypersalivation, due to parasympathetic stimulation of the salivary glands, may be the first sign of muscarinic overdose. Parenteral administration may precipitate a cholinergic crisis and should be avoided. Bethanechol is contraindicated in the presence of functional or mechanical obstruction to urethral outflow. It is frequently used in combination with agents that decrease urethral tone.

Urethral Spasticity

Urethral spasm may accompany infectious, inflammatory, or neurological disorders. α_1 adrenergic antagonists relax urethra smooth muscle. Phenoxybenzamine blocks the α adrenergic response to norepinephrine and epinephrine without interfering with β adrenergic receptors. Phenoxybenzamine is variably absorbed after oral dosing. Its pharmacological effects take several hours to days to become noticeable and may persist for 3–4 days after the drug is discontinued. Persistence is due to the drug's high lipid solubility and thus its accumulation in adipose tissue. Phenoxybenzamine undergoes hepatic metabolism and urinary and fecal excretion. Side effects include hypotension, reflex tachycardia, miosis, increased IOP, weakness, ataxia, and gastrointestinal upset. Prazosin is another α_1 adrenergic antagonist. Its onset of action is faster than that of phenoxybenzamine. It is also used as a balanced vasodilator to control systemic hypertension and decrease afterload in cardiac failure. Prazosin is available orally and circulates highly protein bound. It is metabolized in the liver, generating some active metabolites that excrete in the feces. Terazosin is a close structural analogue that has greater oral bioavailability and a longer duration of action. Use is limited in veterinary patients.

Relaxation of External Urethral Sphincter

Skeletal muscle relaxants such as dantrolene, benzodiazepines, and baclofen have been used to decrease excessive contraction of the external urethral sphincter. Dantrolene has a direct effect on skeletal muscle (inhibition of Ca^{2+} release from the sarcoplasmic reticulum) to cause relaxation. Dantrolene is variably absorbed after oral administration and is highly protein bound. It is metabolized by the liver and excreted in the urine. Hepatotoxicity has been associated with high-dose chronic use in humans. More common side effects include sedation, weakness, and ataxia. Baclofen, a derivative of the inhibitory neurotransmitter GABA, inhibits interneuron neurotransmission within the spinal cord. The drug is rapidly absorbed after oral administration and is excreted unchanged in the urine. Side effects include sedation, weakness, inappetence, and vomiting. The drug should not be used in patients with epilepsy because it can lower the seizure threshold.

21 Respiratory Drugs I

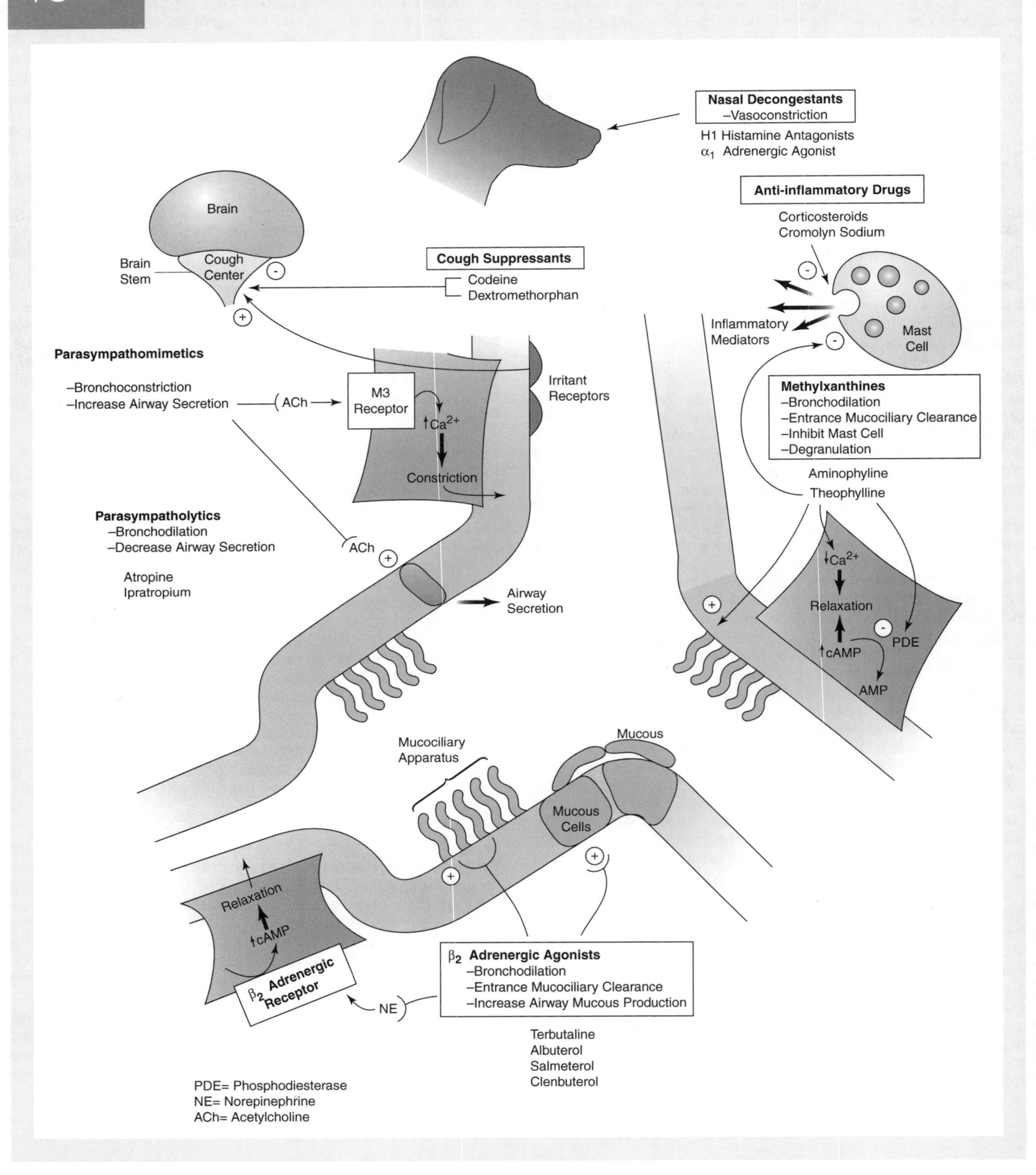

Control of bronchial smooth muscle tone depends on the integration of signals from several mechanoreceptors (irritant receptors) that respond to chemical, mechanical, and physical stimuli. The parasympathetic nervous system is responsible for maintaining baseline airway tone. The sympathetic system balances this by stimulating bronchodilation through activation of β_2 adrenergic receptors. In contrast, α adrenergic stimulation can lead to bronchoconstriction. Respiratory defense mechanisms include coughing, sneezing, clearance action of the mucociliary apparatus lining the airways, and the respiratory mononuclear phagocytic system. The latter is composed of alveolar macrophages and the pulmonary intravascular macrophages, which are important sources for inflammatory mediators including leukotrienes and prostaglandins.

Bronchodilators

β_2 Adrenergic Receptor Agonists

β_2 adrenergic receptor binding results in relaxation of bronchial smooth muscle, stimulation of airway mucus secretion, and enhancement of mucociliary clearance (**Figure**). Activation of β_2 adrenergic receptors results in an increase in intracellular cAMP levels with subsequent activation of cAMP dependent protein kinase A that results in relaxation of bronchial smooth muscle. Epinephrine and isoproterenol are nonselective β adrenergic agonists that stimulate β_1, β_2, and α adrenergic receptors. They are used either parenterally or by aerosol inhalation to achieve rapid bronchodilation in the treatment of allergic reactions or status asthmaticus. Concurrent activation of β_1 and α adrenergic receptors by these agents leads to tachycardia, vasoconstriction, and hypertension, respectively.

β_2-Selective Receptor Agonists

Terbutaline is a β_2-selective agonist used in small animals. It can be administered parenterally or orally. It is rapidly absorbed after parenteral administration and is useful in the treatment of acute asthmatic attacks in feline patients. Rapid first-pass hepatic metabolism results in reduced oral bioavailability. Terbutaline undergoes some inactivation upon hepatic metabolism, but about half is excreted unchanged in the urine. Adverse effects are dose related and include those associated with stimulation of β_1 receptors, including tremors and CNS excitement and sweating in horses.

In equine patients, several short-acting β_2-selective agents are available for aerosol administration. These include albuterol, salmeterol, and terbutaline. The use of the oral short-acting β_2 agonist clenbuterol is illegal in food animals.

Methylxanthines

The methylxanthine theophylline has several mechanisms by which it promotes bronchodilation (*see* **Figure**) these include: 1) inhibition of phosphodiesterase, resulting in increased cAMP levels; 2) competitive antagonism of adenosine, which is a known bronchoconstricting agent; and 3) interference with calcium mobilization. Theophylline also inhibits mast cell degranulation, increases mucociliary clearance, prevents microvascular leakage, and increases the force of respiratory muscle contraction. It is equally effective in dilating large and small airways.

Theophylline is well absorbed after oral administration. It is metabolized in the liver primarily to inactive metabolites that are excreted in the urine. There are marked differences in the elimination of theophylline in humans and the same would be expected in veterinary patients, although such effects have been poorly documented. Ideally, serum levels should be monitored to prevent underdosing or overdosing. The clearance of theophylline is enhanced by concurrent therapy with barbiturates while the administration of macrolide antibiotics, clindamycin, lincomycin, or cimetidine decreases clearance. Because of theophylline's low volume of distribution and low lipophilicity, dosage should be based on lean body weight in obese individuals.

Theophylline is available in a variety of dose formulations with different amounts of active theophylline. Aminophylline is a widely used formulation (theophylline ethylenediamine) and is 86% theophylline. Oxitriphylline is 64% theophylline and the glycinate and salicylate salts, which are available as elixirs, are only 46% theophylline. Although a time-release formulation is available for human patients, it is not as slowly absorbed in veterinary patients. It does permit twice-a-day doses in dogs and once-a-day administration in cats.

The theophyllines are generally considered to have a low therapeutic index. Side effects include CNS disturbances (excitement, seizures, restlessness), gastrointestinal disturbances (nausea and vomiting), and polyuria/polydipsia (associated with a diuretic effect). Cardiac dysrhythmias may be seen when theophylline is administered IV or when it is given along with sympathomimetic agents or inhalational anesthetics.

Anticholinergics

Anticholinergics reduce the sensitivity of irritant receptors and antagonize vagally mediated bronchoconstriction (*see* **Figure**). The predominant action of the anticholinergics is on the large airways. Due to its nonselective action, atropine given either parenterally or by aerosolization is not useful in chronic respiratory conditions. Side effects in horses undergoing chronic therapy with atropine include mydriasis, ileus, dry mucous membranes, restlessness, and ataxia. Atropine is useful in the acutely dyspneic animal since its effects appear to be additive to those of β_2 agonists and corticosteroids. Ipratropium bromide is a synthetic anticholinergic. Its major advantage is that it is not well absorbed following aerosolization, which limits nonselective anticholinergic side effects.

Respiratory Drugs II

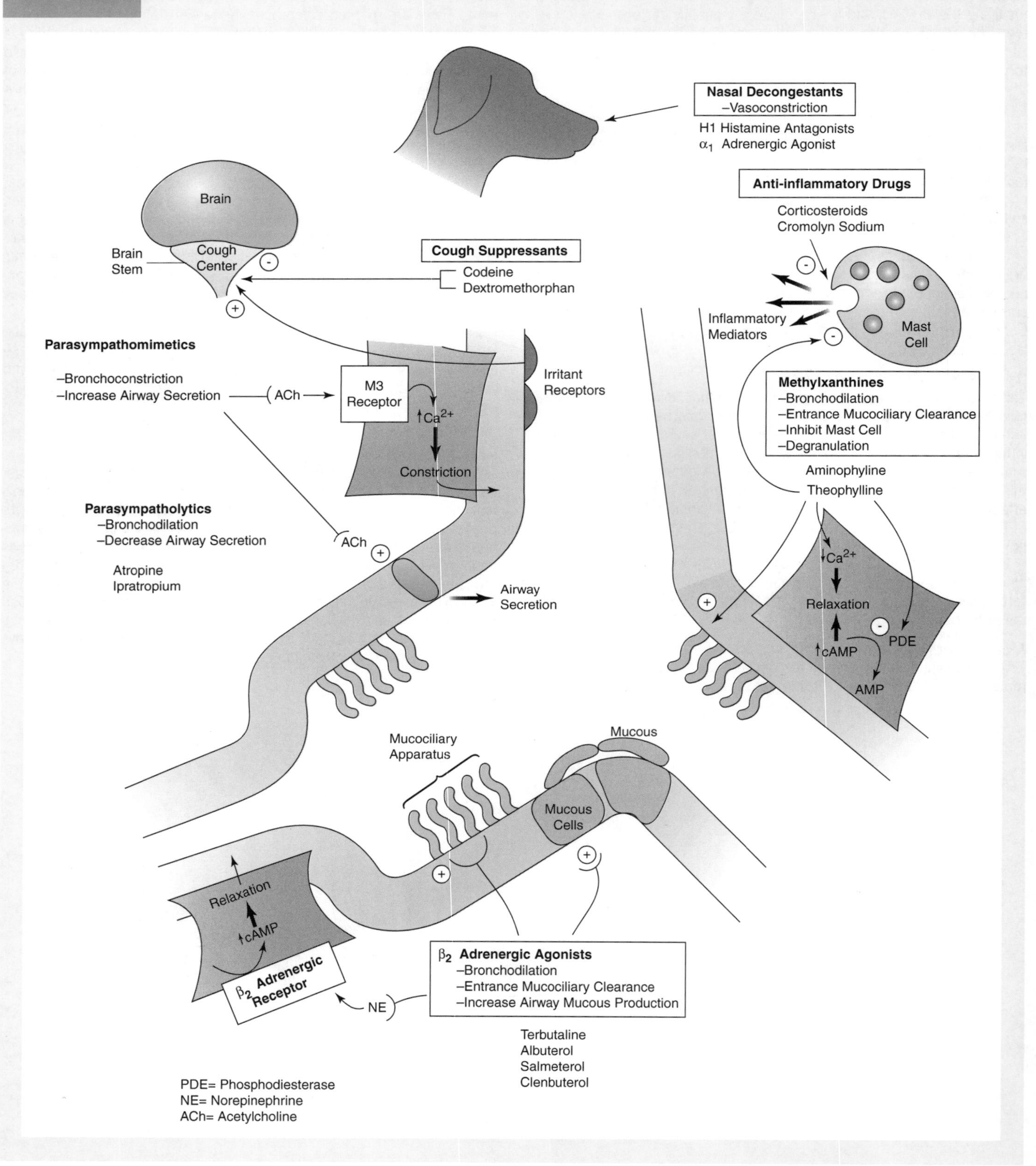

Anti-inflammatory Agents

Corticosteroids have several beneficial actions in the respiratory system including bronchodilation and anti-inflammatory actions. The anti-inflammatory actions of glucocorticoids are described in Chapter 59. Corticosteroids are the mainstay in the treatment of allergic airway disease. Although systemic administration is common and the preferred route in an acute respiratory crisis, inhalation is the preferred method of delivery for the treatment of chronic disorders. The benefits of inhalation therapy include high topical potency and low systemic bioavailability, which limits systemic toxicity.

Mast Cell–Stabilizing Agents

Cromolyn sodium inhibits calcium influx into mast cells, preventing their degranulation and subsequent release of histamine and other inflammatory mediators (**Figure**). It is most effective as a preventative prior to activation of inflammatory cells. It must be given by aerosolization because it is poorly absorbed after oral administration. The drug is used in horses to treat chronic obstructive pulmonary disease. Its pharmacological effects appear to increase over time. Adverse effects are rare.

Antitussives

The goal of antitussive therapy is to decrease the frequency and severity of cough without impairing respiratory defense mechanisms. In many respiratory conditions, especially those caused by infectious organisms, coughing is an important protective reflex. Irritant receptors trigger the cough reflex. The coughing reflex can be blocked peripherally by facilitating removal of the irritant stimuli or blocked centrally at the coughing center in the brain. Since bronchoconstriction is a frequent and important cough stimulus, the use of bronchodilator therapy will often relieve coughing. Expectorants or mucolytics may aid in removing respiratory irritants and may help to control coughing.

Centrally acting antitussives work by directly suppressing afferent input into the coughing center (*see* **Figure**). Narcotic antitussives are associated with significant sedative effects and can lead to respiratory and cardiac depression. Chronic use can cause constipation. They are also controlled substances. Codeine and hydrocodone are the most commonly prescribed narcotic antitussives. Nonnarcotic antitussives include the opioid agonist-antagonist butorphanol and dextromethorphan. Butorphanol is more commonly used as an analgesic, but is also approved as an antitussive in dogs. The drug is well absorbed after oral administration, but is subject to significant hepatic first-pass effect so that oral doses are considerably higher than parenteral doses. The drug is metabolized by the liver and excreted in the urine and bile. It causes minimal cardiac or respiratory depression. Sedation is a common side effect. Dextromethorphan is a semisynthetic derivative of opium. Its onset of action is rapid after oral administration (30 min). Dextromethorphan is as effective as codeine as an antitussive in dogs and cats. It lacks any sedative side effects. Dextromethorphan is available without prescription.

Expectorants

Expectorants stimulate secretion into the tracheobronchial tree and facilitate removal of viscous mucus that can accompany inflammatory and infectious respiratory disorders. Potassium iodide stimulates irritant receptors in the gastric mucosa, which reflexly (via the vagal nerve) stimulate secretion in the respiratory tract. Guaifenesin, a wood tar derivative, is a common expectorant included in over-the-counter cold preparations. Its mechanism of action is not completely understood and its efficacy is questionable.

Decongestants

Decongestants are used in the symptomatic treatment of viral or allergic rhinitis or sinusitis. The 2 major classes of drugs used as decongestants are H1 histamine receptor antagonists (diphenhydramine, dimenhydrinate, chlorpheniramine, hydroxyzine) and α_1 adrenergic receptor agonists (pseudoephedrine, phenylephrine, ephedrine). Antihistamines work by blocking histamine-induced increases in capillary permeability. They are given systemically. Sedation is the common side effect. The α_1 adrenergics cause vasoconstriction and thus reduce the volume of fluid exudation. Systemic administration of α_1 adrenergics should be used cautiously in the setting of hypertension, glucose intolerance, glaucoma, and disorders associated with urinary retention. Use of α adrenergics may also result in CNS excitement. Due to the high incidence of side effects when given systemically, they are best used as topical agents. Antihistamines and decongestants are often used together without enhanced toxicity.

23 Diuretics I

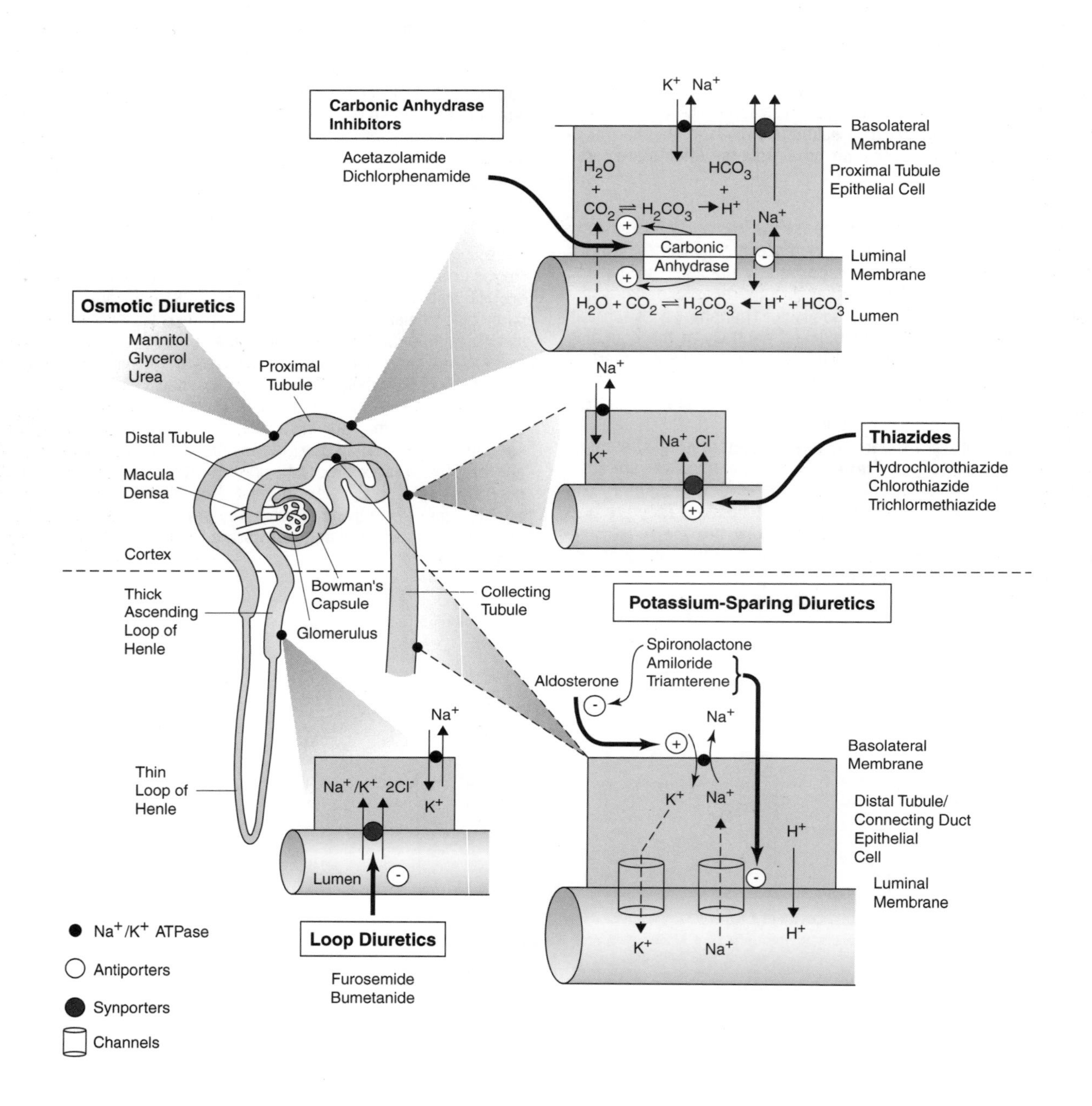

Diuretics increase the urinary excretion of Na^+ and water by altering electrolyte and water reabsorption in the renal tubular epithelium. The net result is to decrease extracellular fluid volume. Diuretics are used to treat conditions associated with hypervolemia.

The ability of the renal tubules to excrete a concentrated urine requires normal tubular transport function, the presence of antidiuretic hormone (ADH), and a concentrated renal medullary interstitium. The first step in urine formation is glomerular filtration, which produces a dilute tubular fluid. In the proximal convoluted tubule, 80% of the Na^+ is passively resorbed by 2 mechanisms: 1) facilitated cotransport with organic solutes and 2) countertransport with H^+. Active transport of Na^+ at the pericapillary membrane by the Na^+/K^+ ATPase keeps intracellular Na^+ concentrations low and maintains the driving force for Na^+ transport across the luminal membrane. The transepithelial movement of Na^+ creates an osmotic gradient for the reabsorption of water. Sodium transport in the proximal tubules is also dependent on HCO_3^- reabsorption. The first step in the reabsorption of HCO_3^-, is protonation to form H_2CO_3. Brush border carbonic anhydrase then catalyzes the formation of H_2O and CO_2 from H_2CO_3. The CO_2 freely diffuses into the tubular cell where it quickly combines with H_2O to re-form H_2CO_3. The H_2CO_3 then ionizes and the resultant H^+ is exchanged for Na^+. HCO_3^-, within the tubular cells is taken up by the peritubular capillaries in a process that is coupled to transport with Na^+. The net result is Na^+ and HCO_3^- resorption and urinary acidification.

Additional Na^+ resorption occurs in the thin loop of Henle. Since these segments are permeable to water, solute-free water moves out into the concentrated medullary interstitium and the tubular fluid becomes concentrated. The thick ascending loop of Henle is the diluting segment of the nephron. This segment is impermeable to water, but actively transports Na^+ and Cl^- out of the lumen into the tubular cells. Intracellular Na^+ is moved into the interstitium by the basolateral Na^+/K^+ ATPase. Since these segments of the nephron are impermeable to water, there is no osmotic movement of water along with the transported ions, and the tubular fluid leaving the loop of Henle is dilute. In the presence of ADH, water channels in the collecting ducts open and as the dilute tubular fluid moves through, it equilibrates osmotically with the concentrated medullary interstitium, resulting in the formation of concentrated urine. In the absence of ADH, the collecting ducts are impermeable to water and dilute urine is excreted.

In the late distal convoluted tubule and collecting ducts, the final qualitative changes in urinary excretion are made. These segments are responsive to the hormone aldosterone that is secreted from the adrenals. Aldosterone promotes Na^+ resorption and K^+ and H^+ excretion.

The ability of a diuretic agent to effect urine concentrating ability is determined by 2 major factors: 1) the amount of Na^+ and water resorption it inhibits and 2) the ability of the nephron segments distal to its site of action to resorb Na^+ and water.

Osmotic Diuretics

Mannitol is the most commonly used osmotic diuretic. Mannitol is freely filtered at the glomerulus and poorly reabsorbed (**Figure**). The presence of unabsorbed solutes in the proximal tubular fluid decreases water reabsorption. Since mannitol also increases renal medullary blood flow and decreases intracranial and intraocular pressures, it is used to treat oliguric renal failure, acute glaucoma, and cerebral edema. Mannitol is administered IV over several minutes. Diuresis begins within 15–30 min and lasts 3–4 hours. Because it is hyperosmolar, mannitol causes water to move out of cells and into the vascular space. As this may cause circulatory overload, mannitol should not be used in patients with heart failure. Glycerol and urea are also used as osmotic diuretics.

Carbonic Anhydrase Inhibitors

The carbonic anhydrase inhibitors (CAIs), acetazolamide and dichlorphenamide, are sulfonamide derivatives. Inhibition of carbonic anhydrase blocks proximal tubular resorption of HCO_3^- and Na^+ (*see* **Figure**). The net effect of CAIs is to reduce the amount of H^+ available for Na^+-H^+ exchange and thus block the formation of carbonic acid from HCO_3^-. The diuretic effect of CAIs is self-limiting since the metabolic acidosis that results from the loss of HCO_3^- in the urine leads to the accumulation of H^+, which permits the exchange process to resume. Side effects of CAIs include vomiting, hypokalemia (due to increase in urine flow to the distal tubules), CNS effects (sedation, depression, excitement), dermatological effects, and bone marrow depression. The primary use for CAIs in veterinary medicine is to decrease intraocular pressure (IOP] in glaucoma. This effect is mediated by inhibition of carbonic anhydrase in the ciliary body where the enzyme participates in the production of aqueous humor. The CAIs are available orally, with an onset of action of about 30 min and a maximal effect in 2–3 hours. These drugs are primarily excreted unchanged in the urine by a combination of glomerular filtration and tubular secretion. The use of CAIs in hepatic disease is contraindicated since the resultant urinary alkalinization may divert ammonia from the urine into the systemic circulation.

24 Diuretics II

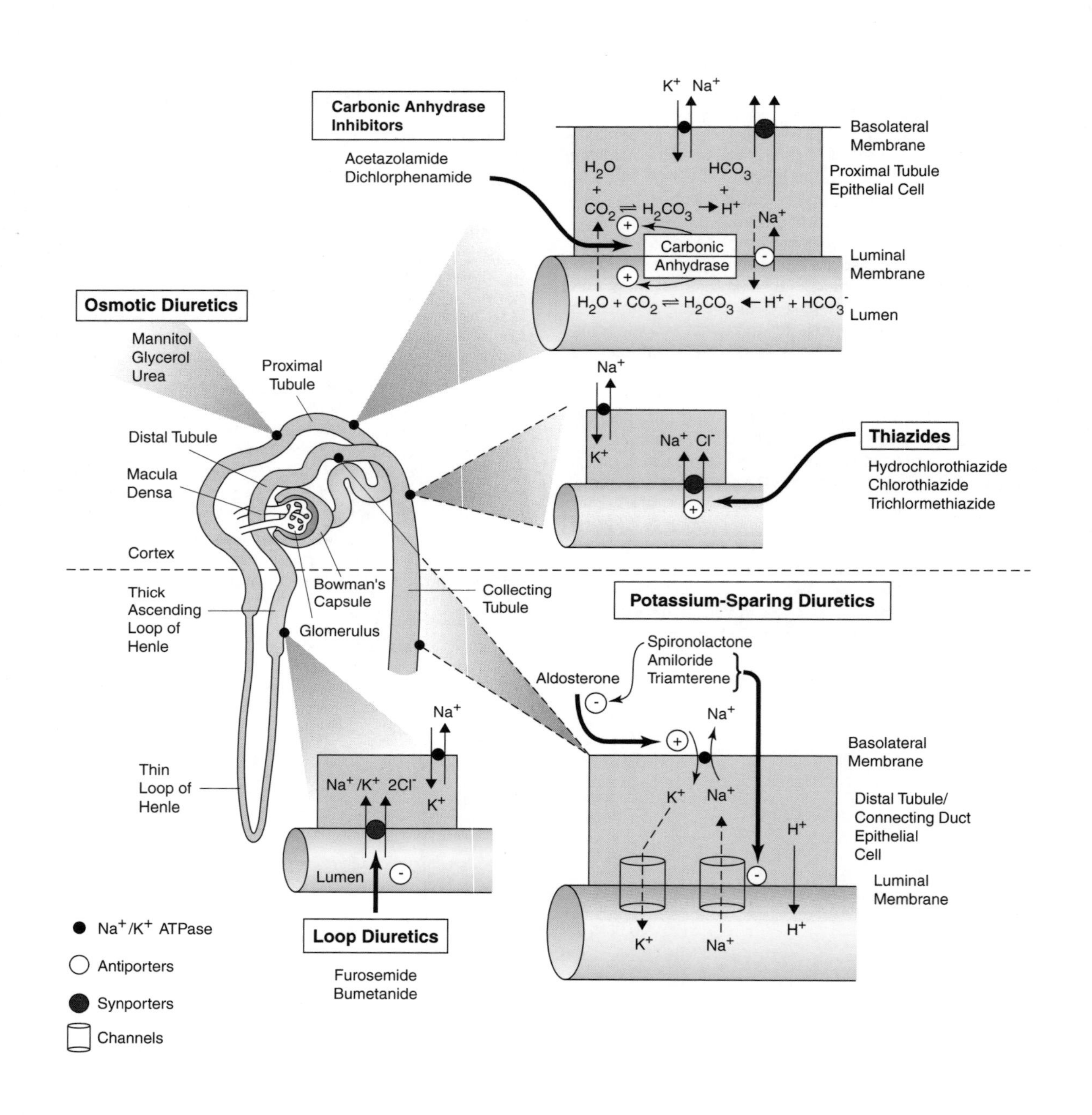

Thiazides

The thiazide diuretics, hydrochlorothiazide, chlorothiazide, and trichlormethiazide are sulfonamide derivatives that work by blocking Na^+ reabsorption in the distal tubule (**Figure**). They are considered weak diuretics since the majority of the filtered Na^+ has already been reabsorbed by the time it reaches the distal nephron. Thiazide diuretics increase Ca^{2+} and decrease K^+ and Mg^{2+} absorption in the distal tubule. They are used to manage edema associated with mild to moderate congestive heart failure (CHF), to control systemic hypertension, and to treat postparturient udder edema in cows. They can also be employed in the management of hypercalciuria in patients with recurrent calcium oxalate stones. The drugs are available orally, circulate highly protein bound, and are largely excreted unchanged by tubular secretion into the urine. Adverse effects include hypokalemia, hypochloremic metabolic alkalosis, hyponatremia, hypomagnesemia, and hyperglycemia (associated with decreased insulin secretion secondary to hypokalemia and stimulation of hepatic glycogenolysis). The latter effect on glucose homeostasis has been exploited in the treatment of insulin-secreting tumors.

Loop Diuretics

Loop diuretics inhibit Na^+ and Cl^- reabsorption from the thick ascending loop of Henle (*see* **Figure**). Since resorption of these solutes in the loop of Henle is essential for urine concentrating ability, the loop diuretics are the most potent of all diuretics. These agents include the sulfonamide derivatives furosemide and bumetanide, and the phenoxyacetic acid, ethacrynic acid. These diuretics also increase the urinary excretion of Ca^{2+}, Mg^{2+}, and K^+. Furosemide is the most commonly used loop diuretic in veterinary medicine. It is a mainstay in the treatment of pulmonary edema associated with CHF. Furosemide is also a mild venodilator that shifts blood volume from the pulmonary to the systemic circulation, an action that augments its effectiveness in treating pulmonary edema. Furosemide is also used to mobilize ascites in hepatic disease and to maintain urine flow in oliguric renal failure (although it does not increase glomerular filtration rate). Since retention of organic anions during renal failure can interfere with the tubular secretion of furosemide, higher doses may be necessary to promote diuresis in this setting. Furosemide is used in horses to treat exercise-induced pulmonary hemorrhage, although its efficacy has been questioned. By decreasing pulmonary capillary pressure, furosemide may prevent pulmonary vascular wall damage and the subsequent hemorrhage that occurs during strenuous exercise. Furosemide is also used to treat hypercalcemia.

Furosemide can be given orally or parenterally. When given IV, its onset of action is within 2–5 min and its effects persist for 2–3 hours. Its onset of action is slower after oral administration (1 hour), but its action is prolonged (6–8 hours). The majority of furosemide is excreted unchanged in the urine after secretion by the organic anion transporter into the proximal tubules. A small portion undergoes glucuronidation in the liver. Adverse side effects include volume depletion, hypokalemia, and development of a metabolic alkalosis. Metabolic alkalosis is a common side effect of therapy with furosemide (and other diuretics that promote Na^+ and Cl^- loss). Alkalosis is due to 1) volume depletion with contraction of the extracellular fluid volume around a constant amount of extracellular HCO_3^- and 2) increased renal loss of H^+ due to the large amounts of Na^+ presented to the distal tubule where it is exchanged for H^+ and K^+. Furosemide may potentiate the ototoxicity of aminoglycosides and promote glucose intolerance. Cats appear to be highly sensitive to the adverse effects of furosemide. Bumetanide and ethacrynic acid are infrequently used in veterinary medicine. Nonsteroidal anti-inflammatory agents inhibit the diuretic and natriuretic action of loop diuretics.

Potassium-sparing Diuretics

The potassium-sparing diuretics include spironolactone, amiloride, and triamterene. Spironolactone is an aldosterone antagonist that blocks aldosterone-mediated Na^+ resorption and K^+ and H^+ excretion in the late distal tubule and collecting ducts (*see* **Figure**). The onset of action of spironolactone is slow, with its maximal effectiveness delayed for 3–5 days. Spironolactone has moderate diuretic action and is often used in combination with other diuretics. It is most effective in pathologic states associated with hyperaldosteronism such as hepatic failure CHF and the nephrotic syndrome. Spironolactone is metabolized in the liver to its active component, canrenone. Adverse effects include hyperkalemia and gastrointestinal disturbances. Gynecomastia has been reported in human males due to inhibition of androgen synthesis. Human patients with congestive heart failure appear to have a survival advantage that treated with low doses of spironolactone.

Triamterene and amiloride inhibit active Na^+ reabsorption in the late distal convoluted tubule and collecting duct. They increase Na^+ and Cl^- excretion without increasing K^+ excretion. They are given orally. They are secreted into the proximal tubule and excreted in the urine. Triamterene undergoes extensive hepatic metabolism whereas amiloride does not. Adverse side effects include hyperkalemia and gastrointestinal disturbances.

Methylxanthines

The methylxanthines include aminophylline, theophylline, caffeine, and theobromide (contained in chocolate). These drugs are weak diuretics that inhibit Na^+ resorption in the proximal tubule and increase renal blood flow. They are not typically used clinically for their diuretic action. Aminophylline and theophylline are used clinically as bronchodilators (*see* Chapter 21).

25 Inhalant Anesthetics

A Structural Formula of Common Inhalant Anesthetic Agents

```
 N
  \
 ||  O
  /
 N
Nitrous Oxide

     Br  F
     |   |
 H - C - C - F
     |   |
     Cl  F
   Halothane

    Cl  F       H
    |   |       |
 H- C - C - O - C - H
    |   |       |
    Cl  F       H
   Methoxyflurane

    F   Cl      F
    |   |       |
 F- C - C - O - C -
    |   |       |
    F   H       F
     Isoflurane

     Cl  F       F
     |   |       |
 H - C - C - O - C - H
     |   |       |
     F   F       F
     Enflurane

     F   H       F
     |   |       |
 F - C - C - O - C - H
     |   |       |
     F   F       F
     Desflurane

                 F
                 |
     H       F - C - F
     |           |
 F - C - O - C - H
     |           |
     H       F - C - F
                 |
                 F
     Sevoflurane
```

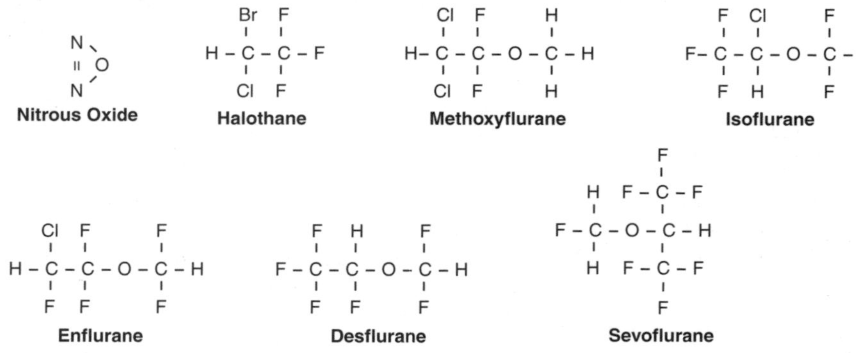

B Pharmacology of Inhalant Anesthetic Agents

Anesthetic Agent	MAC*(%)	Blood-Gas† Partition Coefficient	Dysrhythmias with Catecholamines	Respiratory Depression	Muscle Relaxation	Reduction of Cardiac Output	Metabolism	Intracranial Pressure
Methoxyflurane	0.25	14	++	++	++	++	++++	↑↑↑
Halothane	0.87	2.4	+++	++	+	++	+++	↑↑↑
Isoflurane	1.28	1.4	+	++ (↑Horses)	+++	+	+	↑
Enflurane	2.20	2.0	++	+++	+++	++	++	↑↑
Desflurane	7.2	0.42	+	+++	+++	+	-	↑
Nitrous Oxide	2.00	0.5	-	Mild	-	Mild‡	-	-

*MAC = Alveolar concentration that prevents gross purposeful movement in 50% of patients in response to a standardized painful stimulus. Surgical anesthesia is obtained by alveolar concentrations 1.4-1.8 × MAC.
†Solubility of the agent in blood correlates with speed of induction and recovery.
‡Stimulation of sympathetic nervous system overcomes any NO induced decrease in cardiac output.

Inhalant anesthetics act on the brain and spinal cord to produce unconsciousness and analgesia. The mechanism of action of these agents is unknown, but likely involves direct interference with neurons in the CNS by partitioning into the phospholipid bilayer of the plasma membrane. Inhalant anesthetics are administered to produce the desired depth of anesthesia. The goal is to establish a partial pressure of the agent in the brain so that generalized CNS depression can be achieved. Most of the inhalant anesthetics are liquids at room temperature and must be vaporized and delivered to the lungs. In the lungs they diffuse through the alveoli into the blood. The diffusion of these agents from the alveoli to the blood is affected by the solubility of the anesthetic in blood (blood-gas partition coefficient), blood flow to the lungs (cardiac output), and the concentration gradient (difference in partial pressure) of the agent between the alveolus and the pulmonary venous blood.

The inhalation anesthetics in use in veterinary medicine are enflurane, halothane, isoflurane, methoxyflurane, and nitrous oxide (**Part A**). The potency of each agent is defined by a minimum alveolar concentration (MAC), which is the concentration that prevents gross purposeful movement in 50% of patients in response to a standardized painful stimulus (**Part B**). Anesthetic potency is inversely related to the MAC. The MAC is determined in part by the lipid solubility of the agent. The more lipid soluble the agent, the lower the MAC and the higher the potency. The anesthetic dose required to anesthetize 95% of animals is about 1.2–1.4 times the MAC. The solubility of the anesthetic agent in blood correlates with the speed of induction and recovery and is defined by the blood-gas coefficient. A high blood-gas coefficient indicates that blood can hold a large amount of the agent and predicts that it will take a longer time to raise the alveolar partial pressure. The more soluble in blood, the faster the induction and recovery.

Inhalant anesthetics result in the following effects on body systems: 1) generalized CNS depression; 2) increased cerebral blood flow; 3) depression of alveolar respiration (especially in ruminants); 4) decreased cardiac output due to a decrease in myocardial contractility; 5) increased cardiac automaticity, with sensitization of the myocardium to the arrhythmogenic effects of catecholamines; 6) decreased blood flow to the kidneys and liver; 7) mild skeletal muscle relaxation; and 8) decreased ability to regulate body temperature, resulting in hypothermia.

Halothane

Halothane is a halogenated hydrocarbon (*see* **Part A**). About 80% of the drug is eliminated unchanged by the lungs, with the remainder metabolized by the hepatic microsomal enzyme system to trifluoroacetic acid, inorganic Cl^-, and Br^-. These metabolites undergo renal excretion. Halothane causes a large increase in cerebral blood flow and thus should not be used in patients with increased intracranial pressure. Halothane causes a dose-related depression of the cardiovascular system and sensitizes the myocardium to catecholamine induced premature ventricular contractions. This agent is a moderate respiratory depressant, although it does have some bronchodilatory action. Halothane has been associated with an idiosyncratic acute hepatocellular necrosis. Since hepatotoxicity only occurs with secondary exposures, it is believed to be immunological in origin. Halothane sensitizes swine to the development of malignant hyperthermia. It will potentiate the muscle relaxation induced by nondepolarizing neuromuscular blockers.

Methoxyflurane

Methoxyflurane, a halogenated ether (*see* **Part A**), is very lipid soluble and therefore one of the most potent inhalant anesthetics. Its hydrophobicity is responsible for slow induction and recovery times. About 50% of inhaled methoxyflurane undergoes hepatic biotransformation to Fl^-, dichloroacetic acid, and oxalic acid, which are excreted in the urine. Methoxyflurane's pharmacological effects are similar to halothane's. Idiosyncratic toxic reactions leading to acute hepatic necrosis and acute renal failure have been reported in dogs.

Isoflurane

Isoflurane is also a halogenated ether (*see* **Part A**). Due to relatively low lipid solubility, isoflurane induction and recovery times are rapid. This allows rapid modulation of changes in a patient's level of anesthesia. Isoflurane undergoes little biotransformation and thus there are no reports of liver or renal toxicity in the veterinary literature. Isoflurane causes less cerebral vasodilation and cardiac depression than halothane does, and does not appreciably sensitize the myocardium to catecholamines. It produces the same degree of respiratory depression and mild bronchodilation as halothane. It significantly enhances the effects of nondepolarizing neuromuscular blocking drugs. Isoflurane is the anesthetic agent of choice in critically ill patients and patients with suspected increased intracranial pressure.

Other Inhalant Anesthetics

Enflurane is a chemical isomer of isoflurane that is infrequently used in veterinary medicine (*see* **Part B**). It offers little advantage over isoflurane and has been associated with skeletal muscle twitching and tonic-clonic seizures and rare instances of hepatotoxicity in humans. Desflurane is the newest inhalation anesthetic used in humans. Its chemical structure and pharmacological properties are similar to those of isoflurane (*see* **Part B**). Due to a very high vapor pressure, a special vaporizer is necessary to administer desflurane. Sevoflurane is a polyfluorinated methyl isopropyl ether that has received limited use in veterinary patients in Europe. Its pharmacological properties are similar to those of isoflurane, but it undergoes considerable biotransformation.

Nitrous Oxide

Nitrous oxide is a nonflammable organic gas at room temperature (*see* **Part A**). It has extremely low lipid solubility, making induction and recovery times very quick. The cardiopulmonary and respiratory effects of nitrous oxide are minimal (*see* **Part B**). Although it depresses myocardial contractility, its direct stimulation of the sympathetic nervous system overcomes this depressant action. It is metabolized by intestinal anaerobic bacteria to molecular nitrogen and free radicals. Since its potency as an anesthetic agent is limited, it is often used in combination with another inhalant or injectable agent. Nitrous oxide has no effect on intracranial pressure.

Two unique pharmacokinetic effects known as the second gas effect and diffusional hypoxia are important when considering the pharmacology of inhalation anesthesia with nitrous oxide. During induction with high concentrations of nitrous oxide, the rate of uptake of the gas by the blood and tissue is very high. As this volume of gas disappears from the lung, additional gases must flow into the airway. This second gas effect speeds the onset of action of other inhalant anesthetics. The reverse of the second gas effects happens when nitrous oxide administration is terminated. During anesthesia with nitrous oxide, elimination of this gas from the lung occurs at as great a rate as the uptake. When administration of nitrous is stopped, elimination proceeds that dilutes the alveoli and reduces alveolar oxygen concentration. This effect is called diffusional hypoxia. It can be minimized by oxygenating patients for at least 5 minutes after the nitrous is discontinued.

26 Injectable Anesthetics I

A Mechanism of Action

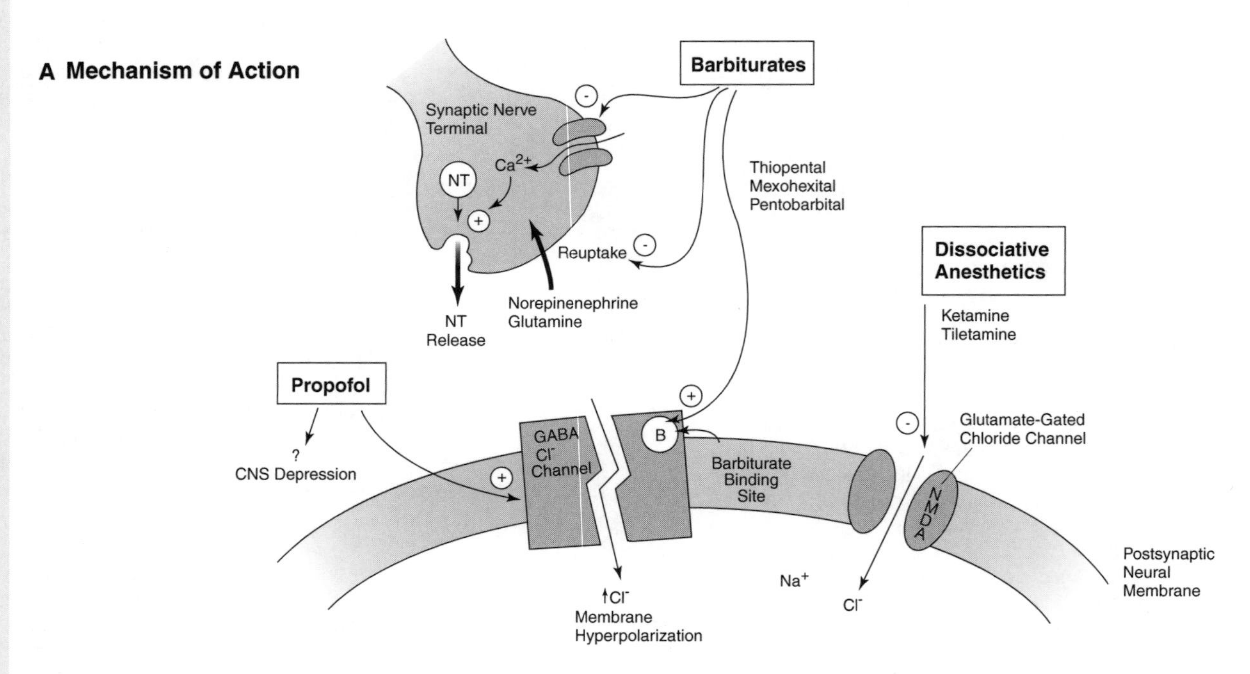

NT = Neurotransmitter
GABA= γ-aminobutyric acid

B Pharmacology of Injectable Anesthetics

Drug	Administration	Induction/Duration of Surgical Anesthesia (min)	Analgesia	Muscle Relaxation	Cardio-vascular Depression	Respiratory Depression	CNS Effects: Induction	CNS Effects: Seizures	CNS Effects: ICP*	Miscellaneous
Barbiturates	IV	Mexohexital: 0.5/30 Thiopental: 1-2/60 Pentobarbital: 3/120	No	Moderate	Mild	Moderate to Severe	Excitement Apnea†	↓	↓↓↓	
Propofol	IV	0.5-1/2-10	No	Mild	Mild → Moderate	Mild → Moderate	Myoclonus Apnea†	↓	↓	May cause histamine release
Etomidate	IV	0.25-0.5/5-10	No	No	Mild	Mild	Myoclonus Excitement	↑	↓	Inhibits steroidgenesis
Dissociative Anesthetics	IV IM	1-2/10-15 10-20/60	Yes	No ↑Tone	Stimulation	Mild	–	↑	↑	Reflexes remain intact Induces profuse salivation

*Intracranial pressure.
†Apnea at induction avoided by slow administration.

Injectable anesthetics permit rapid, easy induction of anesthesia without the need for special equipment. Although it is harder to control the depth of anesthesia with most injectable agents, the recent introduction of agents with very short durations of action has made it easier to control depth by controlling the rate of administration. Recovery is usually rapid and without incidence. Many injectable anesthetics are controlled substances (barbiturates, ketamine) that require special record keeping. Injectable anesthetics undergo extensive metabolism which increases the possibility of an adverse drug reaction.

Barbiturates

Barbiturates are derivatives of barbituric acid. Substitution of the hydrogen at the C5 position with an alkyl or aryl group confers CNS depressant activity on barbituric acid. When the oxygen on C2 is replaced with a sulfur, the thiobarbiturates are created. The thiobarbiturates have increased potency and a shorter duration of action than the oxybarbiturates. The barbiturates, which are weak acids, are formulated as sodium salts, which when dissolved in water form very alkaline solutions. The solutions are quite unstable and will decompose on exposure to light, air, or heat and will cause tissue irritation upon extravasation.

The barbiturates cause a dose-dependent depression in CNS function that progresses from sedation, to hypnosis, to general anesthesia. Although not precisely understood, two mechanisms of action have been proposed (**Part A**). First, they inhibit the release of several neurotransmitters, including norepinephrine and glutamate, an effect that may be modulated by inhibition of calcium reuptake by neuronal tissue. Second, they facilitate inhibitory gabaminergic (GABA) neurotransmission by acting as direct agonists at the GABA receptor and indirectly by decreasing the dissociation of GABA from its receptor.

Barbiturate distribution depends largely on lipid solubility. The thiobarbiturates, which are more lipid soluble than the oxybarbiturates, penetrate into the CNS quickly and induce anesthesia rapidly. Due to their high lipid solubility, they redistribute quickly to highly perfused tissue such as muscle. This redistribution decreases brain concentrations and thus the level of anesthesia. Further redistribution to more poorly perfused adipose tissue results in recovery from anesthesia. Dogs with low body fat stores such as greyhounds or severely cachexic animals have limited redistribution to fat and thus have a prolonged anesthetic recovery. Repeated administration of thiobarbiturates may result in the accumulation of high concentrations in adipose tissue, and subsequent release from these stores once anesthetic administration is terminated may prolong recovery from anesthesia. Barbiturates have poor analgesic activity.

Oxybarbiturates are extensively metabolized in the liver by the hepatic CP450 system, and the metabolites are excreted in the urine. The thiobarbiturates undergo oxidation in hepatic and extrahepatic tissues. Owing to their extensive hepatic metabolism, barbiturates should be used cautiously in patients with impaired hepatic function. Neonates also metabolize barbiturates more slowly, due to immature hepatic biotransformation. The rate of barbiturate metabolism in different species varies, with cats requiring a long time to metabolize barbiturates and rodents and ruminants (especially sheep and goats) metabolizing them quickly. In all species, the barbiturates induce the hepatic CP450 system.

All barbiturates share some common effects (**Part B**). They can induce transient apnea upon initial injection due to a direct depressant effect on the respiratory center in the brain. Slow administration can minimize the incidence of apnea at induction. Barbiturates also cause a dose-dependent depression of respiration. Overdose of barbiturates causes death by depression of the respiratory center and this effect has been exploited for euthanasia. Barbiturates produce transient hypotension and a resultant increase in heart rate. Cardiac arrhythmias may occur with large doses. Barbiturates decrease the sensitivity of the motor end plate to ACh, causing mild to moderate skeletal muscle relaxation. Since barbiturates decrease cerebral blood flow, cerebral blood volume, and intracranial pressure, they are the preferred anesthetic in animals with intracranial disease.

Thiopental

Thiopental is an ultra-short-acting thiobarbiturate. Since it is highly lipid soluble and has a low degree of ionization, it is rapidly taken up by the brain and induction of anesthesia occurs within 1–2 min.

Methohexital

Methohexital is an oxybarbiturate with a methyl addition to one of the nitrogen atoms, which renders it more lipid soluble. Induction of anesthesia is rapid and lasts about 10 min. It is more likely than the thiobarbiturates to cause excitement during induction. Rapid injection minimizes the incidence of CNS excitement during induction. Methohexital can be used safely in greyhounds and other sight hounds, and perivascular injection does not cause tissue irritation.

Pentobarbital

Pentobarbital is a short-acting oxybarbiturate with a duration of action of 30–60 min. Pentobarbital is infrequently used as an anesthetic agent in clinical veterinary practice, but is still used for anesthesia in laboratory rodents. CNS excitement during induction and recovery may occur. In small animals, pentobarbital anesthesia may be used to control status epilepticus. Pentobarbital is also a common component of euthanasia solutions, as high doses lead to predictable paralysis of the respiratory and vasomotor centers in the brain.

Phenobarbital

Phenobarbital is a long-acting barbiturate that is not used for anesthesia, but is an effective anticonvulsant agent (*see* Chapter 33).

Injectable Anesthetics II

A Mechanism of Action

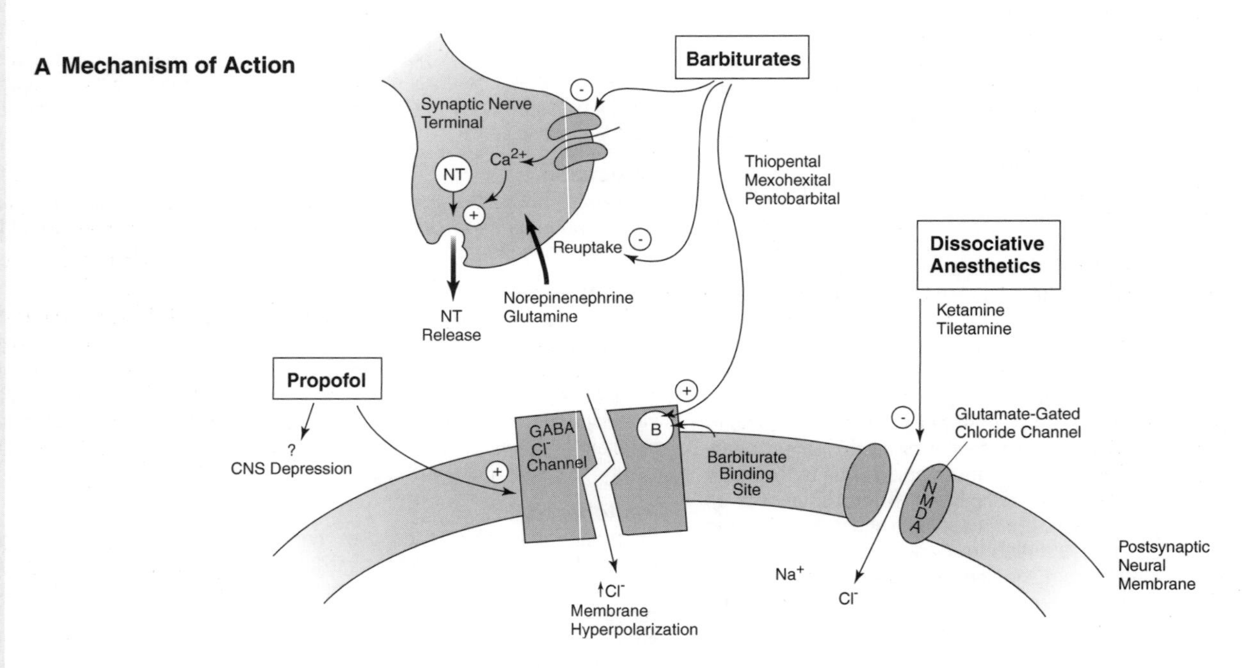

B Pharmacology of Injectable Anesthetics

Drug	Administration	Induction/Duration of Surgical Anesthesia (min)	Analgesia	Muscle Relaxation	Cardio-vascular Depression	Respiratory Depression	CNS Effects: Induction	CNS Effects: Seizures	CNS Effects: ICP*	Miscellaneous
Barbiturates	IV	Mexohexital: 0.5/30 Thiopental: 1-2/60 Pentobarbital: 3/120	No	Moderate	Mild	Moderate to Severe	Excitement Apnea†	↓	↓↓↓	
Propofol	IV	0.5-1/2-10	No	Mild	Mild → Moderate	Mild → Moderate	Myoclonus Apnea†	↓	↓	May cause histamine release
Etomidate	IV	0.25-0.5/5-10	No	No	Mild	Mild	Myoclonus Excitement	↑	↓	Inhibits steroidgenesis
Dissociative Anesthetics	IV IM	1-2/10-15 10-20/60	Yes	No ↑Tone	Stimulation	Mild	–	↑	↑	Reflexes remain intact Induces profuse salivation

*Intracranial pressure.
†Apnea at induction avoided by slow administration.

Propofol

Propofol is a novel alkyphenol anesthetic agent that acts by enhancing the effect of α-aminobutyric acid (GABA) in the CNS (**Part A**). Since it is poorly soluble in water, it is supplied as an aqueous emulsion containing 1% propofol, soybean oil, glycerol, egg lecithin, and sodium hydroxide. The formulation has no preservatives and supports bacterial growth. After IV administration, propofol is taken up by the brain within seconds, resulting in rapid smooth induction of anesthesia. The drug is then quickly redistributed to other tissues, resulting in a short duration of action (2–10 min) and a complete smooth recovery within 20–30 min. Propofol is highly protein bound. It undergoes hepatic metabolism to glucuronide and sulfate conjugates, which are excreted in the urine. Extrahepatic metabolism may also occur. Repeated administration does not result in accumulation in adipose tissue and thus does not prolong anesthetic recovery times.

Propofol causes minimal cardiovascular depression (**Part B**). It induces transient, mild to moderate hypotension and a mild decrease in myocardial contractility. It can also cause bradycardia, especially when used in combination with opioids. It is associated with mild hypoventilation, and short periods of apnea may occur on induction that can be avoided by injecting the drug slowly. Propofol decreases intracranial pressure. It appears to have a dose-related anticonvulsant action. Occasionally dogs may demonstrate myoclonic-like activity at induction that can be treated with benzodiazepines. Propofol may provoke histamine release. It has poor analgesic activity. Greyhounds tolerate propofol well but may have slightly prolonged recovery times.

Etomidate

Etomidate is a hypnotic nonbarbiturate anesthetic of ultra-short duration. It has a wide margin of safety and causes minimal cardiovascular and respiratory depression (*see* **Part B**). It decreases cerebral blood flow and oxygen consumption. Myoclonus and excitement may occur on induction or recovery, but can be minimized with premedication with benzodiazepines. It is available as a hyperosmolar solution in propylene glycol that can cause pain on injection. Etomidate also suppresses steroidogenesis.

Dissociative Anesthetics

Dissociative anesthetic agents used in veterinary medicine include ketamine and tiletamine. These drugs depress the thalamoneocortical system and activate the limbic system. They induce a CNS state of unconsciousness marked by maintenance of many reflexes (photic, pharyngeal, laryngeal, corneal, and pedal). The CNS effects are incompletely understood, but are partially mediated by noncompetitive antagonism of NMDA glutamate-gated chloride channels (*see* **Part A**). Glutamate is a major excitatory neurotransmitter in the CNS. These drugs provide good somatic analgesia but poor visceral analgesia (*see* **Part B**). They actually slightly increase muscle tone. Since they result in the appearance of epileptiform EEG patterns in the limbic system, they are generally not used in patients with seizure disorders. They stimulate salivation and increase tracheobronchial mucous gland secretions. The dissociative anesthetics depress the cardiovascular system, but concurrent stimulation of sympathetic activity compensates for this depression and the overall effect is to stimulate the cardiovascular system. Respiration is mildly depressed. These drugs increase cerebral blood flow and intracranial pressure. Since the eyes remain open under dissociative agents, they are susceptible to corneal ulceration if ocular lubricant is not applied.

Ketamine

Ketamine is approved for use in cats, although it is widely used in other species. Ketamine can be administered IV for fast-acting anesthesia or IM for a slower onset of action with a longer duration (*see* **Part B**). Ketamine is metabolized in the liver by *N*-demethylation and hydroxylation and excreted in the urine. Termination of anesthetic action is also partially the result of redistribution. Ketamine is usually used in combination with a tranquilizer or sedative such as xylazine, acepromazine, or a benzodiazepine to provide muscle relaxation and to decrease the likelihood of seizure activity. Repeated administration of ketamine will not increase the anesthetic intensity, but will prolong the duration of anesthesia. Pain may occur after IM injection or upon extravasation during IV injection. The pharmacological effect of ketamine can be shortened by the administration of yohimbine, a specific α_2 adrenergic receptor antagonist.

Tiletamine

Tiletamine is used in combination with zolazepam (a benzodiazepine) for anesthesia in dogs and cats. Its pharmacological actions are similar to those of ketamine, but it lasts 3 times as long. Recovery in dogs may be accompanied by muscle tremors, paddling, and whining due to more rapid metabolism of the zolazepam. Hyperthermia may result from the increased muscle activity.

28 Tranquilizers and Sedatives I

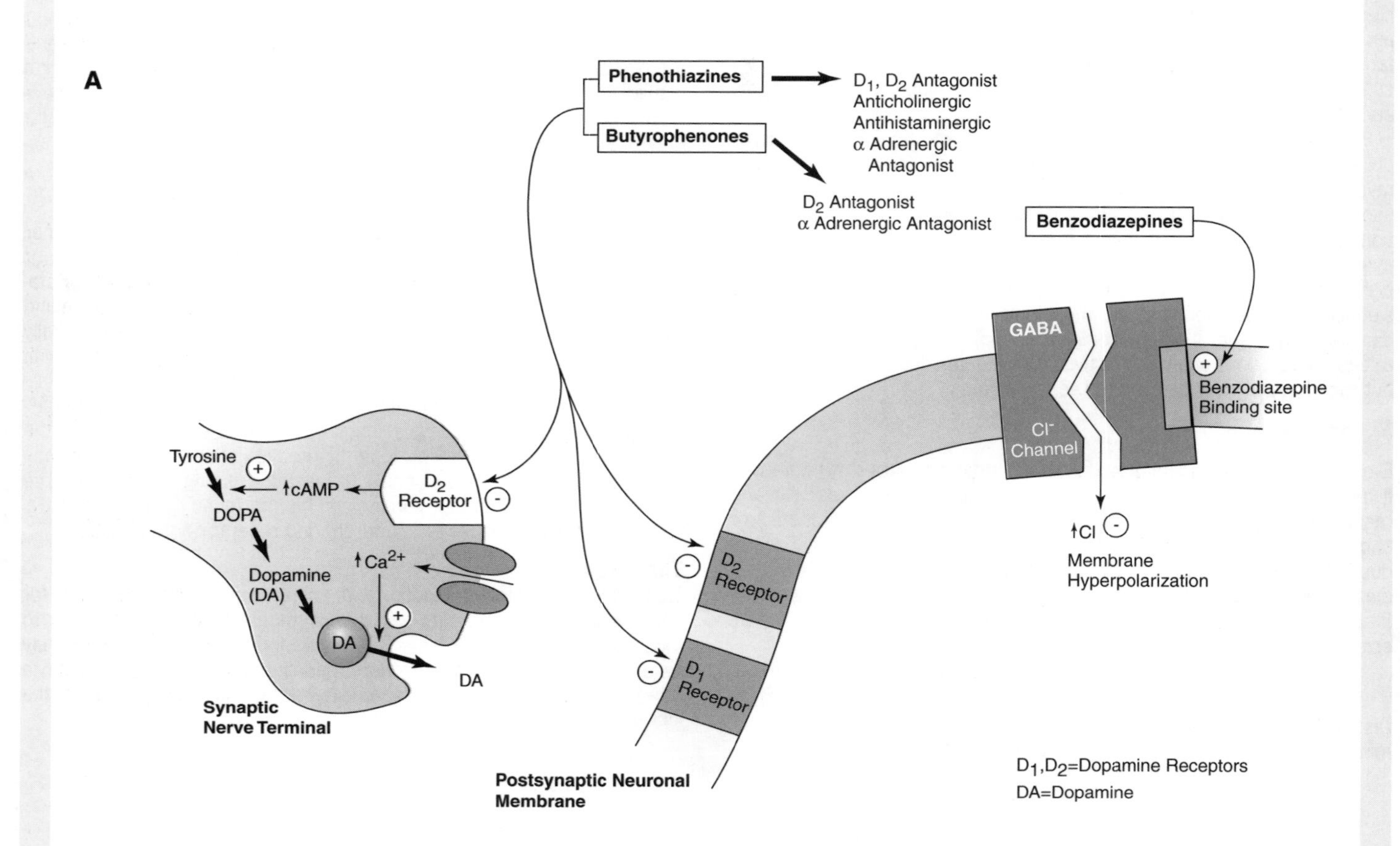

B Characteristics of Tranquilizers

Drug	Mechanism	Analgesia	Muscle Relaxation	Respiratory Depression	Cardiovascular Depression	Seizures	Miscellaneous
Phenothiazines	D_1, D_2 Antagonist	No	No	No	Hypotension Bradycardia SA arrest	↑	Antiemetic
Butyrophenones	D_2 Antagonist	No	No	No	Hypotension	↑	Antiemetic
Benzodiazepines	Bind to GABA Receptor	No	Mild	Mild	Mild	↓↓	Anxiolytic Appetite Stimulant Anterograde amnesia

Sedatives and tranquilizers are used in aggressive animals to facilitate physical examination and diagnostic procedures, to calm animals for frightful procedures such as airplane travel or shipping, and as preanesthetic agents to enable the use of smaller amounts of general anesthetic or to abrogate the adverse effects of some anesthetic agents.

Phenothiazines

Phenothiazine derivatives depress brain stem function by blocking dopamine receptors and increasing the turnover of dopamine (**Part A**). The net result is decreased central nervous system (CNS) arousal. In general, coordinated muscle responses are preserved, but spontaneous muscle activity is decreased. The phenothiazines also have varying degrees of anticholinergic, antihistaminergic, antispasmodic, and α adrenergic receptor antagonist action. One of the most important side effects of treatment with the phenothiazines is hypotension, which is due in part to blockage of α_1 adrenergic–mediated vasoconstriction (**Part B**). This effect can be particularly prominent when high sympathetic tone is present such as in hypovolemic or shocky patients. Systemic hypotension may lead to reflex sinus tachycardia. The phenothiazines may also cause a vagally mediated sinus bradycardia or heart block. They have minimal effects on respiration.

Since phenothiazines block the peripheral effects of catecholamines, they have an antiarrhythmic effect and decrease the incidence of catecholamine-associated arrhythmias seen with ultra-short-acting barbiturates and inhalation anesthetics. By virtue of their action on adrenergic and dopamine receptors in the CNS, phenothiazines can block CNS stimulation associated with catecholamine-like drugs (such as amphetamines) and the locomotor hyperactivity and stereotypical motor behavior evoked by the dopamine agonist activity of opiates. Hypothermia is caused by depletion of catecholamine stores in the hypothalamus. The phenothiazines decrease packed cell volume due to splenic sequestration of erythrocytes. They have no analgesic action. Phenothiazines may lower the seizure threshold and should not be used in animals with seizure disorders. They result in prolapse of the third eyelid in most species and penile prolapse in stallions. Occasionally animals may develop a type of "rage" syndrome after administration of phenothiazines. Since phenothiazines inhibit anticholinesterase, they may potentiate the toxicity of organophosphates and should be used with caution in animals with disorders of the neuromuscular junction.

Acepromazine is the most widely used phenothiazine in veterinary medicine. It can be given parenterally or orally. It is metabolized in the liver and the metabolites are excreted in the urine. Chlorpromazine and prochlorperazine are phenothiazine derivatives primarily used for their antiemetic properties (*see* Chapter 50).

Butyrophenone Derivatives

The pharmacological actions of butyrophenone derivatives include sedation, reduced muscle activity, and antiemesis. Their CNS actions are due to inhibition of dopamine and norepinephrine neurotransmission (*see* **Part A**). Like the phenothiazines, these agents also block peripheral α_1 adrenergic receptors, which may result in hypotension (*see* **Part B**). They cause minimal respiratory and cardiovascular depression and afford protection from epinephrine-induced cardiac arrhythmias. The butyrophenones are metabolized by the liver and excreted in the urine. The butyrophenone, droperidol, is combined with fentanyl for use as an anesthetic agent in small animals. It is a short-acting butyrophenone with a duration of action of about 2 hours. Droperidol has a wide margin of safety in the dog. Azaperone is a butyrophenone derivative that is used in swine for controlling aggressive behavior. It is given IM and has an onset of action of 20 min and duration of 2–4 hours. Transient salivation, piling, and shivering have been reported. The drug may also protect against halothane-induced malignant hyperthermia in pigs.

Benzodiazepines

The benzodiazepines are controlled substances that exert their pharmacological effects (sedation, muscle relaxation, anticonvulsants, anxiolytics, anterograde amnesia) by enhancing α-aminobutyric acid (GABA) neurotransmission in the CNS (*see* **Part A**). The GABA receptor is a Cl^- channel that opens upon activation, permitting Cl^- influx and membrane hyperpolarization. The result of membrane hyperpolarization is inhibition of neurotransmission. Benzodiazepines bind to the GABA receptor, facilitating the action of GABA.

Diazepam

Diazepam is a water-insoluble benzodiazepine. It is available for oral and parenteral administration. Oral absorption is good, with peak levels within 30 min to 2 hours after dosing. When given IV, diazepam has a rapid onset of action (0.5–2.0 min), but a short duration of action (1–2 hours). IM injections are absorbed more slowly than oral doses, due to the diazepam's poor water solubility.

Diazepam is biotransformed in the liver to active metabolites. The major active metabolite is nordiazepam (desmethyldiazepam), with smaller amounts of oxazepam. Both of these metabolites have the same potency as diazepam. These metabolites undergo hepatic conjugation with glucuronide and are excreted in the urine. Adverse effects after IV administration are associated with the propylene glycol vehicle and include venous thrombosis, transient hypotension, and arrhythmias. To avoid these side effects, the drug should be given slowly IV. Diazepam has minimal effects on the cardiovascular and respiratory systems. It may potentiate the respiratory depressant effect of other drugs such as the barbiturates and opioids. An idiosyncratic acute hepatotoxic reaction has been described in cats receiving oral diazepam.

Diazepam is used clinically as an anticonvulsant (*see* Chapter 33), a tranquilizer, and a muscle relaxant. It is also used in conjunction with opioids for neuroleptoanalgesia or with dissociative anesthetics or barbiturates for the induction of general anesthesia. Diazepam's ability to relax skeletal muscles is related to a depressant action on interneuronal reflexes in the spinal cord as well as an inhibitory effect on the presynaptic release of acetylcholine. In general, diazepam is a poor sedative in dogs and cats and may produce excitation in some animals. It does appear to be an adequate sedative in neonatal foals, sheep, and goats. Diazepam has been used in cats for its ability to modify behavior (inappropriate urination) and to stimulate appetite.

Midazolam

Midazolam is water soluble at low pH (<4.0) while at body pH it becomes more lipid soluble. Formulation in a water-soluble base greatly enhances midazolam's absorption from IM sites. Midazolam undergoes hepatic metabolism to inactive metabolites. It is 3–4 times as potent as diazepam, but shares the same pharmacological effects.

Zolazepam

Zolazepam is available in combination with tiletamine as an injectable anesthetic for dogs and cats. Little is known concerning its pharmacology in small animals.

Benzodiazepine Antagonists

Flumazenil is a benzodiazepine receptor antagonist that can be used to reverse the sedative effects of these drugs. It is given IV and is metabolized rapidly by the liver so that the duration of action is only about 1 hour in humans. It has been associated with induction of seizures in humans.

29 Tranquilizers and Sedatives II

Therapeutic Actions and Side Effects of Xylazine

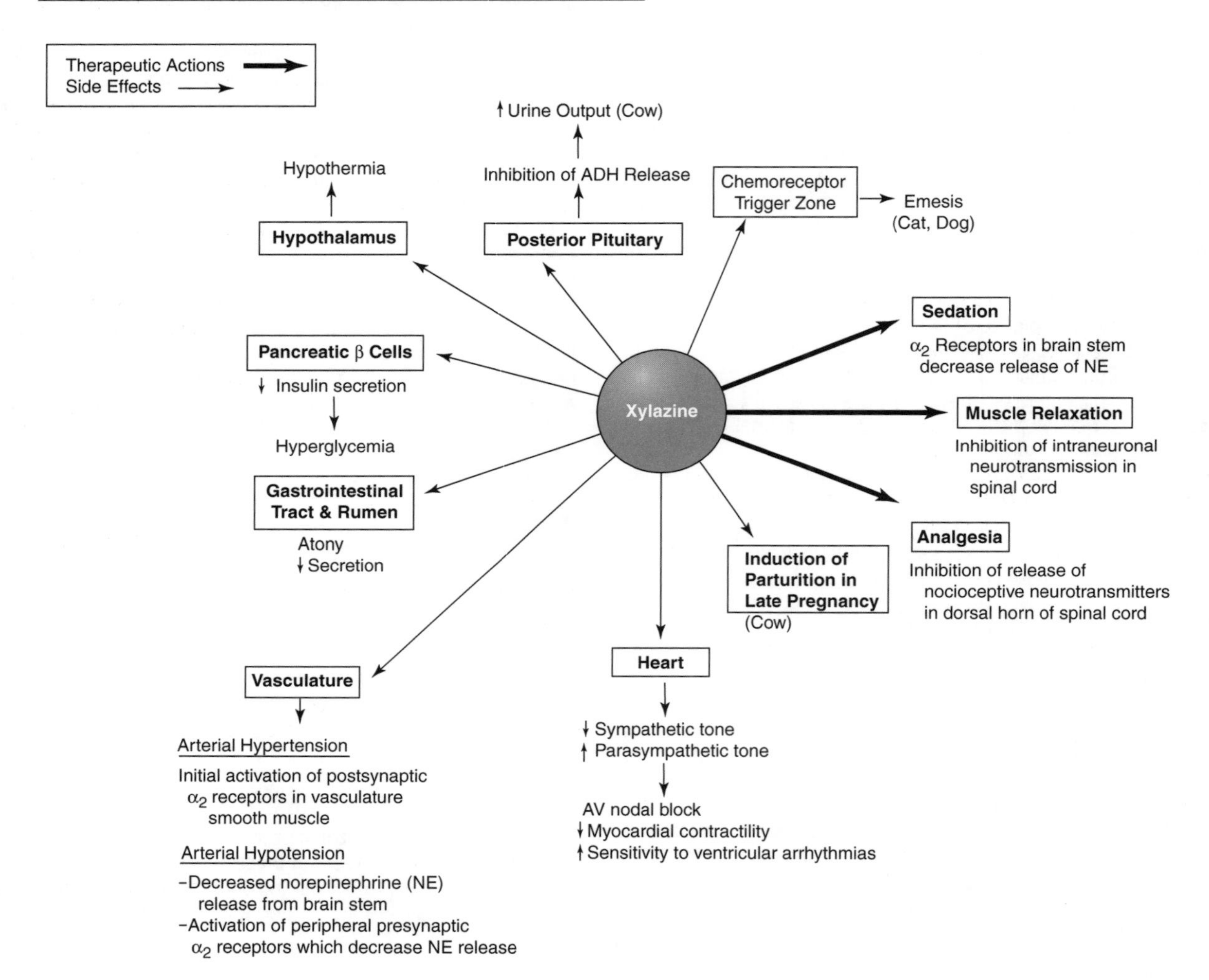

Xylazine

Xylazine is an α_2 adrenergic receptor agonist that results in dose-dependent sedation, analgesia, and muscle relaxation (**Figure**). Activation of α_2 adrenergic receptors in the brain stem and spinal cord decreases the release of neurotransmitters, particularly norepinephrine (NE), from nerve terminals. Stimulation of α_2 adrenergic receptors in the brain stem is responsible for sedation, while stimulation of spinal cord α_2 adrenergic receptors results in analgesia and muscle relaxation. The analgesic action of xylazine is quite potent, especially in horses. The peripheral effects of xylazine are related to blockage of α adrenergic receptors on vascular smooth muscle, or secondary to decreased sympathetic tone related to decreased central nervous system (CNS) release of NE. The cardiovascular effects of xylazine are variable (**Figure**). An initial period of hypertension is followed by hypotension. Initial stimulation of postsynaptic α_2 adrenergic receptors on vascular smooth muscle results in vasoconstriction and increased systemic blood pressure. After longer periods of time, decreases in central NE release result in activation of presynaptic α_2 adrenergic receptors and hypotension. Sinus bradycardia and varying degrees of first- and second-degree heart block may occur due to diminished sympathetic drive and enhanced parasympathetic tone. These conduction disturbances are typically responsive to anticholinergics. The α_2 adrenergic agonist also decreases cardiac contractility. The overall effect on the cardiovascular system is a decrease in cardiac output up to 30%. Xylazine may increase myocardial sensitivity to epinephrine-induced cardiac arrhythmias.

Xylazine has minimal effects on the respiratory system except in ruminants, where marked arterial hypoxemia has been reported secondary to the induction of ventilation-perfusion defects and airway constriction. Xylazine promotes gastrointestinal stasis and may promote gaseous distention of the stomach and rumen and the development of bloat. Xylazine depresses thermoregulatory responses and hypothermia or hyperthermia can result, depending on environmental conditions. Auditory stimuli may provoke arousal with kicking and avoidance responses.

Ruminants are particularly sensitive to the effects of xylazine and require one-tenth the dose required for horses. Hypersalivation and ruminal atony are common in cattle. It may also induce parturition in ruminants. In swine, xylazine produces only mild sedation and is not used. Although it causes penile protrusion in horses, no instances of penile paralysis have been reported.

Xylazine decreases insulin secretion from the pancreatic β cells in cats, dogs, horses, and cattle and may raise blood glucose levels. It inhibits antidiuretic hormone secretion from the pituitary and promotes the development of diabetes insipidus in ruminants and ponies, leading to an increase in urine output. Since xylazine maintains normal micturition reflexes in the dog, it is used when performing cystometry. Xylazine causes CNS excitement and seizures have been reported with overdoses. It stimulates the emetic center and induces predictable emesis in cats.

Xylazine undergoes extensive hepatic metabolism and is excreted in the urine. Its onset of action is within 15 min when given IM and within 3 min when given IV. The sedative effect lasts 1–2 hours, while the analgesic action lasts only 15–30 minutes.

Detomidine

Detomidine is more potent than xylazine, with a greater specificity for the α_2 adrenergic receptors. It has a longer duration of action, but otherwise its pharmacological effects are identical to those of xylazine. Detomidine doses in horses and cattle are the same, although the drug is only approved for use in the former species.

Medetomidine

Medetomidine is approximately 10 times more specific for α_2 adrenergic receptors than xylazine. In dogs it is used IV and IM for sedation and analgesia, with onsets of action of 5 and 15 min, respectively. The drug is poorly absorbed by the SC route. A consistent decrease in heart rate occurs in dogs. Bradycardia should not be reversed with anticholinergics since their use precipitates tachycardia and hypertension. The best treatment for bradycardia is to reverse medetomidine's action with an α_2 adrenergic antagonist. Since medetomidine may cause prolonged sedation or paradoxical excitement, it should not be used in very old or young dogs. Medetomidine causes a predictable respiratory depression in cats and therefore should only be used in healthy animals. Repeat dosing to obtain an adequate response is not recommended.

α_2 Adrenergic Antagonists

The pharmacological actions of α_2 adrenergic agonists can be reversed partially by using the α_2 receptor antagonists, yohimbine, atipamezole, or tolazoline. Yohimbine antagonizes the sedative and analgesic actions of xylazine. It is also capable of partially antagonizing the action of other CNS depressants such as the barbiturates and benzodiazepines possibly by interactions at their receptor binding sites. Yohimbine has also been used to reverse toxicity associated with the insecticidal agent amitraz. When yohimbine is administered IV, its onset of action is within 3 min, with a duration of action of 1–2 hours. It should be given slowly to avoid hypotension and tachycardia. The metabolic fate of yohimbine is unknown. Adverse effects include CNS excitement, muscle tremors, salivation, increased respiratory rate, and hyperemic mucous membranes. It should be used with caution in patients with seizure disorders.

Tolazoline is a nonselective α_1 and α_2 adrenergic antagonist. Although less potent than yohimbine, it has equal efficacy in reversing xylazine anesthesia. The combined action of α_1 and α_2 adrenergic blockade results in vascular smooth muscle relaxation, causing vasodilation and decreased peripheral resistance. Its onset of action is within 5 min, but it has a short duration of action. It is approved for reversal of xylazine in horses. Side effects in this species include tachycardia, peripheral vasodilation causing sweating and injected mucous membranes, hyperalgesia of the lips, piloerection, muscle fasciculations, and nasolacrimal discharge.

Atipamezole is labeled as a reversal agent for medetomidine in dogs. Atipamezole is metabolized in the liver, and the metabolites are excreted in the urine.

30 Opioids

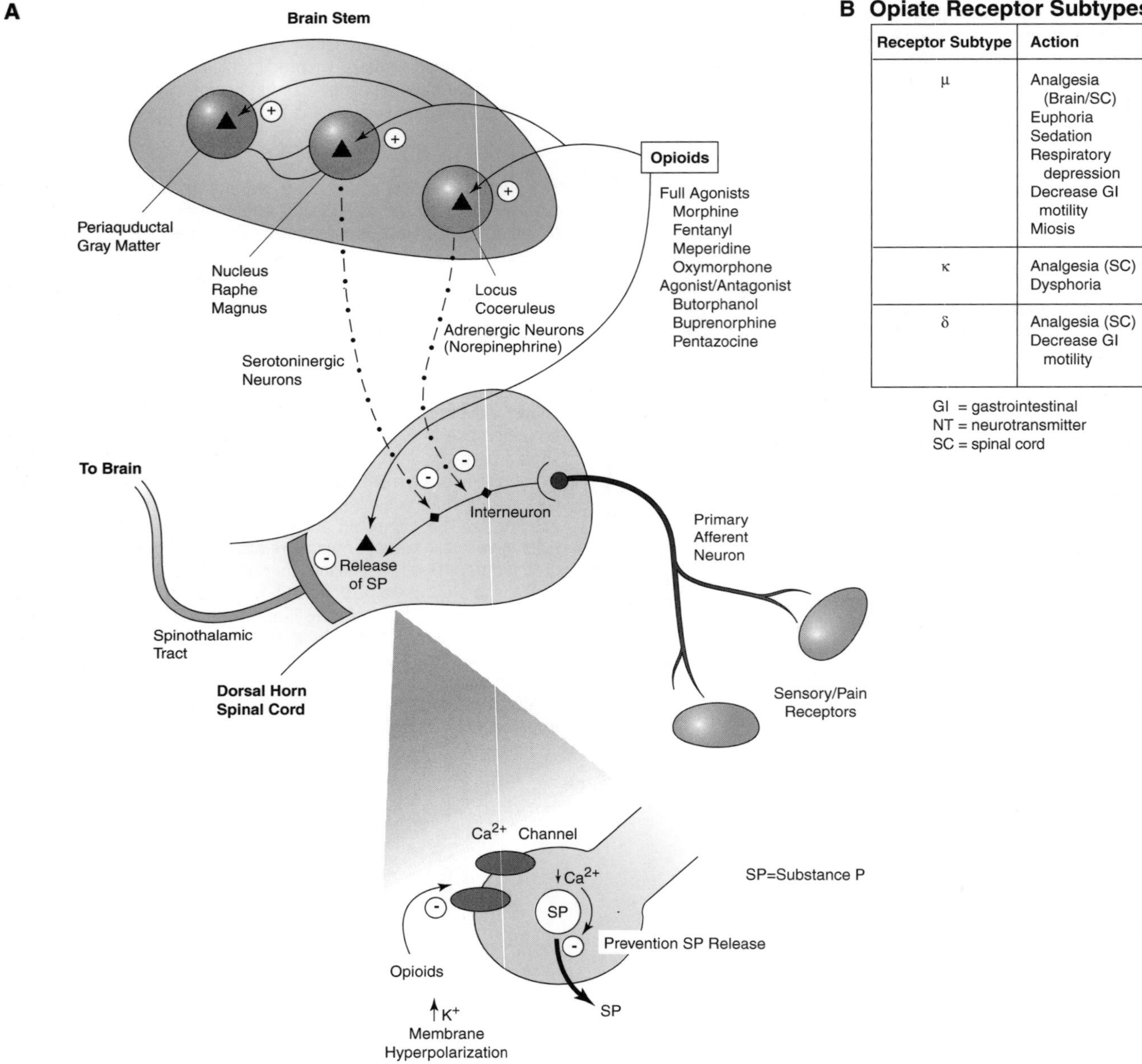

B Opiate Receptor Subtypes

Receptor Subtype	Action
μ	Analgesia (Brain/SC) Euphoria Sedation Respiratory depression Decrease GI motility Miosis
κ	Analgesia (SC) Dysphoria
δ	Analgesia (SC) Decrease GI motility

GI = gastrointestinal
NT = neurotransmitter
SC = spinal cord

Opioids are a diverse group of natural and synthetic drugs that are used in the control of acute pain. In the spinal cord they mediate analgesia by two mechanisms (**Part A**): they inhibit nociceptive neurotransmission from primary afferent neurons 1) by blocking release of substance P and 2) by inhibition of interneurons controlling nocioceptive output via the spinothalamic tract. Their central analgesic action is mediated by enhanced aminergic inhibition of spinal cord interneurons by bulbospinal pathways originating in the brain stem. Analgesia is mediated by the binding of opioids to μ receptors in the brain stem and to μ, κ, and δ receptors in the spinal cord (**Part B**).

The pharmacological effects of opioid receptor binding are mediated by interaction with K^+ and Ca^{2+} channels in the nerve cell membrane. They open inwardly rectifying K^+ channels, resulting in membrane hyperpolarization, or inhibit the opening of voltage-sensitive Ca^{2+} channels to prevent neurotransmitter release.

The full spectrum of pharmacological effects mediated by opioids includes analgesia, sedation, inhibition of gastrointestinal motility, antitussive action, emetic action, alterations in pupillary size, and respiratory depression (*see* **Part B**). Analgesia is mediated primarily by binding to μ receptors although some analgesia is seen with κ and δ receptor binding. Dose-dependent respiratory depression is due to a direct effect on the brain stem and is mediated by binding to μ receptors. There is decreased responsiveness to CO_2. The resultant hypercapnia causes cerebral vasodilation and increases cerebral blood flow. This latter effect limits the use of opioids in patients with evidence of increased cerebral pressure. In general, the opioids have minimal effects on the cardiovascular system. In dogs, opioids may cause bradycardia and hypersalivation. Some opioids induce the release of histamine, causing hypotension secondary to vasodilation and bronchospasm. The emetic action of some opioids is related to stimulation of dopamine receptors in the emetic center. The euphoric effect of opioids is primarily mediated by μ receptor binding and partially mediated by enhancement of dopaminergic neurotransmission. The dysphoric effect of opioids, primarily mediated by κ receptor binding, may be related to inhibition of dopaminergic neurotransmission. Inhibition of gastrointestinal motility and secretion, mediated by μ and δ receptors, has been exploited in the treatment of diarrhea. In cats, sheep, and horses, pupillary size increases with opioids while miosis occurs in dogs and humans.

Central nervous system (CNS) stimulation and excitement may occur upon administration of opioids. This is especially true in cats and horses. Coadministration of a tranquilizer alleviates this response. Convulsions may be seen with high doses of opioids. They may be related to inhibition of GABA neurotransmission in the CNS.

The opioids are primarily metabolized to inactive intermediates in the liver and excreted in the urine. Since most opioids are glucuronidated, excretion time in cats may be delayed. As opioids have low protein binding and a relatively high hepatic first pass, hepatic insufficiency has little effect on their actions. Pathological conditions or drugs that decrease hepatic blood flow (anesthetics, propranolol, congenital or acquired portosystemic shunts), however, may prolong their duration of action.

Drugs that bind at the opioid receptor may be agonists, partial agonists-antagonists, or antagonists at the different subtypes of receptors. All of the opioids are controlled substances.

Agonists

Morphine is a full μ agonist and the standard to which other opioids are compared. It undergoes biotransformation in the liver to a pharmacologically active 6-glucuronide metabolite. Species responses to morphine vary from mild to moderate sedation in dogs to CNS excitement and dysphoria in cats, horses, and ruminants. When given IV, morphine has a high tendency to cause histamine release in dogs.

Fentanyl, a full μ agonist, is a synthetic opioid. Its high lipid solubility accounts for its rapid onset of action and short duration (30 min). It is approximately 100 times more potent as an analgesic agent than morphine. A combination product with the butyrophenone tranquilizer, droperidol is available for IV anesthesia. Fentanyl transdermal patches are available to deliver a continuous dose for the treatment of chronic severe pain. Although limited pharmacokinetic studies have been conducted in dogs and cats, in humans plasma fentanyl concentrations vary from 27% to 99% of the rate of drug delivered. The onset of drug action with the patches is variable but typically is delayed several hours. In dogs, cats, and horses, the patches should be placed 12–24, 6–12, and 3–6 hours preoperatively, respectively. The duration of action is about 72 hours. Side effects include bradycardia, respiratory depression, and rashes at the site of patch application. Some patients may manifest dysphoria that may be controlled with acepromazine. The only approved method of patch disposal is to flush it down the toilet.

Oxymorphone, a full μ agonist, is a derivative of morphine and is about 10 times as potent as the parent drug. It is often used in combination with a tranquilizer or barbiturate for anesthesia. It has little tendency to cause vomiting or histamine release.

Meperidine is a synthetic μ agonist with one-tenth the potency of morphine. Like morphine it should only be administered IM to avoid the histamine release that accompanies IV administration. It has vagolytic and negative inotropic properties. Its short duration of action in small animals (1–2 hours) limits its use as an analgesic agent.

Opioid Agonists-Antagonists

Butorphanol is a synthetic agonist at κ receptors and a weak antagonist at μ receptors. Its analgesic actions are at the subcortical and spinal levels and thus it is better at modulating visceral than somatic analgesia. It is 4–7 times as potent an analgesic as morphine and is associated with only mild cardiopulmonary depression. Butorphanol is not associated with the release of histamine, but it does cause mild sedation. It is less likely to cause dysphoria in cats than other opioids. Butorphanol can be used to reverse the effects of μ agonists such as morphine and oxymorphone. It permits reversal of the sedative effects of these agents without sacrificing analgesia. Butorphanol also has significant antitussive action in dogs.

Buprenorphine is a synthetic opioid that is a partial agonist at μ receptors and an antagonist at κ receptors. It has a high affinity for μ receptors and dissociates slowly from them. These properties account for its slow onset of action and prolonged duration of analgesia (up to 8 hours). Buprenorphine can act as an opioid antagonist by displacing the binding of pure μ agonists. Its effects, however, cannot be antagonized by most μ antagonists due to its strong affinity for the μ receptor. Approximately 70% of the hepatic metabolites of buprenorphine are excreted in the bile.

Opioid Antagonists

Naloxone is a pure opioid antagonist that has a high affinity for μ receptors with a lesser affinity for κ and δ receptors. It has no analgesic activity. Given parenterally, it reverses the CNS sedative and respiratory depressant effects of μ agonists. After IV administration, its onset of action is very rapid (1–2 min) and duration of action is 1–3 hours. It is only minimally absorbed after oral administration, owing to rapid destruction in the gastrointestinal tract.

Naltrexone is an orally available opioid antagonist. It is biotransformed in the liver to the active metabolite 6-β-naltrexol, which is excreted in the urine. Naltrexone may be somewhat more effective in blocking the euphoric effects of opioids as opposed to blocking the respiratory depressant effects. Naltrexone has been associated with hepatotoxicity in humans. It is primarily used in veterinary medicine to treat behavioral disorders (*see* Chapter 35).

31 Nonsteroidal Anti-inflammatory Agents I: General Pharmacology

A Prostaglandins: Production and Clinical Actions

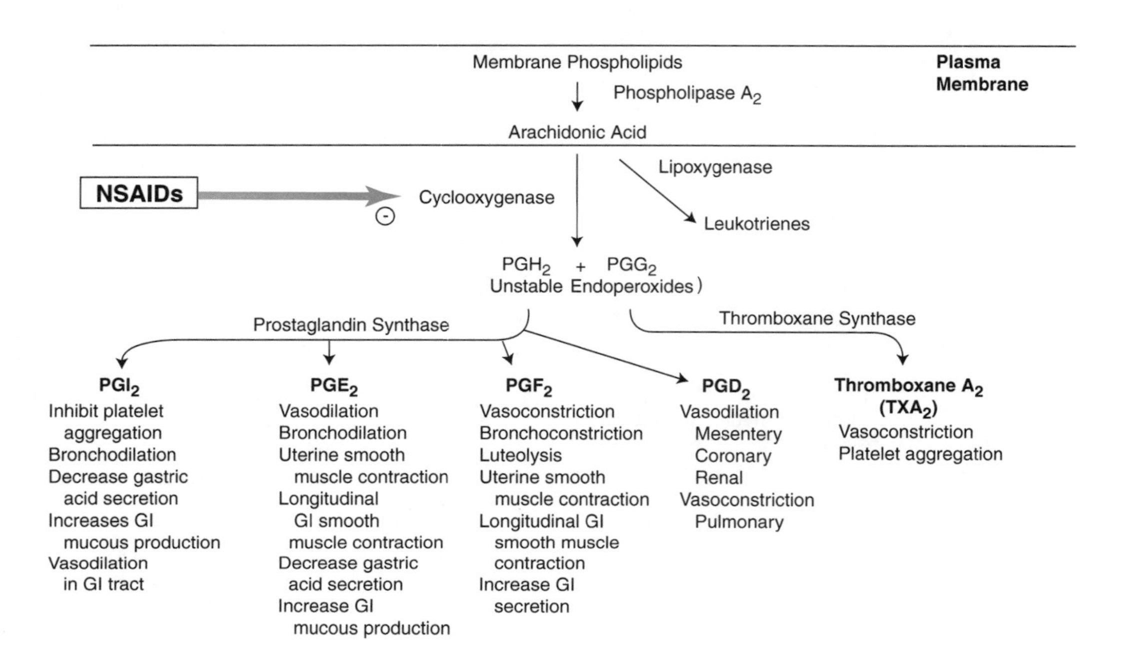

B NSAID Classification

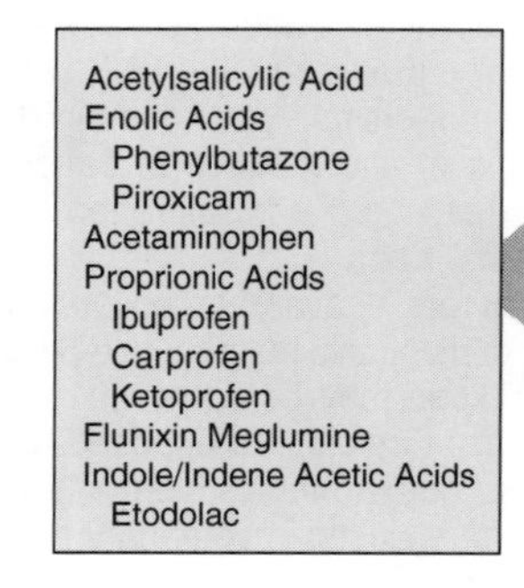

Shared pharmacokinetic properties

Weak acids
Highly protein bound
Hepatic biotransformation
Variable excretion

Shared side effects

Gastrointestinal ulceration and intolerance
Inhibition of platelet aggregation
Inhibition of uterine motility
Inhibition of prostaglandin-mediated renal function

General Pharmacology

Several chemical mediators released from cells during inflammation are responsible for the swelling, redness, edema, hyperthermia, and pain that accompany an inflammatory reaction. Prostaglandins and leukotrienes are 2 of the most important inflammatory mediators. The initial step in the formation of these inflammatory mediators is phospholipase A_2–catalyzed breakdown of the membrane phospholipid arachidonic acid (**Part A**). Arachidonic acid is subsequently acted on by cyclooxygenase or lipoxygenase to yield prostaglandins or leukotrienes, respectively. An important therapeutic strategy in the pharmacological management of inflammation is to control the release or production of prostaglandins or leukotrienes or to inhibit their biological action.

Therapeutic Effects

Nonsteroidal anti-inflammatory drugs (NSAIDs) are a group of related drugs that block the production of prostaglandins by inhibiting cyclooxygenase. Inhibition of prostaglandin synthesis results in anti-inflammatory, antipyretic, antithrombotic, and analgesic activity. Antithrombotic effects due to inhibition of platelet aggregation are seen at low doses; analgesia and antipyretic effects, at intermediate doses; and anti-inflammatory effects, at the highest doses.

Pharmacokinetics

The NSAIDs are weak acids that are generally well absorbed after oral administration (**Part B**). Most are highly protein bound and may displace other drugs from their plasma protein-binding sites. Elimination of NSAIDs is variable and species specific. Most undergo hepatic biotransformation with subsequent renal elimination. Others are secreted in bile and undergo enterohepatic cycling. Still others undergo tubular secretion and renal elimination. Some NSAIDs are able to penetrate and bind to inflammatory tissue better than others and this property may prolong their duration of action.

Adverse Effects

Neonates, geriatric patients, and patients with chronic hepatobiliary disease require smaller doses to obtain therapeutic effects. The most common side effect of NSAID use is gastrointestinal ulceration, which is associated with inhibition of gastrointestinal prostaglandin synthesis. Prostaglandin E_2 (PGE_2) and prostaglandin I_2 (PGI_2) are crucial in the maintenance of gastric mucosal integrity, as they maintain gastric mucosal blood and increase gastric bicarbonate and mucus production. Individual NSAIDs differ in their tendency to cause gastrointestinal ulceration. This variation may be associated with a prolonged duration of action secondary to enterohepatic cycling or with selective inhibition of the 2 cyclooxygenase isoenzymes, Cox-1 and Cox-2. Cox-1 is constitutively expressed in vascular cells, platelets, stomach, and kidneys and is important in maintaining gastrointestinal integrity, normal platelet function, and blood flow to several organs. Cox-2 is an inducible form of cyclooxygenase that is expressed by activated inflammatory cells. NSAIDs that selectively interfere with Cox-2 appear to be less ulcerogenic. Gastrointestinal toxicity of NSAIDs is greatly enhanced by concurrent corticosteroid therapy. This may be related to corticosteroid-induced delays in mucosal healing, increases in gastric acid production, or decreases in gastric mucus production. NSAIDs prolong bleeding times by blocking thromboxane A_2 production in platelets. Thromboxane A_2 is a potent inducer of platelet activation and aggregation. NSAIDs may complicate the management of patients with preexisting renal or hepatic disease, as local production of vasodilatory prostaglandins may be important in maintaining organ function, especially in disease states. NSAIDs also prolong gestation and labor because prostaglandins typically promote uterine motility.

Drug Interactions

Since NSAIDs are highly protein bound, they may displace other highly bound drugs. In humans, increased serum levels and/or duration of action has been noted with concurrent administration of oral anticoagulants, sulfonamides, and sulfonyl urea antidiabetic agents. Since the ability of furosemide and angiotensin converting enzyme inhibitors to promote natriuresis and renal blood flow is related to stimulation of renal prostaglandin protection, NSAIDs can blunt the therapeutic response to these drugs.

32 Nonsteroidal Anti-inflammatory Agents II: Therapeutic Agents

A Structure of Commonly Used NSAIDS

Aspirin

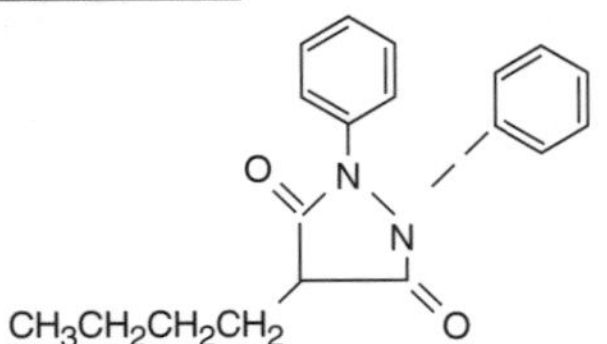

Phenylbutazone

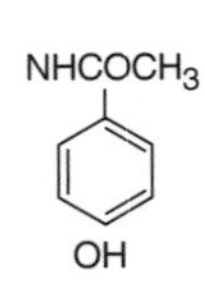

Acetominophen

Meclofenamate

Ibuprofen

Ketoprofen

Carprofen

Piroxicam

Etodolac

B Unique Pharmacologic Features of NSAIDS

Drug	Unique Features
Aspirin	Irreversible cyclooxygenase inhibitor Increased toxicity in cats
Phenylbutazone	Most popular NSAID in horses Induces hepatic CP450 enzymes Idiosyncratic bone marrow toxicity
Piroxicam	Unique adjunctive role in treating transitional cell carcinoma in dogs
Acetominophen	Does not inhibit platelet function Toxic in cats
Ibuprofen	Enhanced ulcerogenic potential in dogs
Naproxen	Enhanced ulcerogenic potential in dogs
Carprofen	Idiosyncratic hepatotoxin
Flunixin meglumine	Potent analgesic agent in the horse

Aspirin

Aspirin, or acetylsalicylic acid, is the prototypic salicylate (**Part A**). It is the only NSAID that irreversibly inhibits cyclooxygenase. In addition to inhibiting cyclooxygenase, salicylates also uncouple oxidative phosphorylation, inhibit the formation and release of kinins, and stabilize lysosomes. Aspirin is well absorbed after oral administration and is rapidly metabolized to the active product, salicylic acid. Salicylic acid is eliminated by glucuronide conjugation, with subsequent renal excretion by glomerular filtration or tubular secretion. Excretion is more rapid in alkaline urine and this fact has been exploited in the treatment of salicylate toxicity. Since cats have a relative deficiency in the hepatic enzyme for glucuronidation, they excrete aspirin much slower than other species ($t^{1/2}$ = 37 hours). In dogs, the $t^{1/2}$ is around 8 hours, but there is evidence of considerable individual variation. Aspirin causes dose-dependent gastric ulceration, although at low doses these lesions may be subclinical. Buffered aspirin, which increases gastric emptying and decreases the time of contact of aspirin with the gastric mucosa, decreases the incidence of gastric irritation. Idiosyncratic hepatic and bone marrow toxicities have been described.

Enolic Acids

Phenylbutazone is the most popular NSAID used in the horse. It is approved for use in the horse and dog and is available for IV or oral administration (**Part B**). It is metabolized in the liver to oxyphenbutazone, which is the active drug. Phenylbutazone has a prolonged duration of action with a long $t^{1/2}$, which may be related to its ability to penetrate inflamed tissue. Idiosyncratic adverse effects include bone marrow hypoplasia, hepatotoxicity, and renal papillary necrosis. Foals and ponies appear to be more sensitive to the ulcerogenic effects than do adult horses. Perivascular administration causes thrombophlebitis. Phenylbutazone induces the hepatic microsomal enzyme system. It can decrease the uptake of iodine by the thyroid gland. Since phenylbutazone promotes Na^+ and water retention, it should be used cautiously in patients with heart disease.

Piroxicam has been used in dogs for its anti-inflammatory activity and as an adjunctive therapy for transitional cell carcinoma of the urethra and bladder. The therapeutic window with this drug is limited, but it has been used safely in dogs at low doses once a day or every other day.

Para-aminophenol Derivatives

Acetaminophen is an effective analgesic and antipyretic agent, but has weak anti-inflammatory activity. It does not inhibit cyclooxygenase but does interfere with endoperoxidase intermediates formed during the metabolism of arachidonic acid. Acetaminophen does not cause gastric irritation or inhibit platelet aggregation. It is rarely used in veterinary medicine and is toxic to cats. In most species, acetaminophen is metabolized in the liver and excreted as a glucuronic acid derivative. Since cats lack the ability to glucuronidate acetaminophen, they metabolize the drug via the hepatic CP450 system, leading to generation of a reactive intermediate, *N*-acetyl-*p*-benzoquinonimine. This reactive molecule is normally inactivated by glutathione conjugation. Because cats typically have low levels of hepatic glutathione, they are unable to handle appreciative quantities of the toxic intermediate and develop oxidant erythrocyte and hepatic damage at low doses. In other species, overdoses of acetaminophen can result in acute hepatic failure.

Fenamates

Meclofenamic acid is a derivative of anthranilic acid, an analogue of salicylic acid. Its onset of action is slow (36–96 hours) and thus clinical efficacy requires 2–4 days. Side effects are limited to gastrointestinal toxicity with only a transient effect on platelet function.

Propionic Acid Derivatives

Since propionic acid derivatives are well tolerated by humans, there has been rapid development of many agents in this NSAID class. These agents include ibuprofen, naproxen, carprofen, and ketoprofen. All of these drugs have the potential to cause gastrointestinal ulceration in veterinary patients.

Ibuprofen is a popular analgesic agent in human medicine, but has a small margin of safety in veterinary patients. Dogs appear to be unusually sensitive to gastrointestinal toxicity of ibuprofen. Naproxen is approved for use in horses and has a wide margin of safety in this species. It undergoes extensive enterohepatic circulation in dogs, resulting in prolonged drug elimination ($t^{1/2}$ = 74 hours) and an increased incidence of gastrointestinal toxicity. Renal and hepatotoxic reactions have also been reported.

Carprofen is approved for use in dogs. Carprofen has some selectivity for the Cox-2 isoform of cyclooxygenase. This effect minimizes its pathological effects on the gastrointestinal tract. There may also be a centrally mediated, poorly understood action that is responsible for its marked analgesic effects. The drug is metabolized by the liver, with most of the metabolites excreted in the feces (80%–90%). Some enterohepatic cycling of the drug occurs. Since serious acute hepatotoxic reactions have been reported in dogs, hepatic function should be monitored prior to and during therapy with carprofen.

Ketoprofen is approved for use in horses and dogs. Besides inhibiting cyclooxygenase, it may also inhibit the formation of some proinflammatory leukotrienes and bradykinins. It may be less ulcerogenic in horses than flunixin meglumine.

Nicotinic Acid Derivatives

Flunixin meglumine is approved for use in horses and cattle. It is one of the most potent cyclooxygenase inhibitors. In horses its analgesic activity is compatible to that seen with opioids. After IV administration, its onset of action is rapid, with a relatively long duration of action (30 hours) due to sequestration in sites of inflammation. Gastrointestinal and renal side effects have been reported in dogs. The drug is better tolerated in horses. IM injections may cause local irritation. Accidental intra-arterial administration in horses can cause CNS stimulation, ataxia, and hyperventilation. The use of flunixin meglumine in cats is contraindicated.

Pyranocarboxylic Acid Derivatives

Etodolac was recently approved for use in dogs. It is a more potent inhibitor of Cox-2 than Cox-1, which makes it less likely to cause gastrointestinal toxicity. The drug is well absorbed after oral administration. It undergoes extensive enterohepatic circulation with a $t^{1/2}$ of 10–14 hours. It is largely metabolized in the liver and the metabolites are excreted in the feces. Side effects include gastrointestinal ulceration and hypoproteinemia.

33 Anticonvulsant Therapy

A Pharmacology of Anti-Seizure Medication

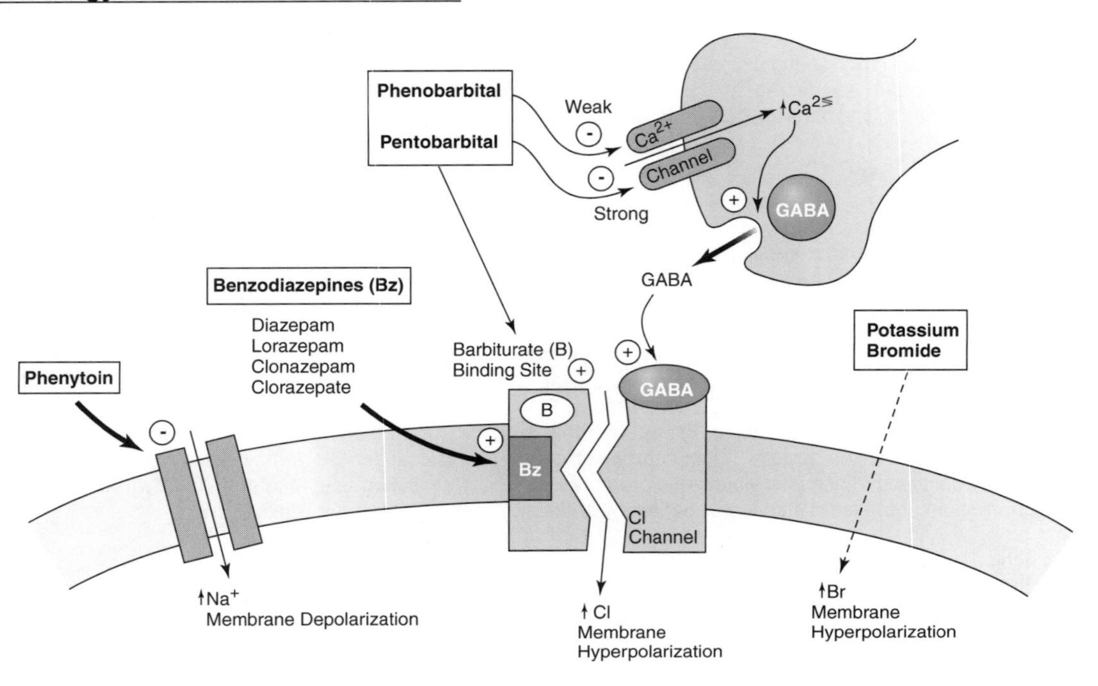

B Management of Seizures

Drug	Route Administration	Anticonvulsant Effect	Indication	Side Effects
Phenobarbital	IM Oral	20Ð30 min 2 weeks	Chronic management	Sedation, ataxia Polyuria/polydipsia/polyphagia Hepatoxicity (dogs) Increased hepatic microsomal enzyme activity
Pentobarbital	IV	Immediate	Refractory status epilepticus	Antiseizure dose associated with marked sedation/general anesthesia
Potassium Bromide	Oral	2-3 months (dogs) 1-2 months (cats)	Chronic management	Vomiting Sedation/ataxia Polyuria/polydipsia Pruritis
Benzodiazepines	IV Oral	Immediate Used to supplement other agents	Status epilepticus Chronic management	Short-acting (10 min) Sedation Tolerance Hepatotoxicity (cat)

Seizures occur when the normal balance between inhibitory and excitatory signals in the brain are shifted toward the latter. They typically start in a focus of hyperexcitable neurons. This excitable focus recruits other neurons in the area and propagates the seizure. Extracellular Ca^{2+} and K^+ are important in the recruitment and propagation of seizures.

Anticonvulsant medications block initiation or propagation of seizures. Many anticonvulsants work by potentiating GABA-mediated neurotransmission in the brain. The GABA receptor is a Cl^- channel, which when open, results in an influx of Cl^- leading to membrane hyperpolarization and suppression of neurotransmission. The GABA receptor has binding sites for 2 common classes of anticonvulsants: benzodiazepines and barbiturates (**Part A**). Benzodiazepine binding to the GABA receptor allosterically modifies the receptor, thereby increasing the number of Cl^- channels that open when GABA binds. Barbiturate binding to the GABA receptor prolongs the amount of time the Cl^- channel remains open after GABA binding. Anticonvulsants also work by inhibiting the action of excitatory neurotransmitters by interfering with Na^+ and Ca^{2+} influxes (*see* **Part A**).

Barbiturates

Phenobarbital increases the seizure threshold (blocks initiation) and prevents the spread of the discharge to surrounding neurons (propagation). These effects are mediated by 3 mechanisms (*see* **Part A**): 1) an enhancement of the inhibitory postsynaptic effect of GABA; 2) a reduction in glutamate-induced depolarization of excitatory neurons; and 3) blockage of the entry of Ca^{2+} into presynaptic neurons, which results in decreased neurotransmitter release into the synaptic cleft.

Phenobarbital is a highly lipophilic weak acid. It can be administered orally or parenterally (**Part B**). It is absorbed well after oral administration with a $t^1/_2$ in dogs of about 2 days. With oral dosing, about 2 weeks is required to obtain steady-state serum concentrations. Phenobarbital is not useful in controlling status epilepticus because its onset of action is delayed for as long as 20–30 min after IV administration. Loading doses can be given to increase serum levels more quickly. Phenobarbital undergoes hepatic metabolism and is primarily excreted as a glucuronide conjugate in the urine. Up to 25% of the drug can be excreted in the urine unchanged.

Phenobarbital is a potent inducer of hepatic microsomal enzymes (*see* **Part B**). This enzyme induction increases the rate at which phenobarbital itself is metabolized and may necessitate increases in doses with chronic therapy. Since phenobarbital increases the metabolism and biliary excretion of thyroid hormone and increases deiodination of T_4, evaluation of thyroid hormone status in animals on phenobarbital may suggest hypothyroidism when the animal is actually euthyroid. Phenobarbital also interferes with performance of the dexamethasone suppression test used to diagnose hyperadrenocorticism.

Since there is individual variability in phenobarbital absorption and metabolism, serum levels should be monitored to ensure they are within the therapeutic range (20–40 μg/mL). Side effects of phenobarbital therapy include dose-dependent sedation and ataxia. These effects may resolve within 1–2 weeks after starting therapy. Some animals become polydipsic, polyuric, or polyphagic. A serious side effect of chronic therapy is hepatotoxicity. This reaction appears to be more common in dogs on high doses of phenobarbital. The typical lesion is a chronic inflammatory hepatopathy that ultimately progresses to cirrhosis. Resolution of hepatic damage can occur if toxicity is detected early and the drug discontinued. Since phenobarbital induces the activity of the major enzymes used to monitor canine liver function, alanine aminotransferase and alkaline phosphatase, without causing morphological damage to hepatocytes, it can be difficult to detect changes in liver function. Periodic assessment of total serum bile acid levels and serum albumin and bilirubin levels can be used to monitor hepatic function. To avoid hepatotoxicity, the minimal dose of phenobarbital that affords seizure control should be used and combination therapy with anticonvulsants that undergo hepatic metabolism (primidone, phenytoin) avoided. Phenobarbital has also been reported to cause a reversible bone marrow dyscrasia marked by neutropenia or pancytopenia.

Pentobarbital is used to control refractory seizures associated with status epilepticus. It is an effective anticonvulsive agent in this setting, but anticonvulsant effects are only seen at doses that produced marked sedation (anesthesia) (*see* **Part B**). The barbiturates are controlled substances.

Primidone and Phenytoin

Primidone is metabolized in the liver to phenylethylmalonic acid and phenobarbital. Although both have anticonvulsant activity, the overall efficacy of primidone as an anticonvulsant agent is correlated with serum phenobarbital levels. The side effects of primidone therapy are similar to those of phenobarbital. Primidone may be a more reliable hepatotoxin. It is also associated with dermatitis and the development of megaloblastic anemia.

Phenytoin is a membrane-stabilizing agent that interferes with Na^+ and Ca^{2+} conductance across neuronal cell membranes. Phenytoin is a very good anticonvulsant in human patients, but its pharmacokinetics in dogs and cats limits its use in veterinary medicine.

Benzodiazepines

Benzodiazepines bind to the GABA channel and enhance the inhibitory action of GABA (*see* **Part A**). IV diazepam is the drug of choice in treating status epilepticus. Since diazepam is highly lipid soluble, it crosses the blood-brain barrier quickly. When given as an IV bolus, it stops seizure activity within seconds (*see* **Part B**). Because of its short duration of action, it may need to be given repeatedly or as a continuous-rate infusion if seizures reoccur. Absorption by the IM route is variable and slow; therefore, this route of administration is not recommended. Diazepam has some efficacy when administered rectally and this route of administration may be useful in managing cluster seizures in dogs at home. Following oral administration, diazepam undergoes extensive first-pass hepatic metabolism. The major metabolites, nordiazepam and oxazepam, are active as anticonvulsants, although less so (25% of the parent compound). Diazepam has limited use in the chronic management of seizures because tolerance to its anticonvulsant action develops rapidly (1–2 weeks). The most common side effect of diazepam therapy is sedation (*see* **Part B**). An idiosyncratic acute fulminant hepatotoxicity has been reported in cats receiving oral diazepam.

Potassium Bromide

Potassium bromide's anticonvulsant action is related to a decrease in neuronal cell excitability. Bromide ions compete for Cl^- transport across cell membranes, resulting in membrane hyperpolarization (*see* **Part A**). Bromide is available as potassium or sodium salt. Bromides are well absorbed after oral administration and are widely distributed. They are not bound to plasma proteins and are excreted in the urine. The $t^1/_2$ is 25 days in dogs and somewhat shorter (10 days) in cats. Steady-state serum concentrations in dogs and cats after oral administration thus take 2–3 months and $1^1/_2$ months, respectively (*see* **Part B**).

Adverse reactions to potassium bromide are dose-dependent sedation and ataxia. Polydipsia, polyphagia, and polyuria may occur. Other side effects include pruritic skin lesions, hyperactivity, and vomiting. Bromide does not increase the hepatotoxicity of phenobarbital, making it a useful adjunct to phenobarbital in animals with refractory seizures. It is also gaining popularity as a first-line drug for canine epilepsy. It is recommended that the drug initially be given at a loading dose due to its long $t^1/_2$. Its anticonvulsant activity is correlated with serum bromide level, with therapeutic serum concentrations reported as 1–3 mg/ml.

34 Behavioral Pharmacology I

A Pharmacology of Behavior Modifying Drugs

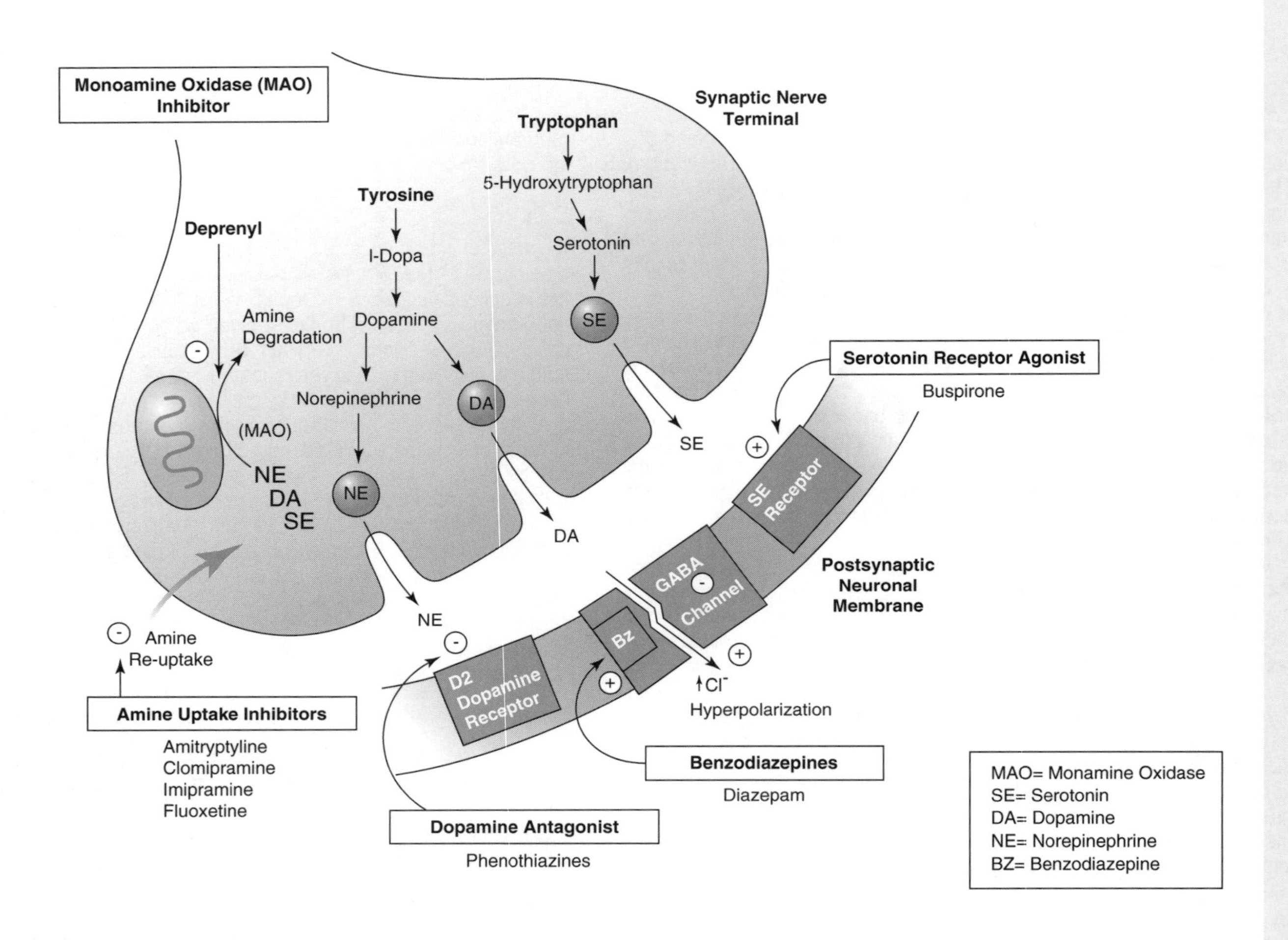

B Pharmacological Properties of Behavioral Drugs

Drug	Amine Effects	Anticholinergic	Peripheral Anti-α_1 Adrenergic	Antihistamine
Amitriptyline	↑NE = SE	+	+	+
Clomipramine	↑SE > NE	+	-	-
Imipramine	↑NE = SE	+	+	+
Fluoxetine	↑SE	-	-	-
Phenothiazine	↓DA	+	+	+
Deprenyl	↑DA ? NE ? SE	-	-	-

Behavioral problems in veterinary patients include aggression, fear, anxiety, compulsive behavior, elimination disorders, and alterations in behavior associated with the aging process. Diagnosis of behavioral disorders requires a comprehensive history, a thorough physical examination, and laboratory tests to rule out physiological causes of aberrant behavior. Treatment of behavioral problems requires environmental modifications. In many cases, the success of behavioral modification techniques is improved by concurrent pharmacological therapy aimed at controlling the biochemical basis of the disorder.

Deprenyl

Cognitive dysfunction disorder is a term used to describe geriatric behavioral changes including alterations in the sleep-wake cycle, decreased social interactions with the owner, disorientation, and impairment of normal housetraining. Alterations in CNS neurotransmission, similar to those seen in Alzheimer's dementia in human patients, may be involved in the pathophysiology. These neurotransmitter abnormalities include depletion of norepinephrine, dopamine, and serotonin.

Monoamine oxidase (MAO) inhibitors are drugs that block oxidative deamination of monoamines (dopamine, norepinephrine, epinephrine, and serotonin) (**Part A**). MAO is a mitochondrial enzyme in nerve terminals that is responsible for the degradation of these neurotransmitters. L-Deprenyl is a selective, irreversible inhibitor of MAO that increases brain dopamine levels and has been used to treat canine cognitive dysfunction. Dopamine is an important neurotransmitter in the prefrontal cortex, a portion of the brain that plays a significant role in cognition. Deprenyl may have neuroprotective properties by acting as a free radical scavenger. It is rapidly absorbed from the gastrointestinal tract after oral administration, but its bioavailability is low (10%). The drug is metabolized in the liver to desmethylselegiline, which is the active MAO inhibitor, and to amphetamine and methamphetamine, which are CNS stimulants. The role of the latter metabolites in the clinical response is unknown. The metabolites are excreted in the urine. Therapeutic responses to deprenyl require up to 30 days. It should not be used in combination with other MAO inhibitors, amitraz, ephedrine, tricyclic antidepressants, serotonin uptake inhibitors, or opioids. A 2-week period of drug withdrawal should be instituted prior to starting these medications. Deprenyl has also been used to control the behavioral abnormalities that accompany canine hyperadrenocorticism, as increasing brain dopamine levels inhibits the secretion of ACTH from the pituitary. Side effects of deprenyl therapy include vocalization, restlessness, pacing, gastrointestinal upset, and anorexia.

Serotonin Uptake Inhibitors

Serotonin plays an important role in facilitating an animal's interaction with its environment and in inhibiting ongoing behavior. Serotonin uptake inhibitors increase brain levels of serotonin and have been used to treat phobias, separation anxiety, aggression, and compulsive behaviors. Serotonin reuptake inhibitors (SRIs) block the neuronal presynaptic membrane amine transporter and thus inhibit synaptic reuptake of serotonin, which is necessary for neurotransmitter inactivation (*see* **Part A**). SRIs include the tricyclic antidepressants and the atypical antidepressant fluoxetine. These drugs have no effect on dopamine levels.

The tricyclic antidepressants, amitriptyline, imipramine, and clomipramine, potentiate the action of serotonin and norepinephrine in the brain and have variable degrees of anticholinergic, antihistaminergic, and antiadrenergic effects (**Part B**). They are incompletely and variably absorbed from the gastrointestinal tract and subject to significant hepatic first-pass metabolism. Typically it takes 5–7 days to establish steady-state serum concentrations and 5–6 weeks to see a therapeutic response to these medications. The tricyclic antidepressants are metabolized in the liver and excreted as glucuronide conjugates in the urine. Side effects include gastrointestinal disturbances, CNS excitement, seizures, and signs of peripheral anticholinergic action (dry mouth and eyes, constipation, tachycardia, urine retention, increased intraocular pressure). Serious cardiovascular toxicity may be manifested by hypotension (due to α_1 adrenergic receptor blockade), prolonged conduction times, and direct depression of the myocardium. In humans, agranulocytosis, rashes, and jaundice have been reported.

Amitriptyline blocks the reuptake of both serotonin and norepinephrine in the CNS and has significant central and peripheral anticholinergic, antihistaminergic, and antiadrenergic actions. It has been used to treat inappropriate urination and interstitial cystitis in cats. Beneficial actions include a tendency to cause urine retention and peripheral analgesic action that may be related to inhibition of norepinephrine uptake. In vitro amitriptyline blocks the release of histamine from mast cells and this action may have a role in mediating its anti-inflammatory–like effect in this disorder. Side effects in cats include somnolence, decreased grooming, and weight gain.

Clomipramine is a rather selective SRI that also has anticholinergic actions. The drug is highly protein bound and is metabolized to several intermediates, one of which, desmethylclomipramine, is active. Two-thirds of the metabolites are excreted in the urine and the rest in the feces. Clomipramine has been used to treat inappropriate urination and psychogenic alopecia in cats and is currently marketed as a treatment for separation anxiety in dogs.

Fluoxetine is a selective SRI that specifically blocks the reuptake of serotonin and has no effect on cholinergic, histaminergic, or adrenergic neurotransmission. The drug is well absorbed after oral administration but undergoes significant first-pass hepatic metabolism to an active metabolite, norfluoxetine. It is eventually excreted in the urine. Therapeutic responses to fluoxetine take 4–5 weeks to be established. After the drug is stopped, 1–2 weeks may be required until all the active drug is eliminated. Fluoxetine has been used to treat obsessive compulsive disorders, separation anxiety, and panic in dogs. Side effects in dogs are mild sedation and gastrointestinal disturbances. This drug has the potential to cause seizure-like activity and paradoxical restlessness and excitement. Fluoxetine should not be given concurrently with MAO inhibitors and should be used cautiously with tricyclic antidepressants because it decreases their metabolism.

Serotonin Receptor Agonists

Buspirone is a serotonin receptor agonist that is used to treat anxiety disorders (*see* **Part A**). Unlike the benzodiazepines, which are the other major class of drugs used to treat anxiety disorders, buspirone has a decreased potential for physical dependence, minimal sedative-hypnotic action, and no withdrawal effects. The drug is well absorbed after oral administration, but undergoes significant hepatic first-pass metabolism to several metabolites including 1-pyrimidinyl-piperazine, which is active. The metabolites are excreted in the urine. The anxiolytic action of the drug is maximal after 3–4 weeks of treatment. Side effects include restlessness, gastrointestinal disturbances, and mild sedation. Cats often become more affectionate and outgoing to people, but intraspecies aggression may occur. Paradoxical excitement occurs in some animals. Although buspirone should not be combined with MAO inhibitors, it can be used with benzodiazepines or tricyclic antidepressants. It has been used with some success to treat separation anxiety in dogs and inappropriate urination in cats.

35 Behavioral Pharmacology II

A Pharmacology of Behavior Modifying Drugs

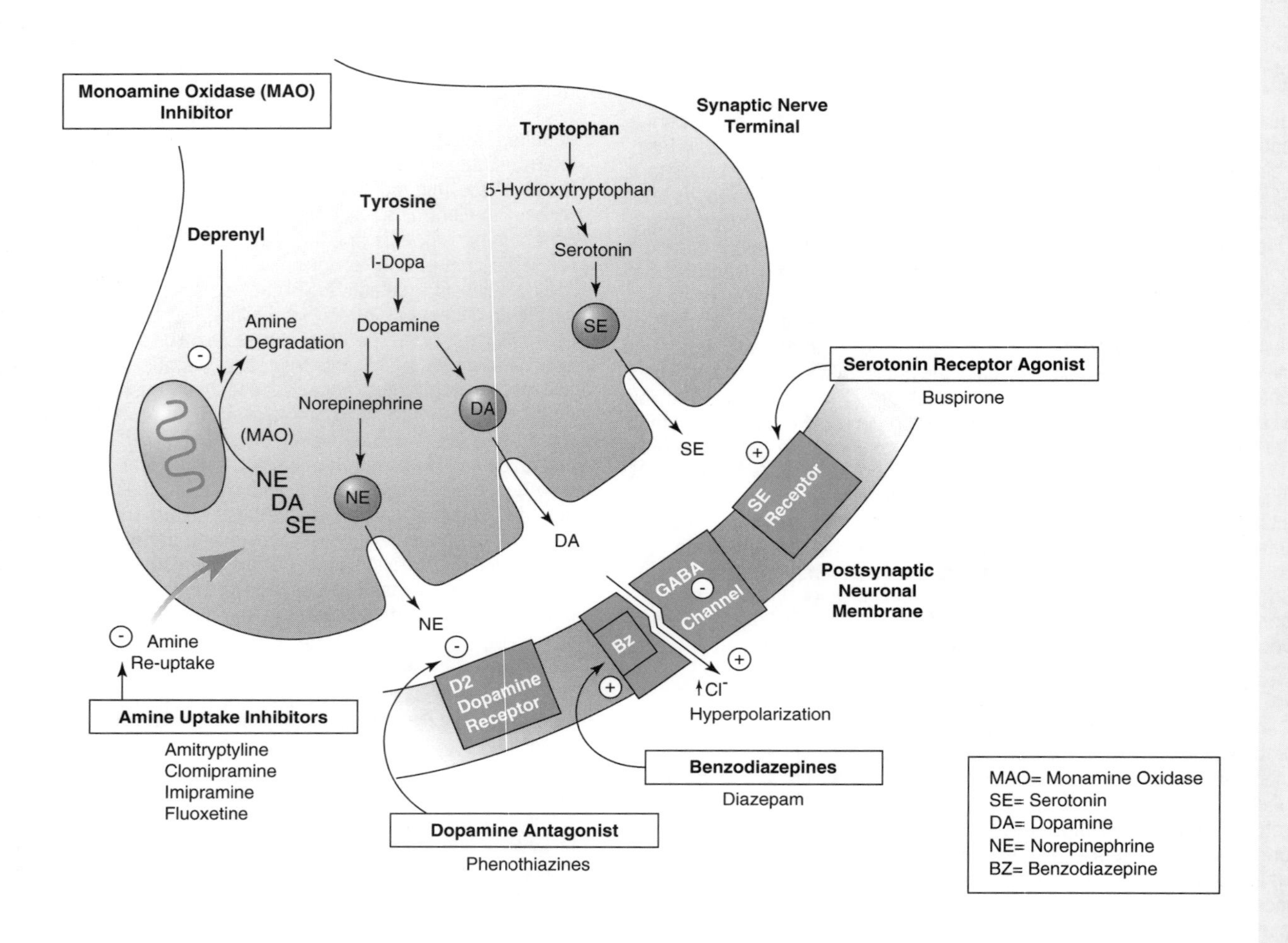

B Pharmacological Properties of Behavioral Drugs

Drug	Amine Effects	Anticholinergic	Peripheral Anti–α_1 Adrenergic	Antihistamine
Amitriptyline	↑ NE = SE	+	+	+
Clomipramine	↑ SE > NE	+	-	-
Imipramine	↑ NE = SE	+	+	+
Fluoxetine	↑ SE	-	-	-
Phenothiazine	↓ DA	+	+	+
Deprenyl	↑ DA ? NE ? SE	-	-	-

Dopamine Antagonists

Phenothiazines are relatively low-potency dopamine antagonists that have anticholinergic, antihistaminergic, antiadrenergic, and antiserotoninergic actions (**Parts A** and **B**). The major disadvantage of using these drugs for behavioral therapy is the occurrence of profound sedation and mild hypotension. Acepromazine has been used in the treatment of situational aggression, anxieties, or phobias. Idiosyncratic rage responses have been reported.

The butyrophenones are associated with less sedation and smaller decreases in blood pressure compared to the phenothiazines, but are associated with a higher incidence of extrapyramidal side effects. Azaperone is a butyrophenone used to control hierarchy fighting in pigs.

GABA Agonists

The γ-aminobutyric acid (GABA) receptor/chloride channel is involved in the modulation of aggressive and social behavior. Benzodiazepines bind to this receptor and enhance GABAergic inhibitory neurotransmission (*see* **Part A**). The benzodiazepines are used to treat anxiety disorders. The major benefit in using them to control anxiety is their rapid onset of action. Use of benzodiazepines can lead to physical dependence, sedation, hyperphagia (in cats), ataxia, increased interspecies aggression, and paradoxical excitement. Acute fulminating hepatotoxicity has been reported as an idiosyncratic reaction to diazepam in cats. Tolerance to the effects of benzodiazepines may develop with time. Abrupt withdrawal of therapy leads to a reaction marked by heightened anxiety or aggression. For this reason, benzodiazepine therapy should be discontinued gradually over 6–10 weeks. Alprazolam may be the benzodiazepine of choice in treating behavioral problems because it produces less sedation. Its shorter $t^{1/2}$, however, requires more frequent dosing and it carries a greater potential for causing withdrawal symptoms.

Opioid Antagonists

Naloxone and naltrexone are opioid antagonists. When administered parenterally these drugs have been used to control stereotypic behaviors in dogs and horses. When administered orally, these drugs have low bioavailability and short $t^{1/2}$s. They are metabolized in the liver and excreted in the urine and feces. Side effects include sedation and diarrhea.

Progestins

Progestins (megestrol acetate, medroxyprogesterone) have nonspecific calming effects. Due to the high incidence of side effects associated with their use, including induction of diabetes mellitus, cyclic endometrial hyperplasia/pyometra complex, and mammary hyperplasia or neoplasia, these drugs should be used only as a last resort for behavioral problems.

β Adrenergic Receptor Antagonists

Catecholamines lower the threshold for aggressive behavior and fear responses. This effect, which is likely mediated by β_2 adrenergic receptors, is associated with a peripheral action on muscle spindle function that attenuates proprioceptive impulses to the reticular activating systems, and a central effect related to an increase in serotonin levels in the brain. The β adrenergic blockers may be useful in the treatment of selected disorders associated with aggression or fear.

36 Antibiotics: General Principles of Use

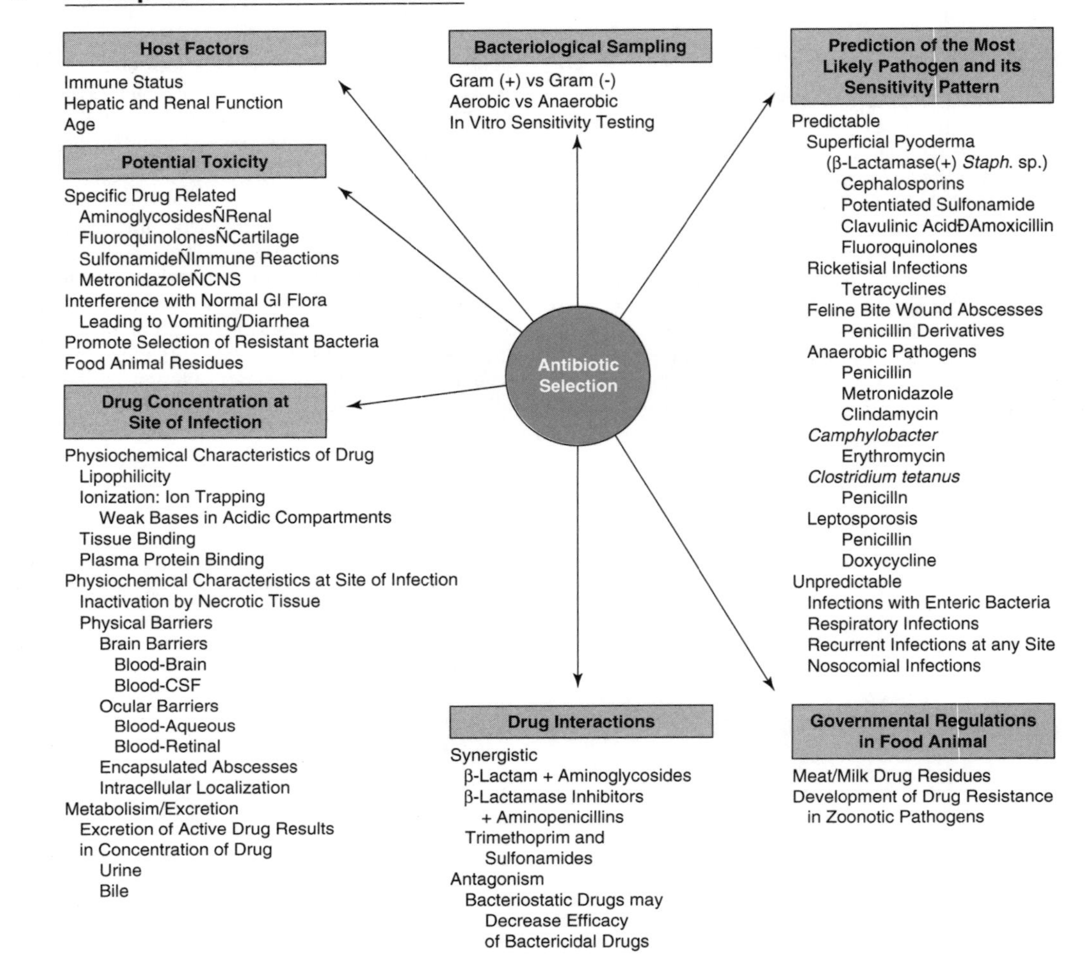

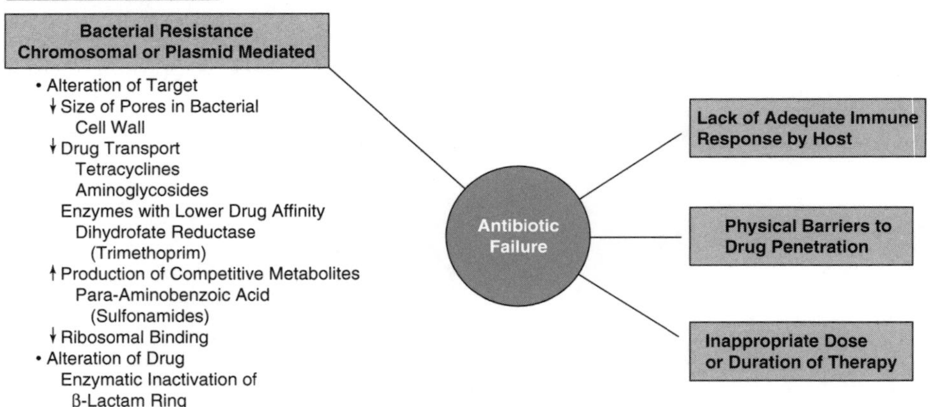

Bacterial Susceptibility to Antibiotics

Gram positive (+) and gram negative (−) bacteria have differential sensitivity to many antibiotics due to differences in cell wall composition. Gram (+) bacteria have a thick outer wall composed of a number of layers of peptidoglycan, while gram (−) bacteria have a lipophilic outer membrane that protects a thin peptidoglycan layer. Antibiotics that interfere with peptidoglycan synthesis more easily reach their site of action in gram (+). Gram (−) bacteria have pores in their outer membrane that allow the passage of small hydrophilic molecules. Some antibiotics, such as the glycopeptides, are excluded from gram (−) bacteria because they cannot pass through these pores. Some gram (−) bacteria, such as *Pseudomonas* sp., can genetically modulate the size of the pores to restrict the entry of antibiotics.

Antibiotic sensitivity differs between aerobic or anaerobic. Anaerobic organisms elaborate a variety of toxins and enzymes that can cause extensive tissue necrosis, limiting the penetration of antibiotics into the site of infection or inactivating them once they are there.

Some bacteria have highly predictable sensitivity patterns because they lack the ability to acquire resistance (e.g., *Pasteurella* and *Rickettsia* spp.) or they have stable resistance [β-lactamase (+) *Staphylococcus* sp.] (**Part A**). Most bacteria, however, have unpredictable sensitivity patterns. For these pathogens, it is important to obtain samples for bacteriological culture and in vitro antibiotic sensitivity testing. In vitro sensitivity testing generates a minimum inhibitory concentration (MIC), which is the concentration of antibiotic that prevents visible growth after incubation for 18 hours. Bacteria are considered susceptible to a particular antibiotic if the MIC is $1/2$ the average clinically obtainable plasma concentration of the drug or $1/4$ of the peak plasma concentration.

Antibiotic Failure

Even when antibiotic choice is based on in vitro susceptibility testing, drug therapy may fail (**Part B**). There may be physical barriers to antibiotic penetration at the site of infection. Of particular concern is the degree of penetration of the blood-brain and blood–cerebrospinal fluid barrier in the treatment of CNS infections. Antibiotics with high lipophilicity and a low degree of ionization (such as chloramphenicol, metronidazole, and clindamycin) penetrate the blood-brain barrier well. Other drugs (e.g., penicillins and cephalosporins) may enter the CNS if the integrity of the barrier is disrupted. Other important barrier sites are the blood-retinal and blood-aqueous barriers in the eye and the presence of poorly vascularized fibrous capsules around abscesses. Another reason for antibiotic failure is inactivation of the drug at the site of infection. Sulfonamides are inactive in the presence of pus, and aminoglycosides are not active in necrotic tissue. Antibiotic therapy may fail due to improper drug dose and/or improper duration of therapy. Antibiotics should be given at least 72 hours before treatment failure is suggested. Unsuccessful antibiotic therapy may also be due to an inadequate host immune response. This is particularly important if the antibiotic is bacteriostatic, in which case an intact immune system is necessary for the removal of remaining organisms.

Bacterial Antibiotic Resistance

By far the most important reason for unsuccessful antibiotic therapy is acquired bacterial drug resistance (*see* **Part B**). Resistance can develop by chromosomal mutation or by horizontal transfer of plasmids between bacteria. Chromosomal mutations are rare. The frequency of mutation is increased by antibiotic use and is correlated with the ratio of the MIC to the antibiotic concentration at the site of infection. Thus, low-level, inappropriate dosing of antibiotics favors the development of resistance. Underdosing also selects for resistant bacteria by killing susceptible bacteria and thus offering a survival advantage to resistant populations that are normally present in small numbers.

The exchange of plasmids is a common way for bacteria to acquire antibiotic drug resistance. Plasmids are circular pieces of extrachromosomal DNA that function independently of the bacterial chromosome. Bacterial plasmids are freely exchangeable among different species of bacteria. Plasmids coding for bacterial drug resistance, so-called *R-plasmids*, may encode resistance to a number of antibiotics.

Plasmid-mediated production of β-lactamases, which can inactivate the β lactam ring containing antibiotics is of major concern in bacterial drug resistance. Plasmids can contain genetic material that encodes for bacterial production of a number of β-lactamases. Gram (+) bacteria secrete these enzymes extracellularly where they are free to inactive antibiotics in the immediate environment. Gram (−) bacteria produce β-lactamases that are secreted and concentrated in the periplasmic space. Since this space is also the site of action of the β-lactam antibiotics, gram (−) β-lactamases are more effective in inactivating antibiotics. Gram (−) bacteria produce a greater variety of β-lactamases capable of destroying both penicillins and cephalosporins.

To avoid bacterial drug resistance, antibiotic agents should be chosen based on in vitro sensitivity testing and administered at therapeutic concentrations for the shortest period of time necessary to eradicate the infection. When a choice of antibiotic agents is available, the drug with the narrowest spectrum of action should be used.

The most important site of plasmid-mediated exchange in bacterial pathogens occurs in the intestine. The use of antibiotics in food animals and the resultant induction of antibiotic resistance to potential human pathogens such as *Salmonella* sp., *Campylobacter* sp., and *Escherichia coli* have led to the establishment of strict rules for the use of antibiotics in food animals.

Empiric Antibiotic Use

Although not ideal, in reality there are many times when antibiotic therapy needs to be started empirically. This occurs when acute life-threatening infections are present, when the site of infection cannot be identified, or when one is awaiting the results of in vitro sensitivity testing. In these cases, one should be confident that bacterial infection is present and attempts should be made to sample infected material and to perform a Gram stain. The selection of an antibiotic should then be based on knowledge of the most likely pathogen involved and its antibiotic susceptibility pattern (*see* **Part A**), knowledge of the factors controlling drug concentration at the site of infection, and knowledge of the drug's toxicity. Other factors affecting the decision are cost of treatment and in food animals, government regulations.

Risks of Antibiotic Use

Some antibiotics have direct dose-related toxic effects (*see* **Part A**). For instance, aminoglycosides cause a dose-dependent renal toxicity. Other drugs are idiosyncratic toxins and have unpredictable dose-independent effects. For instance, the sulfonamides may cause idiosyncratic allergic reactions leading to dry eye syndrome, hepatitis, or immune-mediated blood dyscrasias. Antibiotics may interfere with the normal bacterial flora and thus give a competitive advantage to pathogens. Due to the predominance of gram (+) bacteria in the gut of many small mammals, antibiotics with efficacy against gram (+) organisms destroy gut flora and lead to death. In other animals, antibiotic alterations on gut flora lead to the development of diarrhea. Superinfections with *Clostridium* organisms can cause fatal diarrhea in horses treated with some antibiotics. Inappropriate use of antibiotics also promotes the selection and proliferation of resistant strains of bacteria. In food animals, improper use of antibiotics can lead to the presence of antibiotic residues in the food chain, which can cause toxic reactions in humans.

37 Antibiotics: Drugs that Inhibit Cell Wall Synthesis I

A Structure of Bacterial Cell Wall

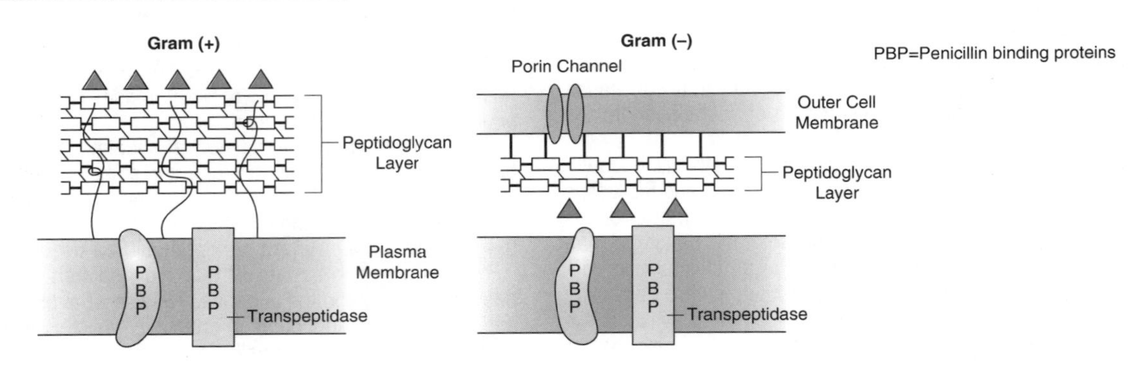

B Classes of β–Lactam Antibiotics

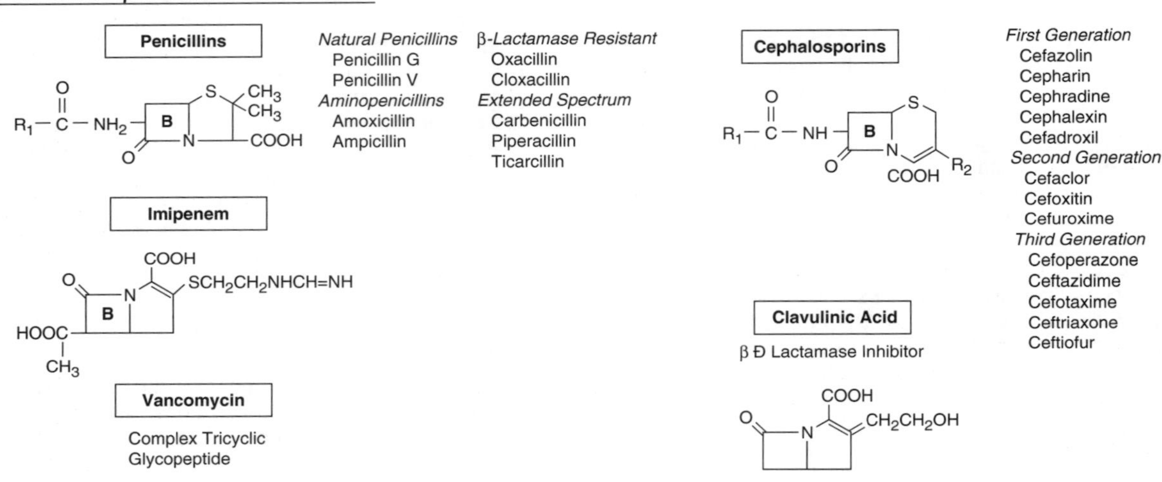

C Mechanism of Action: Disruption Cell Wall Synthesis

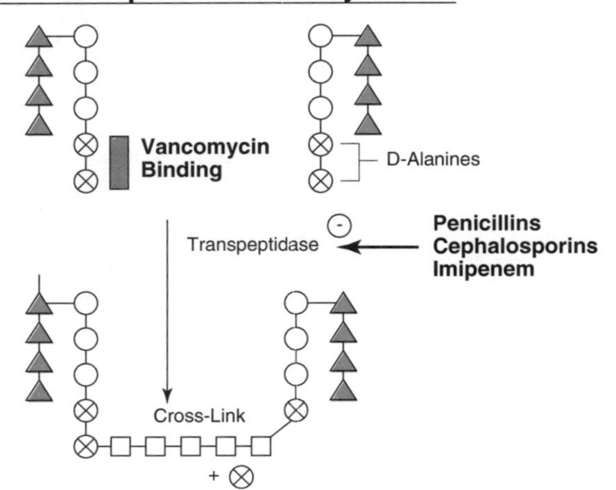

β-Lactam Antibiotics

The β-lactam antibiotics work by preventing bacterial cell wall synthesis. They inhibit transpeptidase, a penicillin-binding protein (PBP), that catalyzes the formation of the peptidoglycan polymers in the bacterial cell wall (**Parts A** and **C**). The β-lactam antibiotics also bind to additional PBPs in the bacterial cell membrane that control cell shape and division. Variation in the activity of different β-lactam antibiotics reflects the relative affinity they possess for these different PBPs. In general, gram (+) bacteria are more susceptible to β-lactam antibiotics, owing to the greater amount of peptidoglycan in their cell wall (*see* **Part A**). Gram (−) bacteria are less sensitive to the β-lactam antibiotics since their peptidoglycan wall is much thinner and is protected by an outer lipophilic cell membrane. The β-lactam antibiotics are bactericidal, but they only work on actively growing cells.

The development of resistance to β-lactams is associated with bacterial production of enzymes called β-lactamases, which catalyze the destruction of the β lactam ring. The β-lactam antibiotics are divided into 3 broad categories: the penicillins, the carbapenems, and the cephalosporins (**Part B**).

Penicillins

There are 4 classes of penicillins: 1) the natural penicillins, 2) the β-lactamase–resistant penicillins, 3) the aminopenicillins, and 4) the extended-spectrum penicillins (*see* **Part B**).

The natural penicillins, penicillin G and penicillin V, have a spectrum of action that includes some gram (+) aerobes [not β-lactamase (+) *Staphylococcus* sp.] and anaerobes. These penicillins are ionized organic acids and therefore do not cross biological membranes well. They are excreted unchanged in the urine by glomerular filtration and tubular secretion. Since they are found in very high concentrations in the urine, their ability to kill gram (−) pathogens in the urinary tract is enhanced. Sodium and potassium salts are available for IV administration. Penicillin G procaine is formulated for slow IM absorption, which results in a prolongation of drug action, although peak serum levels are less than with IV administration. A long-acting penicillin benzathine available for IM administration produces prolonged plasma concentrations (up to 48 hours), but these concentrations are so low that its use can only be recommended for highly sensitive organisms. Both penicillin G and V are available for oral administration, but the latter is preferred because of increased resistance to inactivation by gastric acid. The natural penicillins are considered very safe drugs although hypersensitivity reactions may occur.

The β-lactamase–resistant penicillins (cloxacillin, oxacillin) have enhanced resistance to gram (+) β-lactamases. Their spectrum of action is the same as for the natural penicillins except they are effective against β-lactamase (+) *Staphylococcus* sp. They are only available for oral administration and are generally reserved for treatment of pyodermas.

The aminopenicillins, amoxicillin and ampicillin, have improved activity against gram (−) pathogens, but are not effective against β-lactamase (+) bacteria. The aminopenicillins are available for oral and parenteral use. Amoxicillin is the preferred drug for oral administration since it has a better bioavailability. These drugs are highly disruptive to the gastrointestinal flora and common side effects include vomiting and diarrhea.

The extended-spectrum penicillins, ticarcillin, carbenicillin, and piperacillin, were developed with an expanded gram (−) spectrum including activity against *Pseudomonas* sp. They also have enhanced activity against anaerobes, in particular *Bacteroides* sp. They are only available for parenteral use. Carbenicillin indanyl is available for oral use, but its bioavailability is so low that it is recommended only for treating lower urinary tract infections.

Carbapenems

Carbapenems, such as imipenem, have high stability against the activity of most β-lactamases (*see* **Part B**). They are also capable of penetrating through porin channels in the bacterial cell membrane that usually exclude other drugs. Cilastin is administered along with imipenem to reduce renal metabolism of imipenem, but it has no inherent bactericidal action. Imipenem is effective against most bacteria except some highly resistant *Staphylococcus* sp. and enterococci. The drug must be given parenterally. Side effects include allergic reactions, vomiting, and seizures. The drug is usually reserved for the treatment of mixed infection with multidrug resistant bacteria.

Cephalosporins

The cephalosporins are more resistant to both gram (−) and gram (−) β-lactamases than are the penicillins. Most cephalosporins are excreted unchanged in the urine, although some agents are excreted in bile. Only a limited number of these agents are available for oral use. Most cephalosporins do not cross biological membranes well, unless inflammation is present. The main toxicities are gastrointestinal upset and allergic reactions.

The cephalosporins have been classified loosely based on their chronological discovery, spectrum of action, and chemical structure into 3 generations (*see* **Part B**).

The first-generation cephalosporins [cefazolin, cephapirin, cephradine (parenteral), and cephalexin and cefadroxil (oral)] were developed with increased resistance to gram (+), β-lactamase (+) bacteria. Their spectrum of action for gram (−) bacteria is similar to that of the aminopenicillins, but they are slightly worse with anaerobes.

The second-generation cephalosporins [cefaclor (oral), cefoxitin, and cefuroxime (parenteral)] are less active against gram (+) and more active against gram (−) organisms than are first-generation cephalosporins. Most second-generation cephalosporins are inactive against *Bacteroides* sp. because of the bacterial production of a cephalosporinase. Cefoxitin and cefotetan, which are resistant to this cephalosporinase, have an expanded anaerobic spectrum.

The third-generation cephalosporins [cefoperazone, ceftazidime, cefotaxime, ceftriazone, and ceftiofur (parenteral)] have an enhanced gram (−) spectrum, especially against enteric bacteria, with less activity against gram (+) bacteria. Cefoperazone and ceftazidime have activity against *Pseudomonas* sp., while cefotaxime retains the best gram (+) spectrum. Unlike other cephalosporins, both cefotaxime and ceftazidime obtain good concentrations in the CNS and are considered the drugs of choice in severe life-threatening gram (−) meningitis. The only orally available third-generation cephalosporin is cefixime, which was developed primarily for its activity against resistant *Streptococcus* sp.

Ceftiofur is the only cephalosporin approved for use in food animals. Unlike most third-generation cephalosporins, it has good activity against gram (+) organisms. Ceftiofur is also effective against most gram (−) pathogens, but not *Pseudomonas* sp. It has good activity against the common food animal respiratory pathogens. If used once a day in cows, it has no legal withdrawal time. The drug is approved for once-a-day parenteral dosing in dogs to treat urinary tract infections. Anemia and thrombocytopenia have been reported in dogs treated with doses 3–5 times the approved dose.

When one is interpreting sensitivity disk information, cross-sensitivity between first-generation cephalosporins can be assumed, but due to the variation in spectrum of action in the second- and third-generation cephalosporins, a sensitivity disk for each antibiotic should be tested before the drug is assumed to be effective.

38 Antibiotics: Drugs that Inhibit Cell Wall Synthesis II

A Structure of Bacterial Cell Wall

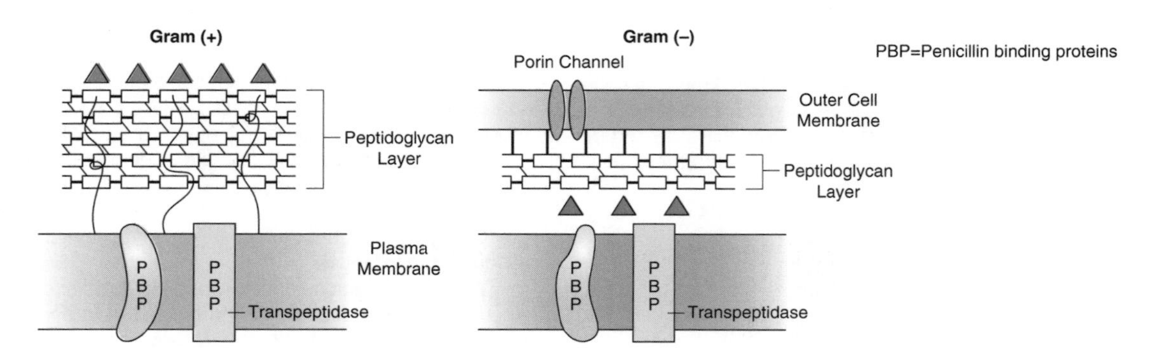

B Classes of β–Lactam Antibiotics

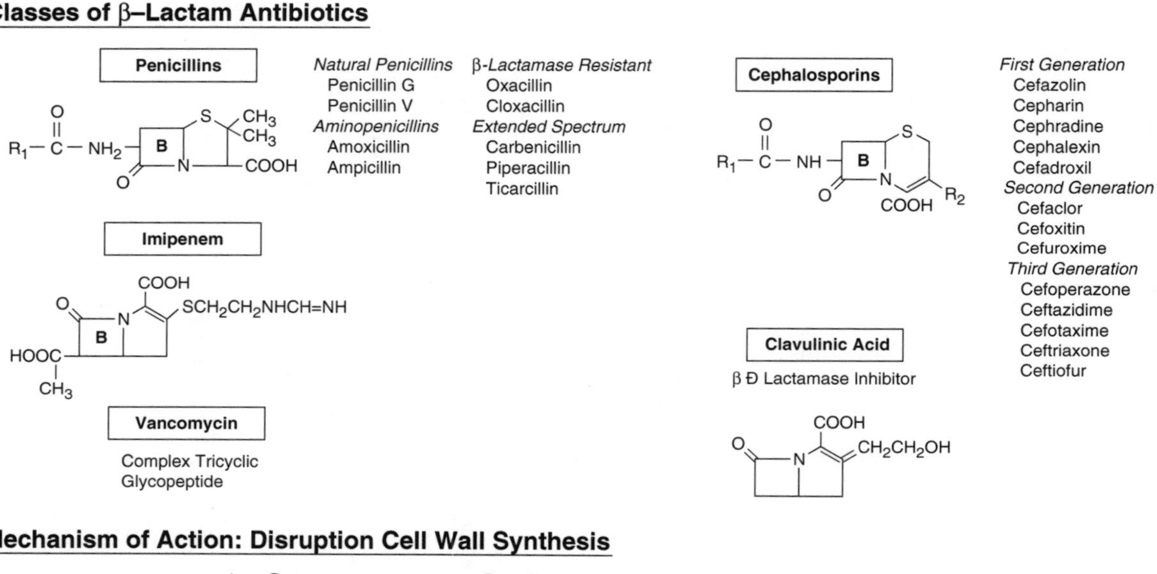

C Mechanism of Action: Disruption Cell Wall Synthesis

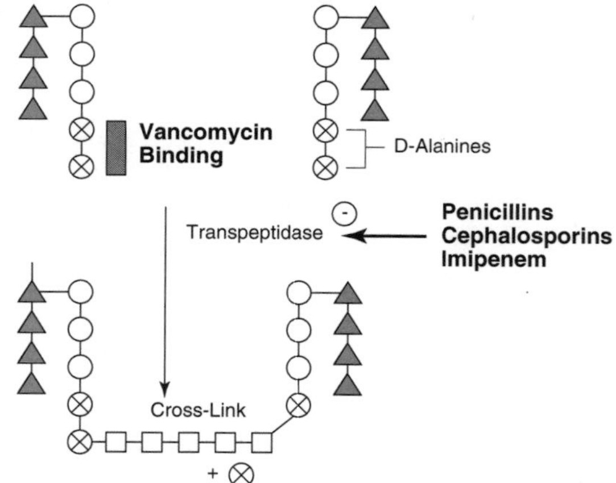

β-Lactam Inhibitors

Clavulinic acid and sublactam are irreversible inhibitors of β-lactamase enzymes (*see* **Part B**). These inhibitors, which have no intrinsic antibacterial activity themselves, are combined with penicillins to enhance their spectrum of action. Both inhibitors are excreted unchanged in the urine. A combination of clavulinic acid and amoxicillin is available for oral use and two combinations, one of sublactam and ampicillin and one of clavulinic acid and ticarcillin, are available for parenteral use. Clavulinic acid–amoxicillin has a spectrum of action similar to the second-generation cephalosporins, but has better activity against anaerobes. The combination is highly moisture sensitive and must be kept in a foil pouch until just prior to administration. The major toxicities are vomiting and diarrhea. The combination is not active orally in herbivores and should not be used in horses, rabbits, guinea pigs, and hamsters, owing to fatal gastrointestinal toxicity.

Polymyxins

Polymyxins disrupt membranes by acting like detergents. They are used topically as they are not available orally and cause neurological and renal toxicity if given parenterally. They are available for topical treatment of ocular, cutaneous, and otic infections. Polymyxins are effective against gram (−) bacteria because of the high lipid content in their cell wall. They are often combined with bacitracin and neomycin as triple antibiotic ointment.

Bacitracin

Bacitracin inhibits peptidoglycan synthesis by blocking phosphorylation reactions that occur during cell wall synthesis. Bacitracin is most commonly applied topically to the skin and eye and is most effective against gram (+) bacteria.

Vancomycin

Vancomycin is a large, complex, tricyclic glycopeptide that inhibits the synthesis of the cell wall peptidoglycan (*see* **Part B**). It binds to the D-alanyl-D-alanine portion of the cell wall inhibiting cross-link formation (**Part C**). Its spectrum of action is primarily against gram (+) bacteria. Its worth lies in the fact that two of the most serious nosocomial multidrug resistant pathogens, *Enterococcus* and *Staphylococcus* sp., remain sensitive to this drug. It must be given IV and is excreted unchanged in the urine. The drug should be administered slowly IV to avoid inducing histamine release. Perivascular administration causes tissue irritation. Vancomycin is ototoxic and nephrotoxic. The drug can be given orally for its local effects in the gastrointestinal tract. It is commonly used in humans to treat pseudomembranous colitis due to *Clostridium difficile*. Vancomycin use in food animals is illegal because of concerns surrounding the emergence of antibiotic resistance.

39 Antibiotics: Drugs that Interfere with DNA Metabolism

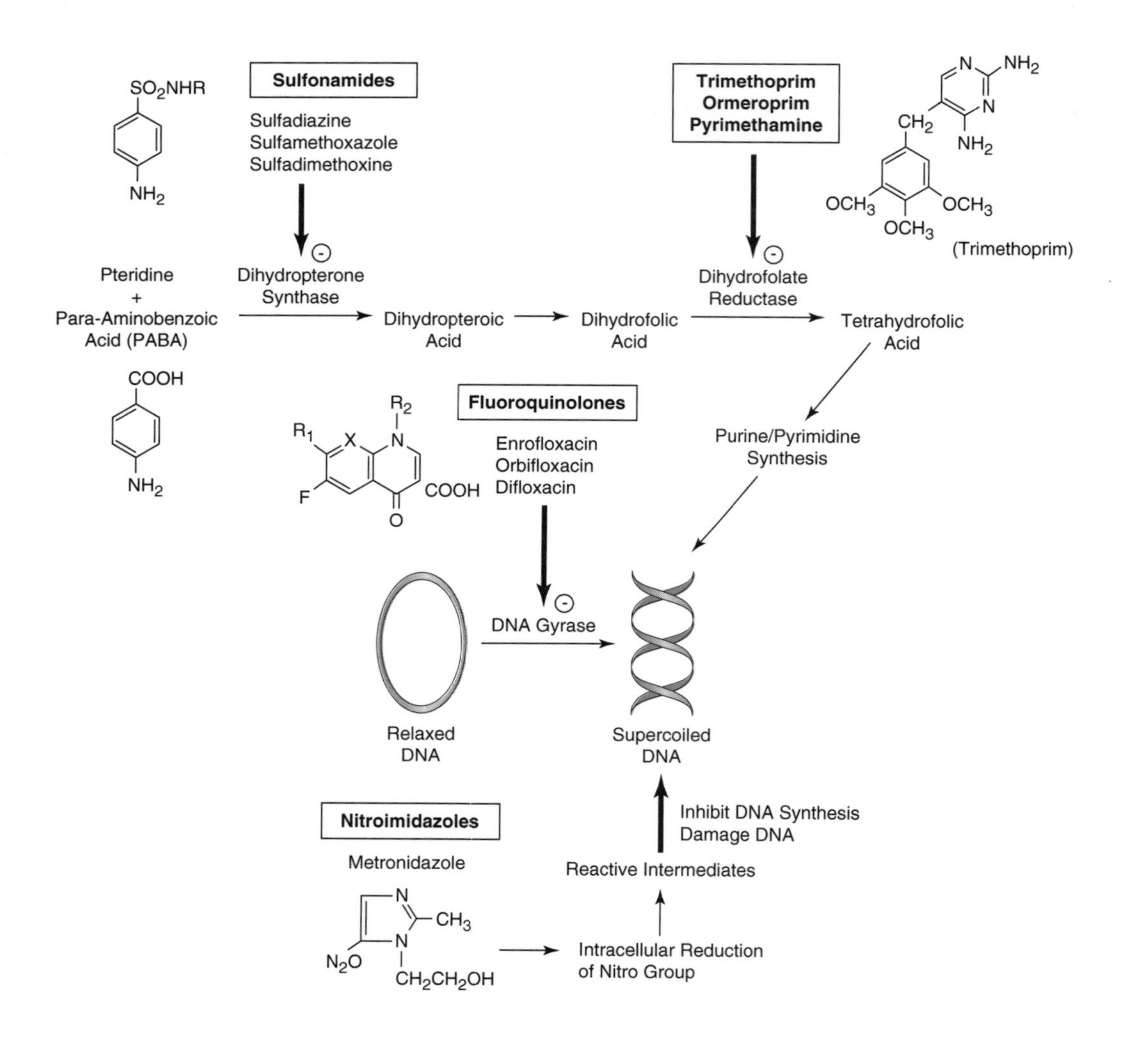

Sulfonamides

Sulfonamides are bacteriostatic antibiotics that interfere with the bacterial production of folic acid from para-aminobenzoic acid (PABA) (**Figure**). Folate is necessary for purine, pyrimidine, and DNA synthesis. Sulfonamides are competitive inhibitors of PABA. Mammals use preformed folate and thus are not sensitive to the toxic effects of sulfonamides.

Sulfonamides are effective against gram (+) and gram (−) aerobes. They have limited action against anaerobes, partially due to the fact that they are readily inactivated in the presence of purulent material. Sulfonamides are active against some protozoans, particularly coccidia, and when used in combination with pyrimethamine, *Toxoplasma* sp. Acquired bacterial resistance to the sulfonamides limits their effectiveness.

The sulfonamides are weak organic acids that are relatively insoluble in water. They are eliminated in the urine by a combination of glomerular filtration and active tubular secretion. They undergo hepatic metabolism by acetylation, glucuronidation, and aromatic hydroxylation. In some dogs, the acetylation pathway may be deficient, which may result in accumulation of toxic intermediates.

Several toxic side effects are associated with the use of sulfonamides. Irreversible keratoconjunctivitis sicca may occur. This toxic reaction may be related to direct toxicity to the lacrimal glands or a hypersensitivity reaction. Several idiosyncratic immune-mediated toxicities have been reported, including cutaneous reactions, arthropathies, hemolytic anemia, thrombocytopenia, and glomerulopathy. Doberman pinschers appear to be uniquely sensitive to these immune reactions. Two hepatotoxic syndromes have been described: an acute hepatic necrosis that may be fulminating and a less severe cholestatic disorder. Renal tubular damage due to precipitation of sulfa crystals has been reported. This is most common with older sulfonamides that have poor solubility in urine. Crystalluria can be avoided by urinary alkalinization and keeping the patient well hydrated. Chronic use of sulfonamides may be associated with folate deficiency. The use of some sulfonamides has been associated with bleeding disorders due to inhibition of vitamin K–dependent activation of coagulation factors and with the development of disorders of thyroid metabolism.

Potentiated Sulfonamides

Trimethoprim and ormetoprim are synthetic folic acid antagonists used in combination with sulfonamides. They bind with a high affinity to bacterial dihydrofolate reductase, but have a very low affinity for the mammalian enzyme (*see* **Figure**). When trimethoprim or ormetoprim is used in combination with a sulfonamide (potentiated sulfonamides), the combination is bactericidal. The potentiated sulfonamides have activity against gram (+) aerobes including β-lactamase (+) organisms, gram (−) aerobes, and some anaerobes. They are also effective against *Chlamydia* sp., and some protozoans including *Toxoplasma* and *Pneumocystis* sp.

Trimethoprim is partially metabolized in the liver and excreted in the urine. It has good tissue penetration into the prostate and cerebrospinal fluid. The combination products available for oral administration include trimethoprim-sulfadiazine and trimethoprim-sulfamethoxazole. The pharmacokinetics of ormetroprim have been poorly characterized. It is available as a combination product with sulfadimethoxine.

Fluoroquinolones

Fluoroquinolones are bactericidal antibiotics that inhibit bacterial DNA gyrase (*see* **Figure**). This enzyme nicks and seals DNA in the process of transcription and reduces DNA size by supercoiling. The fluoroquinolones have a very broad spectrum of action against most gram (−) aerobes including *Pseudomonas* sp. Their gram (+) aerobic spectrum includes β-lactamase (+) organisms, but some *Streptococcus* sp. and *Enterococcus* sp. are resistant. The fluoroquinolones have poor action against anaerobes. They also have some activity against *Mycoplasma*, *Mycobacterium*, *Rickettsia*, *Coxiella*, and *Ehrlichia* sp.

The fluoroquinolones (enrofloxacin, orbifloxacin, difloxacin) are well absorbed after oral administration. Aluminum or magnesium antacids and sucralfate decrease their absorption. Since the fluoroquinolones are highly lipophilic with low protein binding, they penetrate tissues well. High concentrations are obtained in the bile, kidneys, liver, prostate, the reproductive system, and within phagocytes. Concentrations in the CNS are low. The fluoroquinolones are primarily excreted unchanged in the urine by a combination of glomerular filtration and tubular secretion. They undergo some hepatic metabolism to produce a number of metabolites, which are excreted in bile. The fluoroquinolones have a high therapeutic index. Some animals experience gastrointestinal upset and this actually precludes their oral use in horses. In young animals, fluoroquinolones interfere with cartilage maturation, leading to the development of erosions and joint damage. A retinopathy has been seen with high doses (>10 mg/kg/day) in cats. These drugs also have inhibitory action at γ-aminobutyric acid receptor channels in the brain and thus may have neuroexcitatory actions. Concerns over the emergence of fluoroquinolone resistance in enteric pathogens has made the use of these antibiotics illegal in food animals.

The fluoroquinolones are labeled with flexible dosing, which allows for dosing based on bacterial sensitivity, and penetration of the drug at the site of infection. Highly sensitive bacteria on in vitro testing can be treated with lower doses. Since these antibiotics are highly concentrated in the urine (50 × serum level), lower doses can be used to treat urinary tract infections. Bacterial killing with the fluoroquinolones, like the aminoglycosides, is concentration dependent rather than time dependent. Thus, higher once-daily dosing may be more effective than several doses. Higher peak concentrations are also less likely to induce bacterial antibiotic resistance. The fluoroquinolones also exhibit a postantibiotic effect; that is, their effect persists after their plasma concentrations fall below the MIC. This is most likely related to the maintenance of high tissue concentrations.

Metronidazole

Metronidazole is a highly effective agent against anaerobic bacteria and many protozoans (*Giardia* sp., amebae). Under anaerobic intracellular conditions, metronidazole is converted to a variety of unstable intermediates that damage DNA (*see* **Figure**). Metronidazole is not effective against aerobes. The drug is extensively metabolized in the liver to less active intermediates that are excreted in urine (75%) and bile (25%). The drug has excellent tissue penetration, including into the CNS and brain. It is available for oral and parenteral use. The development of bacterial resistance is rare. Metronidazole has poorly understood immunomodulatory properties. The most common side effects are gastrointestinal disturbances. Cats, in particular, object to the very bitter taste. Overdosage can also lead to the development of an acute vestibular syndrome. The drug has been shown to have marginal carcinogenic effects in laboratory animals and this limits its use in food animals.

40 Antibiotics: Drugs that Inhibit Protein Synthesis I

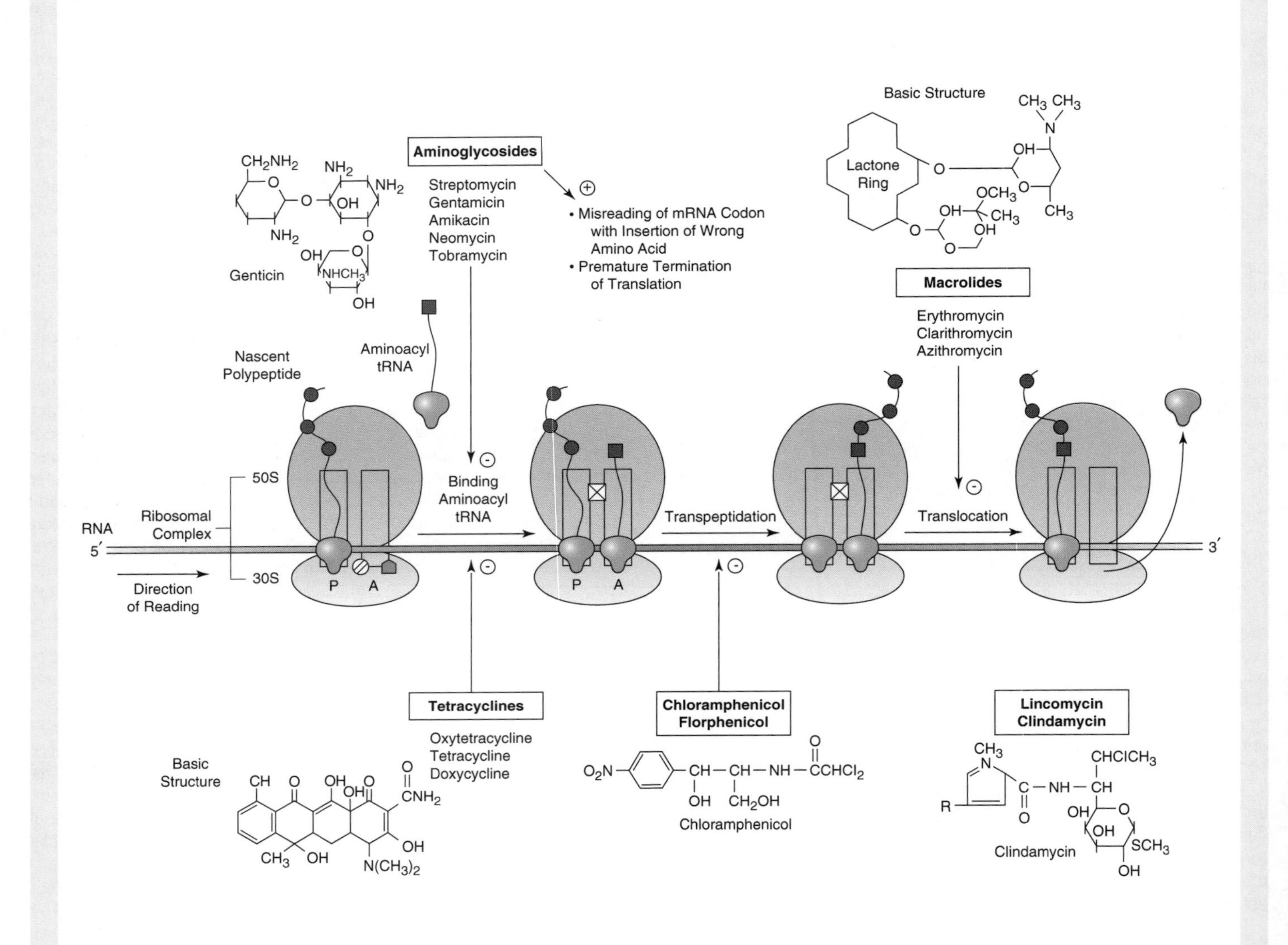

P = Polypeptide Binding Site
A = Aminoacyl tRNA Binding Site
⊘ = Aminoglycoside Binding Site
⊠ = Macrolide, Linomycin, Chloramphenicol Binding Site
⬟ = Tetracycline Binding Site

Mechanism and Spectrum of Action

Aminoglycosides are bacteriocidal antibiotics that bind to the bacterial 30S ribosome (**Figure**). Ribosomal binding blocks initiation of protein synthesis and results in premature termination of mRNA translation and the incorporation of the wrong amino acid into the growing peptide. The aminoglycosides diffuse through pores in the outer membrane of gram (−) bacteria and are then actively transported intracellularly. Since they enter into the cell by active transport, their spectrum of action is limited to aerobic bacteria, particularly gram (−) organisms including *Pseudomonas* sp. They have poor activity against gram (+) bacteria. Aminoglycosides must be given parenterally because they are not absorbed after oral dosing. They are polar organic bases whose distribution is limited to extracellular fluids. Low concentrations are found in secretions and tissues including the CNS, although penetration is improved in the presence of inflammation. The aminoglycosides are actively accumulated in the renal cortex and the inner ear. They are not active in acidic environments such as necrotic tissue or in the presence of purulent material. They are eliminated unchanged in the urine.

Toxicity

The major side effect of aminoglycosides is nephrotoxicity. The exact mechanism by which the aminoglycosides cause renal proximal tubular necrosis is not known. The nephrotoxicity is believed to be related to interference with phospholipid metabolism in the lysosomes of renal proximal tubular cells that results in leakage of proteolytic enzymes. Risk factors for nephrotoxicity include high doses, long duration of treatment, preexisting renal dysfunction or dehydration, concurrent administration of nephrotoxic drugs or diuretics, and very young or old age. Nephrotoxicosis is also enhanced by persistently elevated trough concentrations. Drug dosing that minimizes the dosing interval is associated with a decreased nephrotoxicity. Animals on aminoglycosides must be monitored closely for renal damage. Early signs of tubular damage include increased urinary concentrations of the renal brush border enzyme, γ-glutamyltranspeptidase, isosthenuria, and the presence of urinary casts. Later proteinuria, glucosuria, and azotemia develop. The renal damage is usually reversible, although in advanced stages short-term peritoneal dialysis may be necessary to provide time for the proximal tubules to regenerate.

The aminoglycosides can cause ototoxicity by damaging cranial nerve VIII. The damage may be to the cochlear and/or vestibular nerve resulting in loss of hearing or balance, respectively. Auditory toxicity is more common with amikacin and neomycin, while vestibular toxicity is more common with streptomycin and gentamycin. Cats are more sensitive than other species to vestibular toxicity. The aminoglycosides also inhibit the prejunctional release of acetylcholine and decrease postsynaptic acetylcholine receptor sensitivity. They can potentiate the action of neuromuscular blocking agents and are relatively contraindicated in patients with disease at the neuromuscular junction (e.g., myasthenia gravis).

The aminoglycosides are synergistic in vivo with β-lactam antibiotics and clindamycin and antagonistic with chloramphenicol, tetracyclines, and macrolides. In vitro, β-lactam antibiotics inactivate the aminoglycosides.

Optimal Dosing Schedules

Although original dosing schedules for aminoglycosides indicated the need to give the antibiotics every 8 hours, today once daily dosing schedules for aminoglycosides are preferred for several reasons. First, since aminoglycosides bind irreversibly to ribosomes, they inhibit protein synthesis long after the plasma concentrations have decreased. Second, aminoglycosides kill bacteria rapidly. Lastly, bacterial killing by the aminoglycosides is concentration, not time, dependent. Unlike the β-lactam antibiotics whose efficacy is related to the amount of time that serum concentrations are maintained above the MIC, aminoglycoside effectiveness is linked to the magnitude of peak concentrations. Once daily dosing of aminoglycosides is appropriate as long as the peak drug concentrations are 4 times the MIC. An added benefit of once daily dosing is a reduction in the emergence of bacterial resistance and a decrease in nephrotoxicity.

Bacterial Resistance

Plasmid-mediated drug resistance is a major problem with the aminoglycosides. The major mode of resistance is bacterial production of inactivating enzymes. The fear of induction of bacterial resistance in human pathogens has limited the use of aminoglycosides in food animals. Although not illegal, their use is prohibitive because of extremely long withdrawal times.

Individual Aminoglycosides

Streptomycin is the prototypical aminoglycoside, but is of limited availability. It is less active than other aminoglycosides, owing to acquired resistance. Ototoxicity is the major side effect. Neomycin is available for oral or topical use. Nephrotoxicity precludes its use parenterally. It is not absorbed orally to any appreciable extent (3%), but absorption can be increased in the presence of severe enteric disease. Neomycin is frequently used for its local actions on the gastrointestinal tract. It is a mainstay of therapy for hepatic encephalopathy since it reduces the population of ammonia producing bacteria in the colon.

Gentamycin has the broadest spectrum of action of the aminoglycosides. It is effective against gram (−) bacteria and some gram (+) bacteria. Bacteria acquire plasmid-mediated resistance to gentamycin quickly in the presence of a selective pressure. Amikacin is most often used to treat gentamycin-resistant infections because it has a unique resistance to bacterial aminoglycoside-inactivating enzymes. Tobramycin is another aminoglycoside which is used primarily to treat gentamycin-resistant gram (−) infections.

41 Antibiotics: Drugs that Inhibit Protein Synthesis II

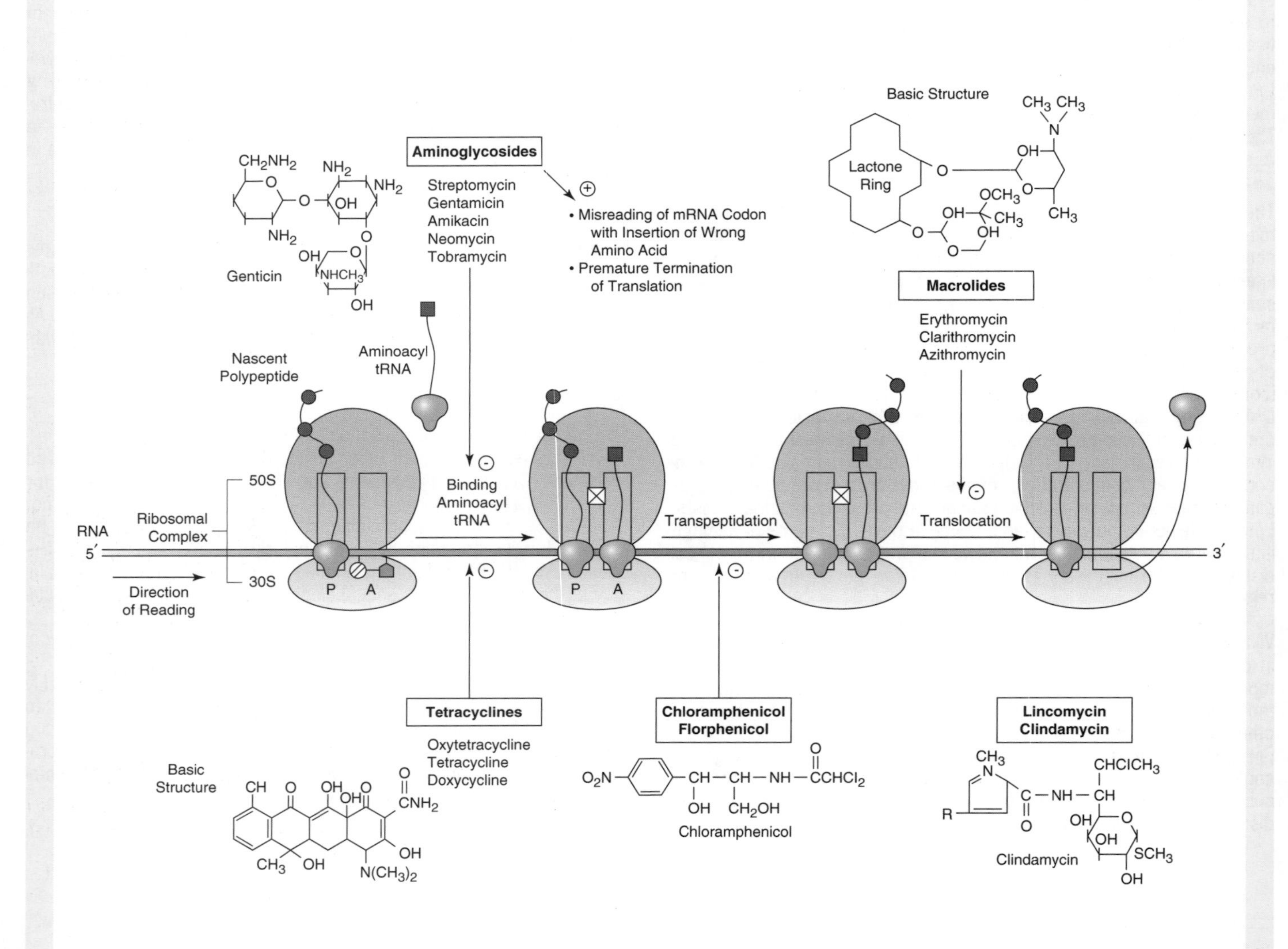

P = Polypeptide Binding Site
A = Aminoacyl tRNA Binding Site
⊘ = Aminoglycoside Binding Site
☒ = Macrolide, Linomycin, Chloramphenicol Binding Site
⬟ = Tetracycline Binding Site

Lincomycin/Clindamycin

These bacteriostatic antibiotics reversibly bind to the bacterial 50S ribosome and inhibit protein synthesis (**Figure**). Their spectrum of action includes gram (+) aerobes including β-lactamase (+) *Staphylococcus* sp. They are highly effective against anaerobes including *Bacteroides fragilis*. They have poor activity against gram (−) aerobes. Clindamycin has better activity against gram (−) organisms than lincomycin does, with some efficacy against *Mycoplasma* sp. and *Toxoplasma* sp. These drugs are available for oral or parenteral administration. They undergo hepatic metabolism to active and inactive metabolites that are excreted in the urine and bile. Clindamycin is actively concentrated in white blood cells and penetrates well into respiratory secretions, pleural fluid, soft tissues, bone, and joint fluid. It does not cross the blood-brain barrier well. Since they are organic bases, these antibiotics are trapped in the udder and prostate.

The most common side effect is gastrointestinal toxicity. They can cause fatal diarrhea in horses, rabbits, hamsters, guinea pigs, and ruminants. They are much less toxic to the gastrointestinal tract in dogs and cats.

Macrolides

Macrolides (erythromycin, clarithromycin, azithromycin) are bacteriostatic antibiotics that reversibly bind to the bacterial 50s ribosome (*see* **Figure**). These drugs interfere with translocation of the newly synthesized peptidyl tRNA molecule from the acceptor site to the peptidyl site on the ribosome (*see* **Figure**). Their spectrum of action includes gram (+) bacteria including β-lactamase (+) organisms and *Mycoplasma* sp. They are moderately effective against anaerobes, but not *B. fragilis*. The macrolides have limited action against gram (−) organisms, but do have activity against some important pathogens including *Bordetella* sp. and *Camphylobacter* sp. They are primarily excreted unchanged in the bile. They are available for oral or parenteral administration. These drugs are contraindicated in adult horses and most ruminants because they can cause fatal diarrhea. They can, however, be safely used in foals.

Erythromycin is administered orally as an enteric coated tablet, as it is subject to degradation by gastric acids. Formulations containing acid-stable salts (ethylsuccinate, estolate) are also available. A common side effect of therapy is vomiting. The estolate formulation has been associated with a cholestatic hepatopathy in humans.

Clarithromycin and azithromycin are derivatives of erythromycin that have better oral absorption, are better tolerated, and have a longer $t^{1/2}$ than the parent drug. These derivatives are more active against gram (−) bacteria, particularly *Haemophilus* sp. and *Bordetella* sp. and also have efficacy against *Chlamydia* sp. and *Toxoplasma* sp. Clarithromycin is used to treat *Helicobacter* sp. Both derivatives have an extraordinary ability to concentrate in tissues, particularly in leukocytes and macrophages. They undergo some hepatic metabolism to active intermediates that are excreted in urine and bile.

Chloramphenicol

Chloramphenicol is a bacteriostatic antibiotic that reversibly binds to the bacterial 50S ribosome. It prevents binding of the amino acids with the aminoacyl tRNA (*see* **Figure**). Chloramphenicol has a very broad spectrum of action against gram (+) organisms including β-lactamase (+) bacteria, gram (−) aerobes, *Rickettsia* sp., *Mycoplasma* sp., *Chlamydia* sp., and virtually all anaerobes. Chloramphenicol undergoes hepatic glucuronidation with subsequent renal and biliary excretion. Its $t^{1/2}$ is increased in neonates and feline patients due to poor hepatic metabolism. The drug is very lipophilic and readily passes through cellular barriers into the cerebrospinal fluid, brain, and aqueous humor. It is available for oral or parenteral administration.

Chloramphenicol causes an irreversible aplastic anemia in humans. This toxicity is associated with the compound's *p*-nitro group (*see* **Figure**). Reversible dose-related disturbances in erythrocyte maturation are seen and may be associated with interference with mitochondrial protein synthesis. The use of chloramphenicol is illegal in food animals for fear of antibiotic residues leading to the development of aplasia anemia in humans. The most common side effects in veterinary patients are vomiting and diarrhea. Chloramphenicol is a potent inhibitor of the hepatic CP450 enzyme system.

Florphenicol is a thiamphenicol derivative that is approved for parenteral use in cattle to treat bovine respiratory pathogens. Florphenicol lacks the *p*-nitro group and has never, despite widespread use in other countries, caused aplastic anemia. Adverse effects include pain at injection sites, anorexia, and decreased water consumption.

Tetracyclines

Tetracyclines (oxytetracycline, tetracycline) and doxycycline are bacteriostatic antibiotics that bind to the bacterial 30S ribosome. They prevent access of the aminoacyl tRNA to the receptor site on the ribosomal complex (*see* **Figure**). They enter bacteria by an active transport. The tetracyclines have a wide spectrum of activity, which includes gram (+) bacteria [not β-lactamase (+) organisms], some gram (−) organisms including *Brucella*, *Leptospora*, *Bordetella*, *Borrelia*, *Mycoplasma*, *Rickettsia*, *Chlamydia*, and *Coxiella* sp. Their activity is limited by widespread plasma-mediated resistance.

Tetracyclines are incompletely absorbed after oral administration and therefore concentrations in the gastrointestinal tract are high. These high concentrations are highly disruptive to the normal bacterial gastrointestinal flora. Food and aluminum- or magnesium-containing antacids interfere with the absorbance of tetracyclines, but not doxycycline. Tetracyclines are eliminated unchanged in the urine. Doxycycline is secreted into the intestine where it is inactivated and eliminated in the feces. A small portion of doxycycline undergoes renal (25%) or biliary (5%) excretion. Since doxycycline is more lipophilic than tetracycline, it penetrates biological membranes such as the blood-brain barrier better. Doxycycline is highly protein bound, while the tetracyclines are not. The major toxicity seen with both antibiotics is gastrointestinal upset. Older formulations of tetracycline may cause renal damage due to the presence of degradation products. Both antibiotics should not be used in young animals because they can cause yellow discoloration of teeth. The tetracyclines may inhibit mammalian protein synthesis, leading to catabolism and the development of azotemia or hepatic lipidosis. Both antibiotics are available for IV administration. Rapid IV administration can cause hypotension or cardiac arrhythmias due to chelation of calcium ions. This is particularly true in equine patients.

42 Antifungal Agents

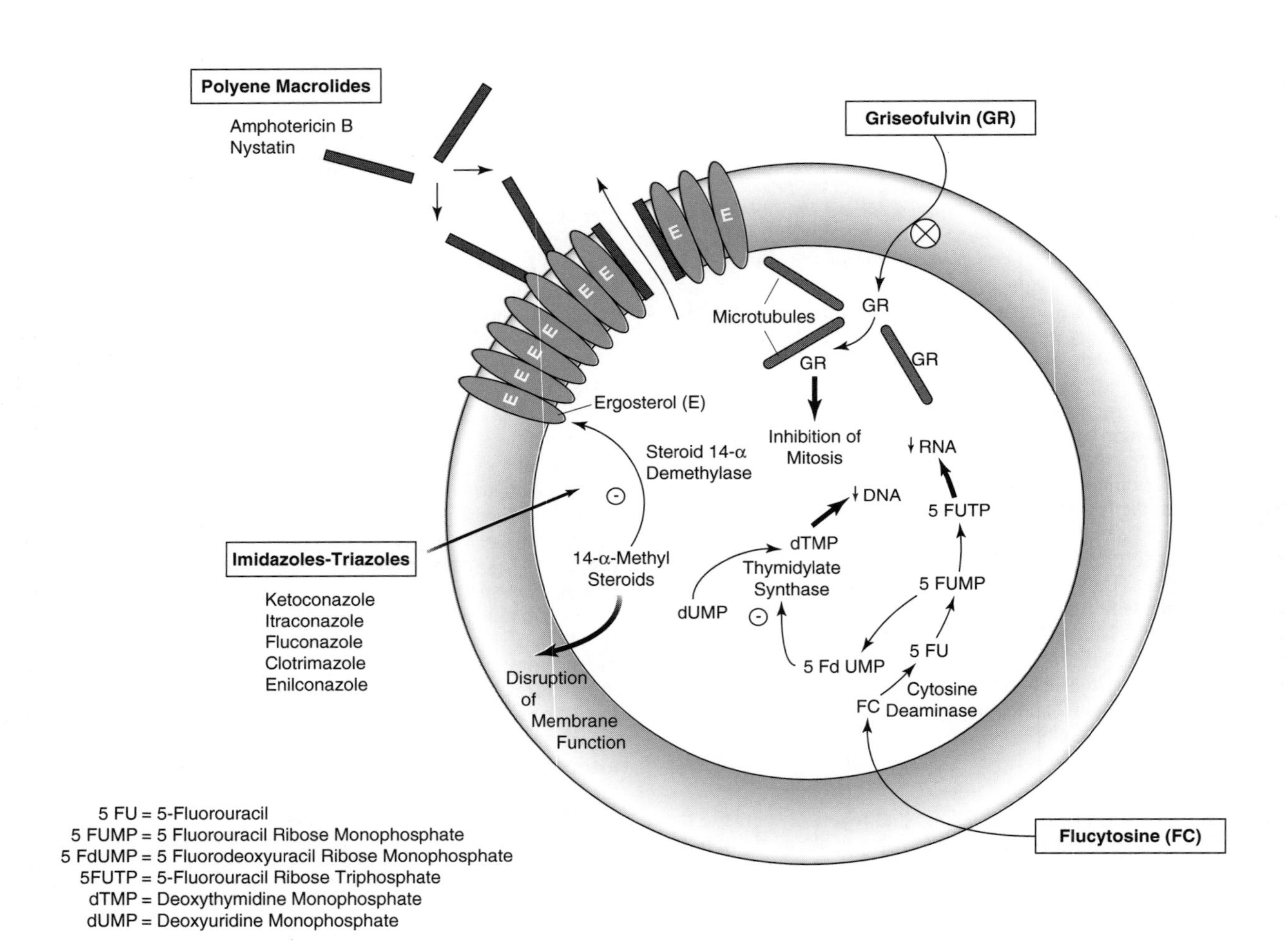

Drugs Affecting the Nucleus

Griseofulvin

Griseofulvin, a fungistatic agent, enters the fungi by an energy-dependent mechanism, binds to polymerized microtubules, and disrupts mitosis (**Figure**). Griseofulvin is active against the dermatophytes, *Microsporum* and *Trichophyton*. The microsized formulation of griseofulvin has variable bioavailability (25%–70%) while the ultramicrosized formulation is almost 100% bioavailable. The drug is deposited in keratin precursor cells and remains in mature keratinocytes until they are shed. Griseofulvin is metabolized by the liver and eliminated as a glucuronide conjugate in the urine. It induces the hepatic microsomal enzyme system.

Although griseofulvin is used quite safely in most species, cats (particularly kittens) have an increased risk of developing anorexia, vomiting, bone marrow suppression, neurotoxicity, and hepatotoxicity. The drug is teratogenic in cats.

Flucytosine

Flucytosine is a fluorinated pyrimidine that fungi deaminate to 5-fluorouracil (*see* **Figure**). Fluorouracil is an antimetabolite that interferes with thymidylate synthase and disrupts fungal RNA and DNA synthesis. Selective fungal toxicity is due to the fact that mammalian cells lack the enzyme cytosine deaminase to convert flucytosine to 5-fluorouracil. Flucytosine is well absorbed from the intestinal tract and is distributed to most tissues including the CNS and aqueous humor. It is largely excreted unchanged in the urine. It is effective against *Cryptococcus* and *Candida* spp. but the occurrence of resistant fungal strains is high. It is almost always used in combination with amphotericin B as these 2 agents are synergistic. Side effects include bone marrow suppression, gastrointestinal disturbances, cutaneous eruptions, and oral ulcerations. Renal failure and cholestatic liver disease have been reported. Aberrant behavior and seizures have been reported in cats.

Cell Membrane–Directed Agents

Amphotericin B

Amphotericin B is a polyene macrolide antibiotic. Amphotericin B selectively binds to ergosterol, a component of the fungal cell membrane that is not present in mammalian cell membranes (*see* **Figure**). This interaction results in the formation of membrane pores that alter membrane permeability and result in the leakage of many small proteins. Most fungal pathogens, except the dermatophytes, are sensitive to amphotericin B.

Since amphotericin B is water insoluble, it is either formulated as a suspension with the bile acid deoxycholate or complexed with lipids. The major advantages of lipid-complexed amphotericin B are enhanced uptake by the mononuclear phagocytic system and better access to the site of infection. Since the lipid-complex formulation also limits kidney accumulation, it decreases the incidence of nephrotoxicity. Amphotericin B is not absorbed from the gastrointestinal tract and must be given IV. The drug is highly protein bound, which limits its distribution to the CNS, bone, brain, and aqueous humor.

The metabolic fate of amphotericin B is unknown. It exhibits a biphasic elimination. An initial $t^{1/2}$ of 24–48 hours is followed by a longer terminal $t^{1/2}$ of up to 15 days. Amphotericin B is slowly excreted unchanged into the urine, although some excretion into bile occurs.

The major toxicity of amphotericin B is nephrotoxicity. Toxicity is associated with both a direct toxic effect on the renal tubular epithelium and renal vasoconstriction leading to decreased glomerular filtration rate. Nephrotoxicity is enhanced by concomitant therapy with other nephrotoxic agents. Nephrotoxicity may be decreased by sodium loading before treatment or by concurrent administration of mannitol. The drug also causes renal tubular acidosis with loss of potassium and magnesium. Supplemental potassium is required in most patients receiving long-term treatment. Administration of amphotericin B can also be associated with development of fever, vomiting, myalgia, and occasionally anaphylaxis. These latter signs may be related to induction of cytokine release from reticuloendothelial cells. A normocytic, normochromic anemia develops in many patients, possibly associated with decreased production of erythropoietin.

Imidazoles and Triazoles

Azoles exert their antifungal effect by inhibiting sterol 14-α-demethylase, a fungal microsomal enzyme, necessary for ergosterol synthesis (*see* **Figure**). Inhibition of ergosterol synthesis also leads to the accumulation of 14 α methyl sterols. These methyl sterols impair membrane function, leading to alterations in energy metabolism and growth inhibition. Some azoles inhibit mammalian microsomal enzyme systems, which accounts for their ability to inhibit the synthesis of mammalian cortisol and reproductive hormones.

Ketoconazole

Ketoconazole is effective against yeast and dimorphic fungi such as *Candida, Malassezia, Coccidioides, Histoplasma*, and *Blastomyces* spp. and dermatophytes. It has less activity against *Cryptococcus* and *Aspergillus* spp. Oral absorption of ketoconazole is variable and requires an acidic environment. Therapy with H2 receptor blockers or antacids or the presence of food in the stomach decreases absorption. Ketoconazole is metabolized in the liver and excreted in feces. It is highly protein bound and does not penetrate the CNS, although in the presence of inflammation CNS levels may increase. It takes 14–21 days to reach steady-state serum concentrations.

The most common side effects of therapy are anorexia and vomiting. Administration of the drug with food or in divided doses may improve these signs. Mild transient asymptomatic elevations in hepatic enzymes may occur. If they persist, hepatotoxicity should be suspected and the drug discontinued. Cats appear to be more susceptible to hepatotoxic symptoms. The drug is teratogenic. A reversible lightening of the hair coat may occur with chronic therapy. Ketoconazole inhibits mammalian cortisol synthesis and has been used to treat canine hyperadrenocorticism.

Itraconazole

Itraconazole is a fungistatic triazole. Compared to ketoconazole, itraconazole is absorbed better (especially if given with a fatty meal), is associated with fewer side effects, has a longer $t^{1/2}$, and has a slightly expanded spectrum of action to include *Aspergillus* sp., with better activity against *Candida* sp. and dermatophytes. The drug is highly lipophilic and is well distributed, but high plasma protein binding limits penetration of the CNS and aqueous humor. The drug is extensively metabolized in the liver and excreted primarily in the bile. Itraconazole has increased specificity for the fungal CP450 system and does not inhibit mammalian steroid hormone synthesis. Itraconazole is teratogenic and associated with gastrointestinal disturbances, hepatotoxicity, and cutaneous reactions.

Fluconazole

Fluconazole is water soluble and therefore readily absorbed from the gastrointestinal tract independent of food intake or stomach acidity. It is not highly protein bound and distributes to the aqueous humor and CNS. Effective concentrations are reached in 5–10 days. Fluconazole is eliminated primarily by the kidneys in an unchanged form. Fluconazole is generally well tolerated, causing less gastrointestinal upset and less frequent elevations in hepatic enzymes. There is no evidence of teratogenicity. Its spectrum of action is similar to ketoconazole and itraconazole, with the exception of poor activity against *Aspergillus* sp.

43 Antiprotozoal Agents

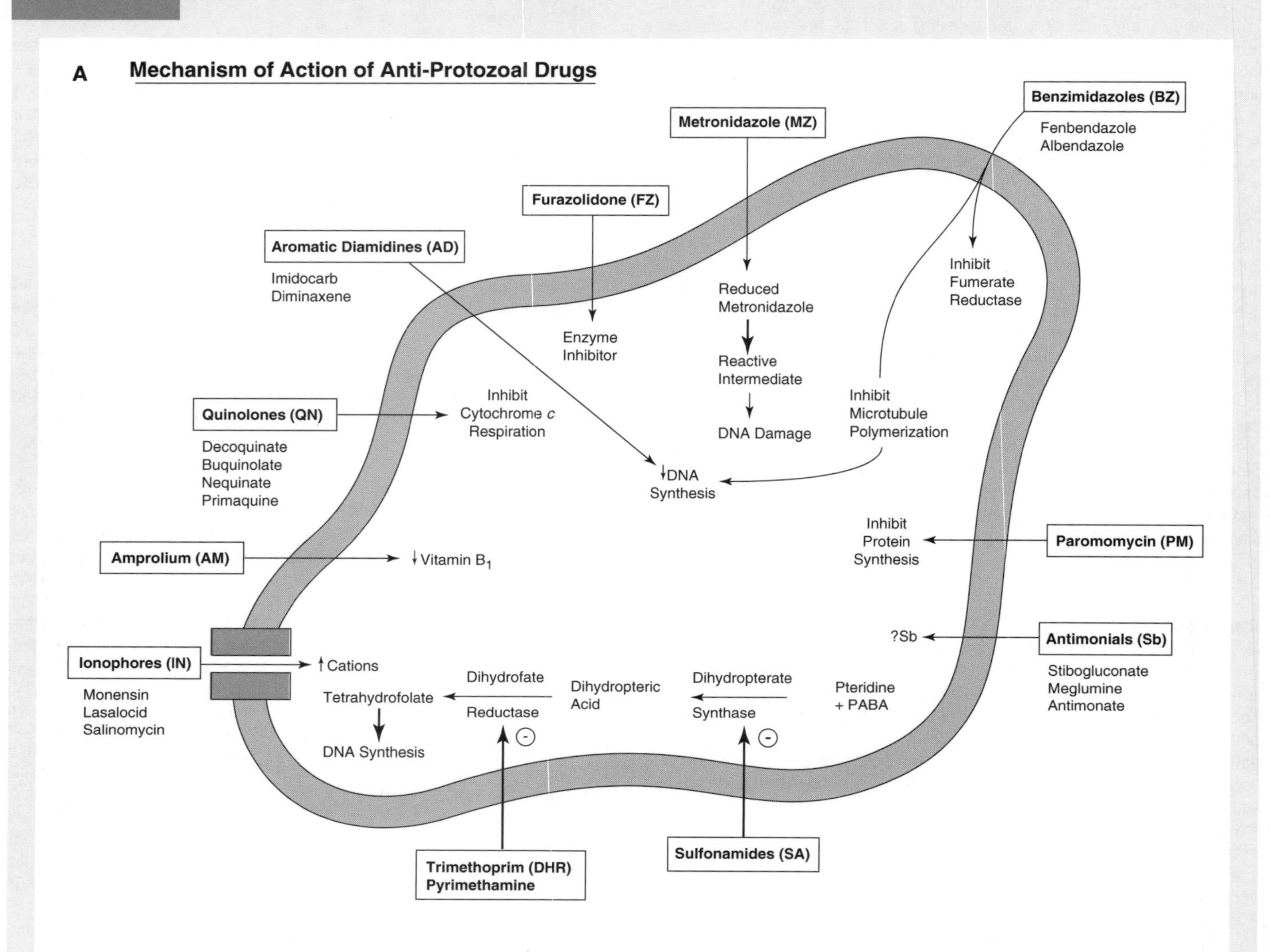

B Anti-Protozoal Drug Susceptibility

Protozoa	Drug Treatment
Babesia	AD
Coccidia	SA, FZ, AM, IN, QN
Cryptosporia	PM
Cytauxzoon	AD
giardia	MZ, BZ, FZ
Hepatozoan	AD
Leishmania	Sb
Neospora	SA, DHR, CLD
Sarcocystis	SA, DHR, CLD
Toxoplasma	SA, DHR, CLD
Trypanosoma	AD

CLD = clindamycin

Metronidazole

Metronidazole is well absorbed from the gastrointestinal tract and reaches high tissue concentrations, making it effective against tissue and gastrointestinal protozoans. Once metronidazole enters the protozoan, its nitro group is reduced which results in the generation of toxic intermediates that damage the parasites' DNA (**Part A**). Metronidaozole is metabolized by the liver and excreted in the urine. It is generally well tolerated, but may cause gastrointestinal upset, and at high doses, neurotoxicity manifested by ataxia, tremors, and weakness may become apparent. Metronidazole is effective against *Giardia* sp (**Part B**).

Furazolidone

Furazolidone is a nitrofuran derivative that interferes with parasitic enzyme systems (*see* **Part A**), although its exact mechanism of action is unknown. It is effective against *Giardia* sp., coccidia, trichomonas, and some enteric bacteria. It is available as a liquid suspension, which facilitates administration to puppies and kittens. Furazolidone is minimally absorbed from the gastrointestinal tract. Side effects include anorexia and vomiting. The nitrofurans are suspected mutagens and carcinogens and their use in food animals is prohibited.

Paromomycin

Paromomycin is an aminoglycoside antibiotic used in the treatment of amoebiasis and cryptosporidiosis. It is poorly absorbed orally and has little activity against intestinal bacteria. Due to the potential to cause nephrotoxicity, it should not be used in patients with renal disease.

Sulfonamides

Although sulfonamides were the first effective anticoccidial agents developed for use, acquired drug resistance, especially in the poultry industry, has limited their use as coccidiostats. They are still effective anticoccidial drugs in small animals and ruminants. Sulfonamides work by interfering with protozoal folate synthesis (*see* **Part A**). They competitively inhibit the bacterial enzyme dihydropteroate synthase, which catalyzes the formation of dihydropterate from PABA and hydromethyldihydropterine. Inhibition of folate synthesis results in decreased production of the nucleotides needed by the developing coccidial stages for DNA synthesis. Mammalian cells use preformed folate.

Dihydrolfate Reductase Inhibitors

Dihydrofolate reductase inhibitors (trimethoprim, ormetoprim) are often combined with sulfonamides to inhibit sequential steps in folate metabolism and increase their antiprotozoal action (*see* **Part A**). These potentiated sulfonamides are effective against *Toxoplasma* and *Neospora* spp.

Pyrimethamine is another dihydrofolate reductase inhibitor used in the treatment of toxoplasmosis in small animals and for equine protozoal myelitis (*see* **Parts A** and **B**). It is often combined with other drugs that share some antiprotozoal action such as clindamycin or sulfonamides. It is structurally similar to trimethoprim and is well absorbed after oral administration. The drug crosses the blood-brain barrier and also accumulates in the liver, lungs, and spleen. It is highly protein bound. The metabolites are excreted in the urine. The effects of the drug can remain for 1–2 weeks after the last dose. It is associated with a higher incidence of gastrointestinal disturbances than trimethoprim particularly in cats. Bone marrow suppression may occur. Concurrent administration of folinic acid prevents the development of mammalian folate deficiency. Very high doses of pyrimethamine are teratogenic in laboratory animals.

Quinolones

Decoquinate, buquinolate, and nequinate are quinolone coccidiostats. These agents permit the coccidial sporozoites to enter the host, but interfere with further development. They work by inhibiting coccidial respiration by interfering with cytochrome *c*–mediated electron transport (*see* **Part A**).

Primaquine is an 8-aminoquinolone used to treat hepatozoonosis. It is less effective against *Babesia* and *Pneumocystis* spp. It binds to protozoal DNA and alters mitochondrial function (*see* **Part A**). Primaquine is available for oral administration and is highly concentrated in the brain, liver, lungs, and muscle. It undergoes hepatic metabolism followed by renal excretion. Side effects include myelosuppression and methemoglobinemia.

Amprolium

Amprolium is structural analogue of vitamin B_1, thiamine (*see* **Part A**). It acts on the first-generation coccidial schizonts to prevent further development to the merozoite stage. It has some activity against the sexual stages and the sporulating oocyst of coccidial organisms. Prolonged use can lead to the development of thiamine deficiency and the development of CNS dysfunction (ventroflexion of neck, anisocoria, seizures). Therapy should be limited to 2 weeks.

Ionophores

The anticoccidial action of polyether ionophores is related to their ability to form lipophilic complexes with cations and to facilitate their transport across biological membranes. The intracellular accumulation of these ions interferes with parasite metabolism. These drugs include monensin, lasalocid, salinomycin, narasin, maduramicin, semduramicin, and arprinocid (*see* **Part A**). Ingestion of monensin causes fatal cardiotoxicity in horses.

Aromatic Diamidines

Imidocarb dipropinate has activity against *Babesia*, Hepatozoon, *Cytauxzoon* sp, and *Ehriichia* sp. (*see* **Part B**). The drug interferes with DNA formation (*see* **Part A**). It must be injected IM or SC. Since the drug is concentrated in organs such as the liver and brain and is eliminated slowly, one injection has antiprotozoal activity for several weeks. Imidocarb has a low therapeutic index. Parenteral administration consistently causes vomiting. Other side effects are pain on injection and signs of cholinergic stimulation (hypersalivation, tachycardia, vomiting, restlessness, and diarrhea). These latter signs may be due to inherent anticholinesterase activity and can be controlled with atropine. Imidocarb should not be used concurrently with other cholinesterase inhibitors. Imidocarb is also associated with dose-dependent increases in liver enzyme activity and hepatic damage.

Diminazene diaceturate is available for parenteral administration for treatment of trypanosomiasis, babesiosis, and cytauzoonosis. The drug undergoes hepatic metabolism and gradual urinary excretion over several weeks. Side effects include CNS signs, acute hemorrhage, diarrhea, and cardiomyopathy.

Antimonials

Sodium stibogluconate and meglumine antimonate are used in the treatment of leishmaniasis. Their exact mechanism of action is unknown. They are seldom effective in curing the disease. Side effects include gastrointestinal disturbances, nephrotoxicity, and cardiac arrhythmias.

Benzimidazoles

The antihelmintics, fenbendazole and albendazole, have activity against *Giardia* sp. Fenbendazole is not appreciably absorbed from the gastrointestinal tract and is not associated with toxicity. Since up to 40% of the dose of albendazole is absorbed from the gastrointestinal tract, it has been associated with more systemic toxicity. Bone marrow suppression has been reported in dogs and it has been shown to be teratogenic in rats, rabbits, and sheep.

44 Antiparasitic Drugs: Endoparasites I

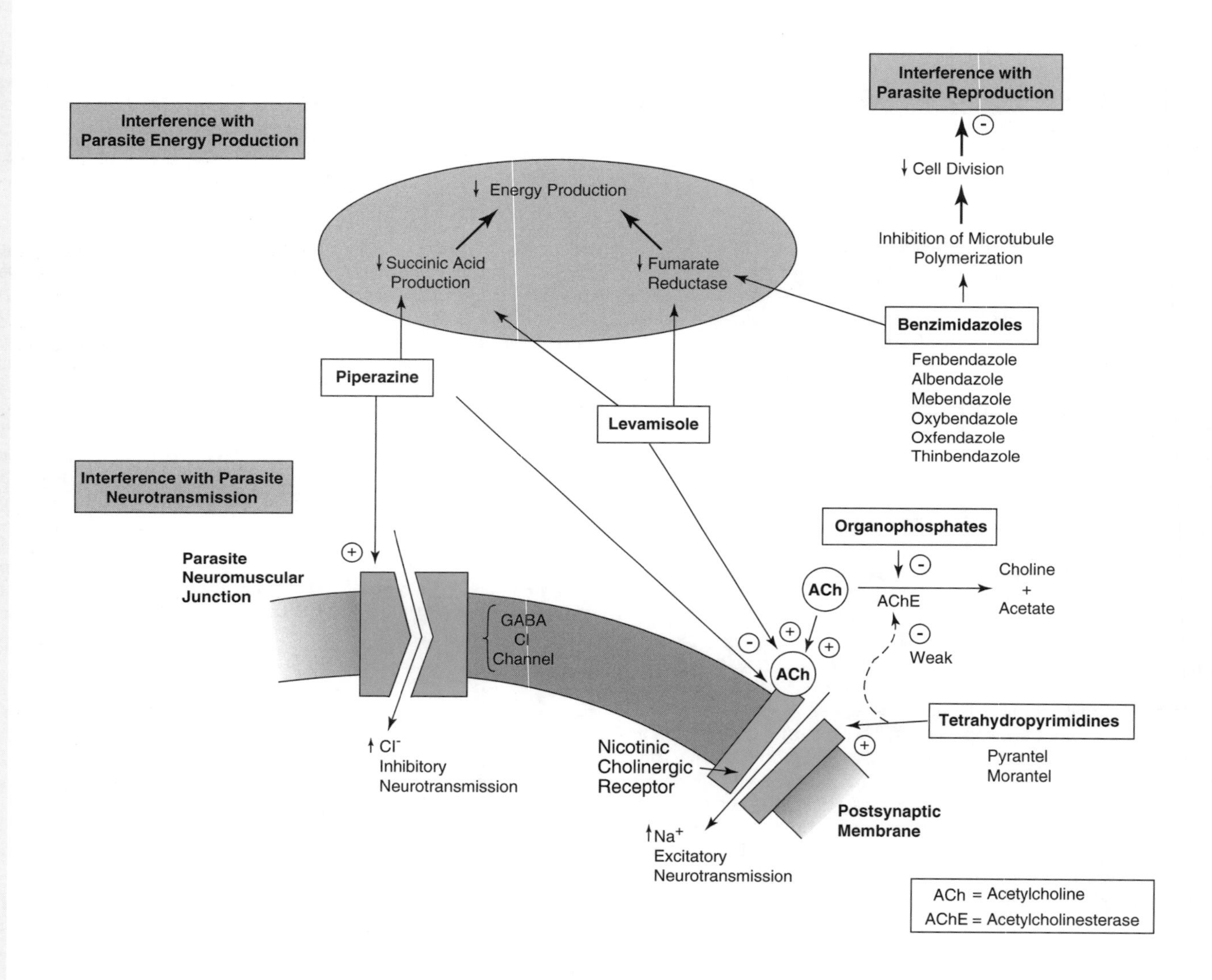

Piperazine

Piperazine blocks excitatory cholinergic transmission and/or stimulates inhibitory GABA transmission at the parasite neuromuscular junction, resulting in worm paralysis (**Figure**). In ascarides, succinic acid production is also inhibited. Piperazine is readily absorbed after oral administration, undergoes hepatic biotransformation, and is excreted in the urine. Piperazines are considered safe in all species. Occasionally neurological side effects (ataxia, tremors, and aberrant behavior) have been reported, most often in cats or with overdosage. Piperazine is only effective against ascarides and some pinworms. It is marginally effective against equine strongyli.

Diethylcarbamazine citrate, a piperazine derivative, is most commonly used as a daily heartworm preventative for dogs. It is effective against the infective larvae stage of heartworm. Since diethylcarbamazine is also effective against circulating heartworm microfilariae, dogs should be free of microfilariae prior to treatment as the sudden death of microfilariae can cause cardiovascular collapse. The drug also has some benefit in the prophylaxis of ascarid infestations.

Benzimidazoles

The benzimidazoles are broad-spectrum antinematodal agents with a wide margin of safety and a high degree of efficacy. Several derivatives of the prototypical benzimidazole thiabendazole (mebendazole, fenbendazole, albendazole, and oxibendazole) are available. The drugs bind to nematode tubulin, preventing its polymerization during microtubule assembly and thus disrupting cell division (*see* **Figure**). They also inhibit parasite fumarate reductase, an enzyme important in energy generation (*see* **Figure**).

The benzimidazoles vary in their degree of absorption from the gastrointestinal tract after oral dosing. The drugs in order of increasing absorption are fenbendazole, mebendazole, thiabendazole, and albendazole. Mebendazole and fenbendazole are primarily excreted unchanged in feces. Thiabendazole and albendazole undergo hepatic metabolism and are excreted primarily in urine.

The benzimidazoles are effective against roundworms, whipworms, and hookworms in most species and have some activity against tapeworms. In horses they are effective against large and small strongyli, and pinworms. Migrating larvae to strongyli are not susceptible. All of the major gastrointestinal parasites are eliminated in ruminants. Albendazole and fenbendazole are effective against lungworms in all species. In dogs and cats, fenbendazole is effective against *Taenia* sp. but not the more common tapeworm *Dipylidium caninum*. Albendazole and fenbendazole are effective against *Giardia* sp.

The benzimidazoles are generally well tolerated and safe for administration to young, pregnant, sick, or debilitated animals. Mebendazole and oxibendazole have been associated with idiosyncratic hepatotoxic reactions in dogs. Bone marrow suppression has been reported following treatment of dogs with albendazole. Albendazole is contraindicated in cattle and sheep during early pregnancy because it is teratogenic and embryotoxic. Thiabendazole has been associated with idiosyncratic toxic skin reactions in dogs.

Febantel is another broad-spectrum benzimidazole. It is metabolized in the gastrointestinal tract to fenbendazole and oxfendazole, which accounts for its antihelmintic activity. Its spectrum of action is similar to fenbendazole alone, but it is also effective against *Echinococcus* sp. The drug is safe for use during pregnancy and in young animals.

Imidazothiazoles

Levamisole works as a ganglionic stimulant (cholinomimetic) causing worm paralysis (**Figure**). At high doses it also interferes with nematode carbohydrate metabolism by blocking fumarate reductase and succinate oxidase (*see* **Figure**). Levamisole undergoes systemic absorption upon oral administration. It is metabolized in the liver and excreted in the urine. In food animals, levamisole has a broad spectrum of action against roundworms, hookworms, and lungworms. It is not effective against whipworms. Levamisole has a relatively low therapeutic index. Signs of toxicosis are related to stimulation of mammalian nicotinic and muscarinic cholinergic synapses. These signs are similar to organophosphate intoxication and include salivation, defecation, respiratory distress, and seizures. Due to its low margin of safety, it is rarely used in horses or small animals.

Tetrahydropyrimidines

Pyrantel is a depolarizing neuromuscular blocker that has an irreversible cholinergic action on the musculature of the parasite (*see* **Figure**). It is available as either the tartrate or pamoate salt. Pyrantal tartrate is more water soluble and better absorbed from the gastrointestinal tract. The pamoate salt is less water soluble and more poorly absorbed. This accounts for its increased efficacy against pinworms in the large intestine. Absorbed pyrantel is metabolized in the liver and excreted in the urine and feces.

Pyrantel has a large margin of safety and is considered safe to administer to young, sick, or pregnant animals. Because of its cholinergic properties, it should not be administered concurrently with other medications that have a similar mechanism of action such as levamisole or organophosphates. It is antagonistic with piperazine.

In small animals, pyrantel is effective against hookworms and roundworms, but has no action on tapeworms or whipworms. In horses, it is effective against roundworms, large strongyli, and pinworms. It is somewhat less effective against small strongyli. At twice the recommended dosage, it is effective against equine tapeworms. The drug is also effective against the fourth-stage larval infections of large strongyli.

Morantel tartrate is a broad-spectrum antihelmintic for cattle and sheep and swine.

Organophosphates

Organophosphates inhibit nematode acetylcholinesterase (AChE), leading to interference with nematode neuromuscular transmission and paralysis (*see* **Figure**). Since the margin of safety of organophosphates is generally less than that of other broad-spectrum antihelmintics, they are infrequently used. They are no longer used in small animals, but have some application in swine and horses. Dichlorvos is an organophosphate that is incorporated into polyvinyl chloride resin pellets, which are slowly released in the gastrointestinal tract. This permits therapeutic concentrations of the drug throughout the length of the intestinal tract. In pigs it is effective against most nematodes including whipworms. In horses it is used primarily for its activity against stomach bots. Trichlorfon is another organophosphate that is used to treat bots in horses. Signs of organophosphate toxicity are outlined in Chapter 46, where their more common use as insecticidal agents is discussed. Organophosphates should not be used in young, sick, or pregnant animals.

45 Antiparasitic Drugs: Endoparasites II

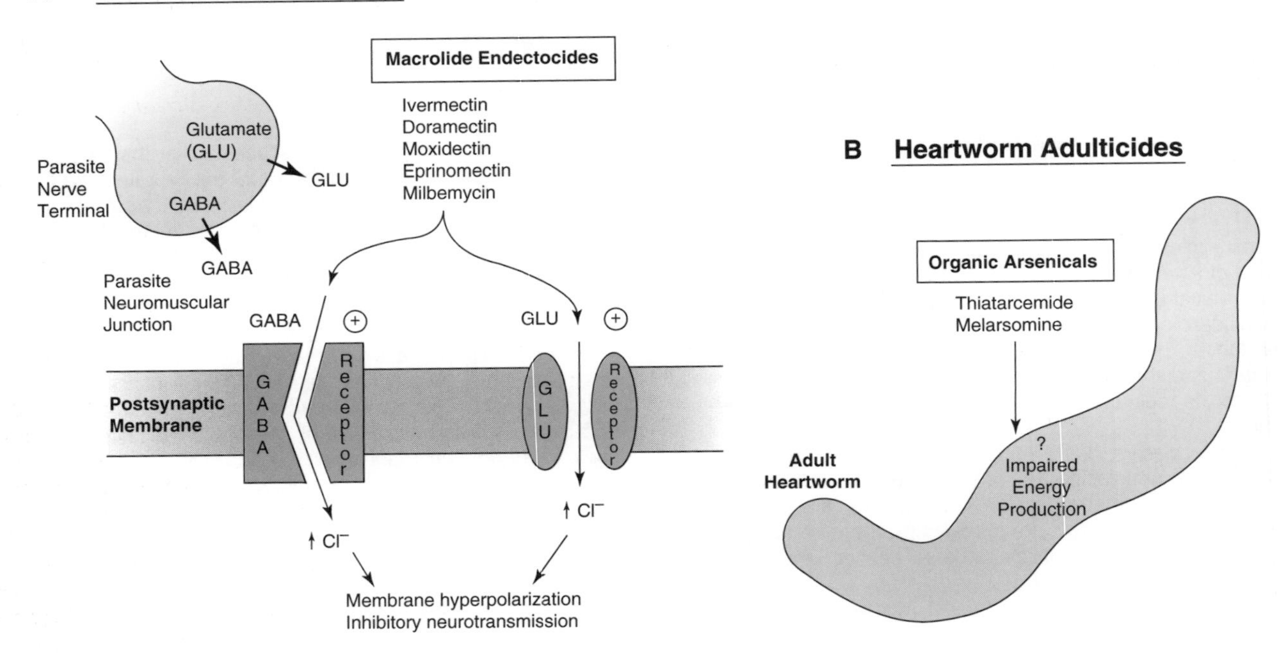

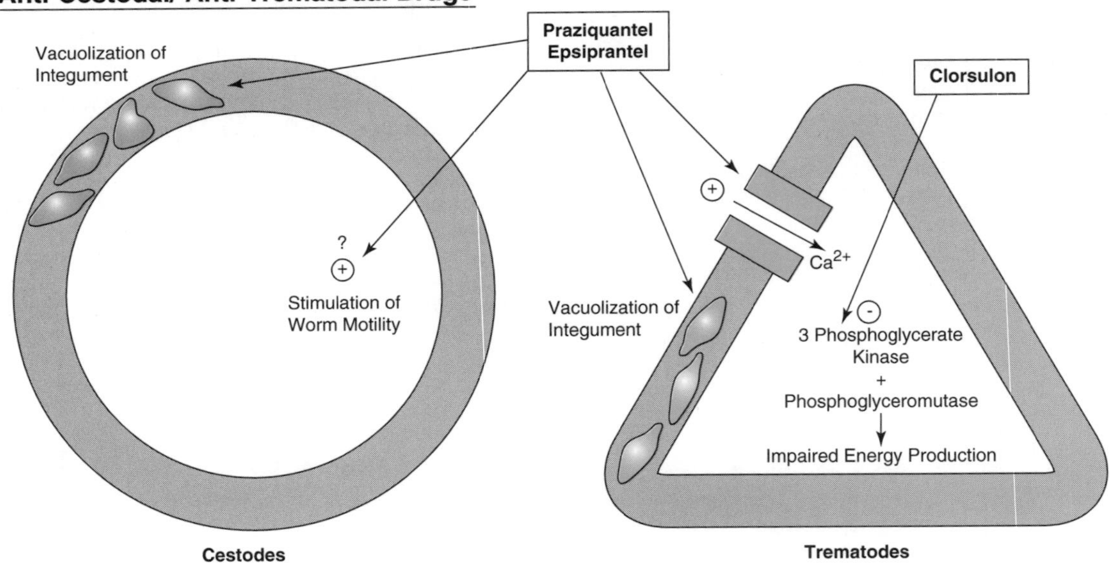

Macrolide Endectocides

The macrolide endectocides are divided into 2 groups: the avermectins (ivermectin, moxidectin, doramectin, and eprinomectin) and milbemycin. These drugs have a dual mechanism of action (**Part A**). First, they potentiate the action of the inhibitory neurotransmitter GABA. Binding of GABA to its receptor, a Cl^- channel, results in a rapid influx of Cl^- ions, membrane hyperpolarization, and inhibition of neurotransmission. Since nematodes and arthropods have GABAergic neurotransmission in their peripheral nervous system, potentiation causes worm paralysis. Selective toxicity to the worm is due to a much higher affinity of the drugs for the parasite GABA receptor and the fact that GABAergic neurotransmission is confined to the CNS in mammals. Since the macrolide endectocides are excluded from the brain, they have no effect on mammals. In addition to their action on GABA neurotransmission, the macrolides may also open glutamate-gated Cl^- channels. The macrolide endectocides have a wide margin of safety and can be used in young, sick, or pregnant animals.

Ivermectin

Ivermectin is a semisynthetic derivative of avermectin with broad-spectrum activity against nematodes and arthropods. It has efficacy against all major gastrointestinal and pulmonary nematodes, many ectoparasites (*see* Chapter 47), and infective-stage heartworm larvae in dogs and cats. The drug is available for oral and SC administration. After oral administration, 95% of the drug is absorbed. It is highly protein bound. Ivermectin undergoes extensive hepatic metabolism, with the majority of the inactive metabolites excreted in the bile. There is a delayed terminal $t^1/_2$ of several days in most species.

Ivermectin is approved for use as an antihelmintic at a dose of 200 μg/kg in horses, cattle, swine, reindeer, and bison. In cattle, sheep, and horses, ivermectin is effective against all important nematodes including grubs in cows, early- and late-fourth-stage larvae of *Strongylus vulgaris* migrating in the walls of the cranial mesenteric arterial system in horses, and the intestinal stage of *Trichinella* sp. in swine. It is also approved for use as a heartworm preventative in dogs and cats at a dose of 6 μg/kg and 24 μg/kg, respectively. Ivermectin is also effective as a heartworm microfilaricide at 50 μg/kg but it is not approved for this use. It has no activity against adult heartworms.

Ivermectin has an excellent safety profile except when administered to collie breed dogs. Signs of toxicosis in this breed or in animals receiving an overdose are mydriasis, depression, ataxia, and coma. In cattle, the drug is so effective against *Hypoderma* grubs, that animals with migrating larvae in their spinal cord or esophagus at the time of ivermectin treatment have developed paresis or edematous esophagitis.

Other Avermectins

Doramectin is a novel avermectin prepared by mutational biosynthesis. In cattle, it is effective against nematodes, lungworms, eyeworms, lice, grubs, ticks, mites, and screwworms. Moxidectin is a broad-spectrum antinematodal and antiarthropod drug marketed for use in dogs, cattle, sheep, and horses. Moxidectin is more lipophilic than other avermectins and tissue concentrations may persist longer than those of other members of this family. It has the same indications and spectrum of activity as ivermectin. Eprinomectin is available in a pour-on formulation for the treatment of endoparasites in cattle.

Milbemycin

Milbemycin oxime is marketed as a heartworm preventative for dogs. At heartworm preventative doses (0.5 mg/kg), it also controls infection with hookworms, roundworms, and whipworms in dogs. Milbemycin is also an effective heartworm microfilaricidal agent at this dose. Administration of milbemycin at higher doses (1–4 mg/kg) has some activity against canine demodicosis, but is not approved for this use. The drug has an excellent safety profile and heartworm preventative doses can be used safely in collies.

Heartworm Adulticides

Thiacetarsamide is an organic arsenical used as a heartworm adulticide in dogs and cats. The drug's exact mechanism of action is unknown but may be related to reaction of the arsenic with sulfhydryl groups in essential parasite energy systems (**Part B**). The drug is only effective against adult heartworms and has no activity against microfilariae. It is given IV. Extravascular administration must be avoided since the drug is a severe vesicant. It is metabolized in the liver and eliminated equally in feces and urine. The drug concentrates in the liver and kidneys and can be associated with nephrotoxicity and hepatotoxicity. Other side effects include gastrointestinal disturbances. The drug has a very narrow therapeutic index and seems to be more toxic when administered to cats.

Melarsomine is another organic arsenical used to treat adult heartworms. Compared to thiatarcemide, it is more efficacious and less likely to be associated with hepatotoxicity. It is administered by deep IM injection in the lumbar epaxial musculature in dogs. It should not be given IV or SC. About one-third of dogs experience muscle pain at the injection sites, which resolves within 1–2 weeks. Rarely, severe myositis has been reported. The drug should not be given to cats. Other side effects include depression, anorexia, fever, and vomiting.

Anticestodal and Antitrematodal Drugs

Praziquantel

Praziquantel is an isoquinolone whose mechanism of action is not totally understood. In cestodes it impairs the function of the parasite's suckers and stimulates worm motility. It causes irreversible vacuolization and eventual disintegration of the worm's integument (**Part C**). In trematodes it increases calcium ion flux into the worm and also facilitates phagocytic ingestion of the worm by causing focal vacuolization of the integument. Praziquantel is completely absorbed after oral administration and is widely distributed. It is metabolized in the liver into inactive metabolites, which are excreted primarily in the urine. Praziquantel is effective against all species of tapeworms in all animals. It is effective against the adult and juvenile forms of *Echinococcus* sp., but has limited activity against the larval stage (hydatid cyst). It also has some antitrematodal activity. It is not effective against nematodes. The drug has a high margin of safety. It can safely be used in pregnant, breeding, sick, and young animals.

Epsiprantel is a pyrazino-benzazepine with activity against intestinal tapeworms. It is not appreciably absorbed from the gastrointestinal tract and therefore has no activity against extraintestinal cestodes.

Benzimidazoles

Some of the benzimidazoles including fenbendazole, mebendazole, and albendazole have limited activity against cestodes. In dogs they are effective against *Taenia* sp., but not the more common tapeworm *Dipylidium caninum*. Albendazole is approved for beef and nonlactating cattle to treat the adult form of *Fasciola hepatica*.

Clorsulon

Clorsulon is a benzosulfonamide that works by inhibiting 3-phosphoglycerine kinase and phosphoglyceromutase, which are necessary for energy production in trematodes. It is considered a safe drug and can be used in breeding and pregnant animals. It is approved for the treatment of immature and adult forms of *F. hepatica* in beef and nonlactating dairy cows.

46 Antiparasitic Drugs: Ectoparasites I

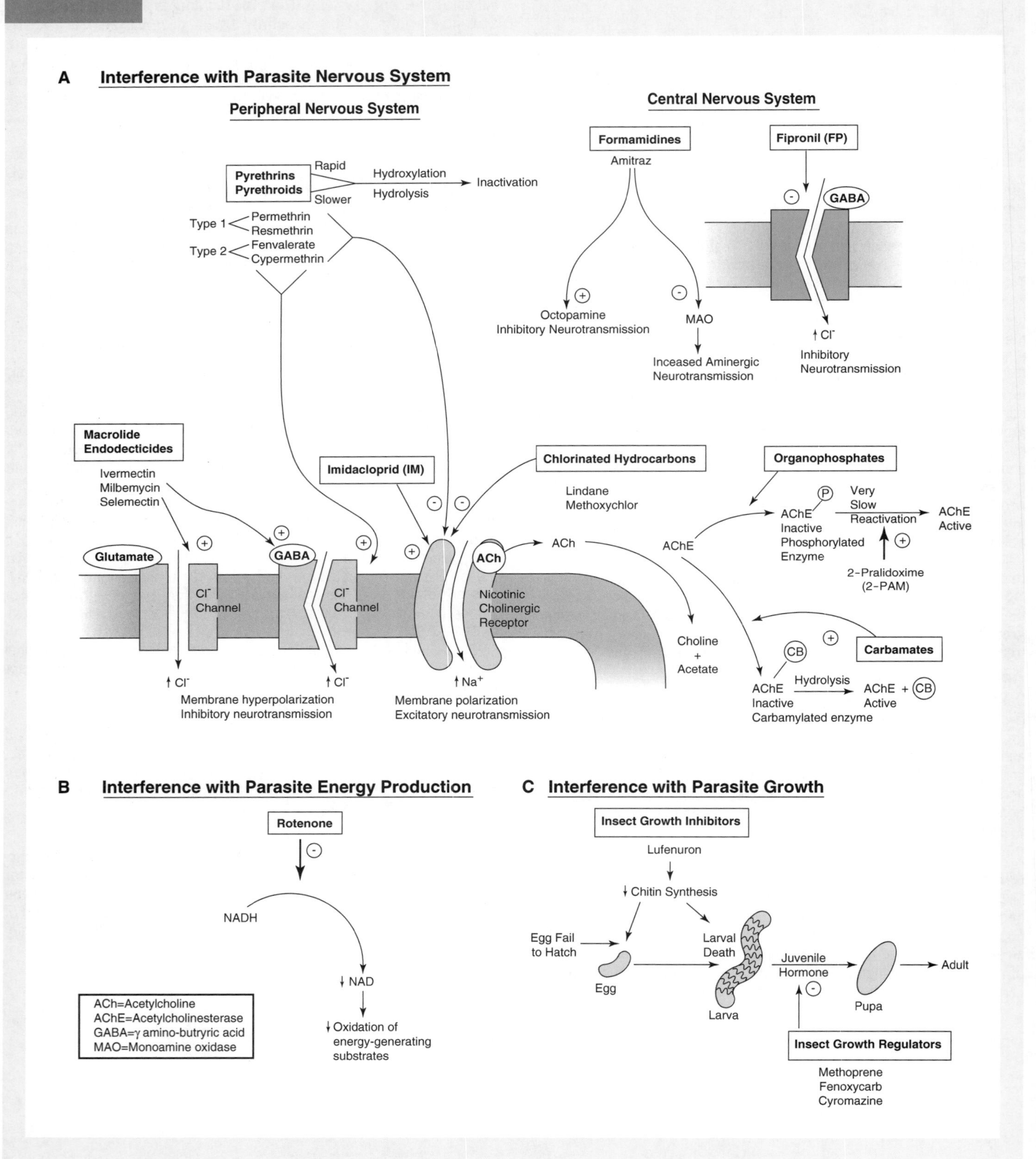

Ectoparasites of veterinary importance include insects (fleas, lice, flies) and acarines (mites and ticks). Drugs used to control these parasites include the pyrethrins and pyrethroids, chlorinated hydrocarbons, organophosphates, carbamates, the formamidines, insect growth inhibitors and regulators, and the avermectins and milbemycins.

Botanical Compounds

Pyrethrin insecticides are natural botanicals derived from the chrysanthemum flower. The pyrethroids (permethrin, cypermethrin, fenvalerate, resmethrin) are synthetic analogues of pyrethrins. These drugs exert their action by slowing conduction in voltage-sensitive sodium channels (**Part A**). The pyrethroids are classified as type 1 or type 2 based on their neurophysiologic effects. Type 1 pyrethroids (permethrin, resmethrin) prolong individual sodium channels long enough to cause repetitive discharges, leading to tremors. The type 2 pyrethroids (fenvalerate, cypermethrin) cause a more prolonged opening of sodium channels, leading to eventual suppression of neurotransmission. They also increase release of the inhibitory peripheral neurotransmitter γ-aminobutyric acid (GABA).

The pyrethrins and pyrethroids have efficacy against flies, fleas, lice, and ticks. These drugs exert a rapid knockdown effect, but have little residual activity. They are commonly used in combination with synergists such as piperonyl butoxide or *n*-octyl bicycloheptane dicarboximide to increase their potency. These synergists inhibit the enzymes responsible for degradation of the pyrethrins or pyrethroids.

The pyrethrins and pyrethroids are relatively safe insecticides. Adverse reactions, which are rare, include depression, hypersalivation, muscle tremors, vomiting, ataxia, dyspnea, and anorexia. Products containing synergists should be used with caution in cats because they increase the likelihood of a toxic reaction. Permethrin is the most widely used pyrethroid. The superior stability of permethrin results in some residual activity for up to 28 days. Permethrins have a high incidence of toxic reactions in cats and should not be used in this species.

Rotenone, another botanical insecticide, inhibits the respiratory system of the parasite by blocking NADH oxidation and subsequent ATP generation (**Part B**). Although it is more toxic than the pyrethroids or pyrethrins, it is still used in dogs and cats. Signs of toxicosis include gastrointestinal upset, respiratory stimulation, seizures, coma, and death.

d-Limonene and linalool are natural insecticides found in the oils of citrus fruit rinds. They are quick-acting agents without any residual action. Both are considered safe, although toxic reactions (hypersalivation, ataxia, and muscle tremors) have been reported in cats.

Chlorinated Hydrocarbons

The chlorinated hydrocarbons in use today are lindane and methoxychlor (*see* **Part A**). These agents lack the environmental persistence that was characteristic of earlier chlorinated hydrocarbons such as DDT. Although the exact mode of action is unknown, these drugs may target the nervous system by affecting sodium channels or potentiate GABAergic neurotransmission. Clinical signs of toxicosis may be immediate or, since the drugs are stored in body fat, they may be delayed for days. These signs include apprehension, exaggerated response to stimuli, vomiting, muscle twitching, tremors, and seizures. Lindane is used in tick sprays for horses and methoxychlor is a common ingredient in products to treat fleas, ticks, and lice.

Organophosphates

Organophosphates are a large group of drugs with insecticidal, acaricidal, and helminthicidal properties. These drugs work by binding to and inhibiting the enzyme acetylcholinesterase (AChE), which degrades acetylcholine (ACh) (**Part A**). ACh thus accumulates at the neural synapse, resulting in continuous neuronal stimulation and paralysis of the parasite. The binding of AChE with organophosphates results in phosphorylation and inhibition of the enzyme. The phosphorylated enzyme is capable of reactivation, but this reaction proceeds so slowly that the organophosphates are considered irreversible AChE inhibitors. After a period of time, the phosphorylated enzyme complex ages and reactivation is not possible. Commonly used organophosphates include chlorpyrifos, coumaphos, cythioate, diazinon, dichlorvos, fenthion, malathion, phosmet, tetrachlorvinphos, and trichlorfon. These products are effective on flies, fleas, ticks, mites, and lice.

The organophosphates have a low margin of safety. They should not be used with other agents that inhibit AChE or block neuromuscular transmission, and care should be taken when used in combination with CNS depressants. Organophosphates should not be used in young, sick, or pregnant animals or in greyhounds, whippets, Persian cats, and certain breeds of cattle (Chianina, Charolais, Simmental, Brahman) due to unique sensitivity.

Clinical signs of organophosphate toxicosis are due to interference with muscarinic or nicotinic neurotransmission. Muscarinic side effects include miosis, lacrimation, salivation, diarrhea, frequent urination, bradycardia, and hypotension. Nicotinic effects are manifested as muscle twitching and fasciculations followed by severe weakness and paralysis. CNS signs such as depression and seizures may also be noted. Toxicosis is treated with anticholinergics (atropine) and antihistamines. Pralidoxime chloride (2-PAM) can be given in an effort to promote the reactivation of AChE (*see* **Part A**).

Carbamates

Carbamates are slowly reversible AChE inhibitors that carbamylate and inhibit enzyme action. When the bond joining the carbamate insecticide to the enzyme is hydrolyzed, the fully active enzyme is released (*see* **Part A**). The 2 most commonly used carbamates are carbaryl and propoxur. The clinical signs of toxicosis are similar to those seen with organophosphates.

Formamidines

Formamidines are acaricides that have a direct effect on the parasite's CNS. Although their exact mechanism of action is unknown, they may interfere with the enzyme monoamine oxidase, which is responsible for metabolism of neurotransmitter amines, and/or activate the inhibitory neurotransmitter octopamine (*see* **Part A**). The formamidine amitraz is used to treat canine demodicosis and is present in tick collars. It is also marketed for use in cattle and swine to control ticks, mange mites, and lice. Signs of toxicosis include sedation, bradycardia, hypotension, mydriasis, and vomiting and diarrhea. Since some of the side effects are associated with amitraz's ability to activate α_2 adrenergic receptors, yohimbine, an α_2 antagonist, is used as an antidote.

Antiparasitic Drugs: Ectoparasites II

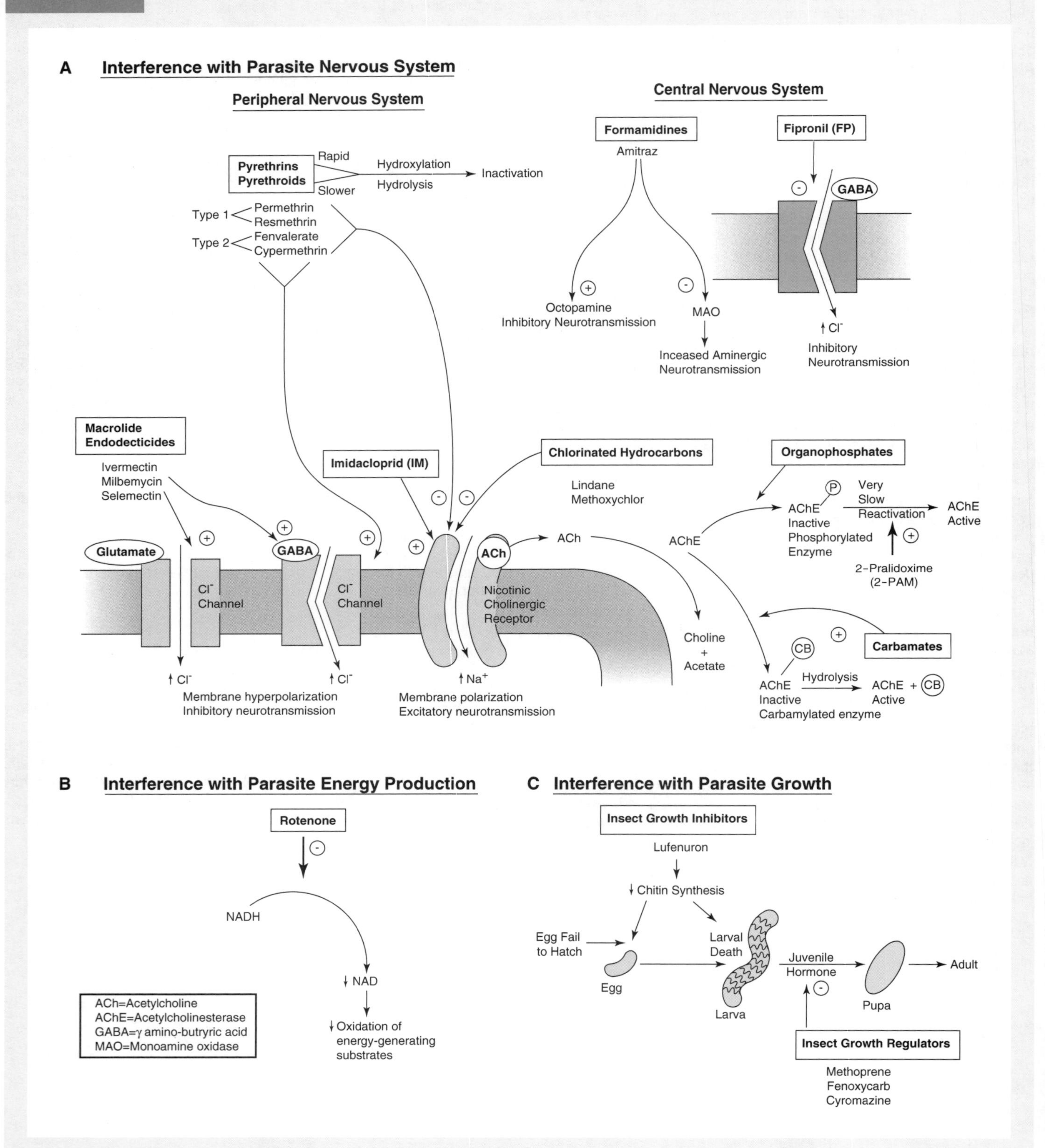

Insect Growth Regulators and Development Inhibitors

Insect growth regulators mimic the effects of an endogenous insect growth hormone, juvenile hormone. Juvenile hormone maintains the larval stage of the insect and inhibits maturation to the pupal and adult stages (**Part C**). These drugs include methoprene, fenoxycarb, and cyromazine. Insect growth regulators have very low toxicity for mammals. They are used in dogs and cats to control fleas and to control flies in cattle.

Insect development inhibitors interfere with the development of the insect's exoskeleton by inhibiting chitin synthesis or deposition (*see* **Part C**). These drugs include diflubenzuron and lufenuron. Lufenuron is marketed as an oral and parenteral flea-control product for dogs and cats. Lufenuron, a very lipophilic compound, accumulates in adipose tissue and is subsequently released slowly into the bloodstream. Adult fleas ingest lufenuron when they feed on the host and the drug is passed transovarially to the flea egg. Most flea eggs exposed to lufenuron fail to hatch and the ones that do die during the first molt. The drug is very safe and can be used in nursing, breeding, and pregnant animals and can be given to animals as young as 6 weeks. The drug is excreted unchanged in the feces. Lufenuron can be safely combined with other insecticidal products.

Imidacloprid

Imidacloprid is a member of a new class of insecticides that binds to the invertebrate nicotinic acetylcholine receptor and blocks neurotransmission (**Part A**). It is not degraded by acetylcholinesterase. Imidaclopramid is applied topically and distributes over the entire coat. It kills fleas, but not ticks. No systemic absorption of the drug occurs after topical administration. The product is applied once every 30 days, although it can be washed off by frequent bathing or swimming. No toxicity has been reported with the drug. Imidacloprid should not be used in animals <16 weeks old and its use in pregnant and sick animals is not recommended.

Fipronil

Fipronil is a phenylpyrazole that works by binding to the GABA receptor and blocking neurotransmission in the insect's CNS (*see* **Part A**). It is active against fleas and ticks for up to 1 month when applied topically. Fipronil dissolves in the oils of the skin and is deposited in the sebaceous glands, sebum, and hair follicles. The compound wicks out of the hair follicles continuously to cover the skin and fur. Reapplication after swimming or bathing is not necessary. No toxicity has been reported with the drug. Fipronil should not be used in pregnant, sick, or debilitated animals or in dogs or cats <10 or 12 weeks old, respectively.

Macrolide Endectocides

Ivermectin (*see* Chapter 45) possesses a broad spectrum of activity against ectoparasites. It is used to control grubs, lice, mites, and flies in horses, cattle, swine, and sheep. The ectoparasitical dose of ivermectin is 200 μg/kg. Ivermectin is only approved for heartworm prevention in small animals at much lower doses. However, it is frequently used off label at the higher ectoparasitical dose to treat sarcoptic and Otodectic mange mites in dogs and cats, respectively. High doses of milbemycin oxime (0.5–1.0 mg/kg) or ivermectin (300–400 μg/kg) for 30 days have been used to treat canine demodicosis.

Selamectin is a semi-synthetic avermectin developed as a topical agent with activity against endoparasites and ectoparasites. It is applied topically, systematically absorbed, and distributed from the blood stream into the sebaceous glands of the skin which then provide a continuous reservoir of drug. In cats, the topical bioavailability is greater than in dogs and the drug is thus effective in cats against intestinal parasites. Selamectin is indicated in the prevention of heartworm disease in the dog an cat, killing of adult fleas and prevention of hatching of flea eggs in the dog and cat, control of ticks (*Dermacentor* sp. only) in dogs, treatment and control of sarcoptic mange in dogs, treatment and control of ear mites in dogs and cats, and treatment and control of hookworms and roundworms in cats. It provides protection against ectoparasites and heartworms infestation for up to 30 days. It is safe to use in cats and dogs 6 weeks of age or older and in avermectin-sensitive breeds of dogs.

48 Antineoplastic Agents I: Alkylating Agents, Antimetabolites, and Mitotic Inhibitors

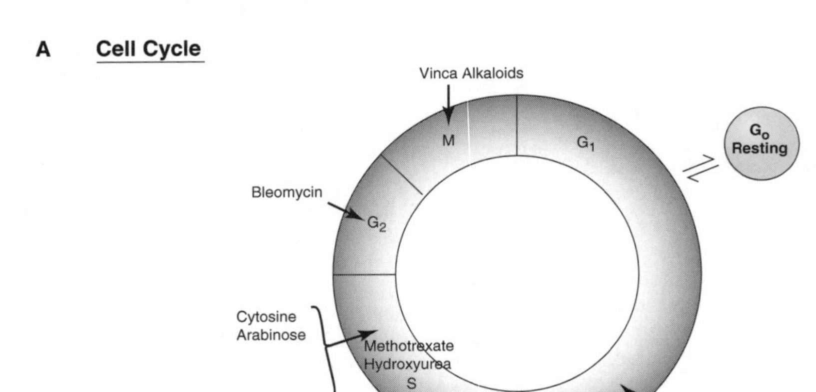

B Mechanisms of Action

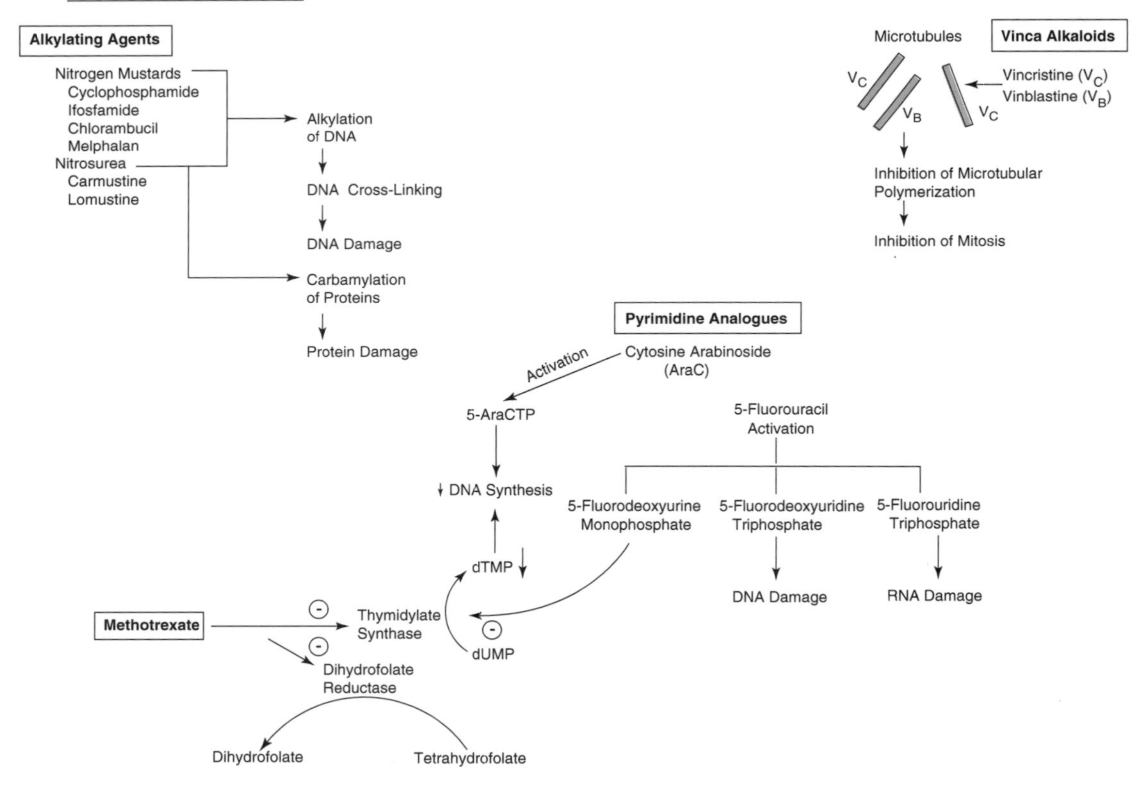

dUMP = Deoxyuridine Monophosphate
dTMP = Deoxythimidine Monophosphate
5-AraCTP = Ara-Cytidine Triphosphate

The goal of cancer treatment in veterinary patients is often remission or palliation and not clinical cure. The type of cancer and its clinical stage dictate which antineoplastic agents are utilized. Since cancerous cells and normal cells are biologically similar, the selectivity of antineoplastic agents is due to quantitative differences between the two cell types. Most drugs exploit the fact that neoplastic cells divide more rapidly than normal cells. An understanding of the kinetics of the cell cycle is essential for understanding the use of antineoplastic agents because most antineoplastic agents have activity only against cells that are in the process of cell division. The cell cycle is divided into five stages: G1, the phase of protein and ribonucleic acid synthesis; S, the phase of DNA synthesis; G2, the phase of additional protein and RNA synthesis; M, the phase of mitosis; and G_0, the resting phase (**Part A**). Nonproliferating cells in G_0 are usually resistant to chemotherapy. Many antineoplastic agents work at only one point in the cell cycle and are called *cell cycle–specific drugs*, while others are not cell cycle specific. Since some normal cells, particularly those in the bone marrow and gastrointestinal tract, divide rapidly, they are highly susceptible to toxicity from antineoplastic drugs. Often combinations of antineoplastic drugs with slightly different mechanisms of action, cell cycle specificity, or nonoverlapping toxicity are used to minimize toxicity to normal cells and maximize tumor kill.

Intrinsic or acquired resistance to antineoplastic agents is the major reason for chemotherapy failure. Resistance may develop due to a number of factors including: 1) enhanced drug efflux due to the expression of membrane drug transporters; 2) increased production of enzymes by the tumor cells that degrade the drug; 3) increased capacity of the tumor cell to repair DNA damage; and 4) decreased binding of drug to receptors or target enzymes in tumor cells.

Alkylating Agents

Alkylating agents are not cell cycle specific but are more effective against rapidly dividing cells. They work by substituting an alkyl group (alkylating) for a reactive hydrogen atom in DNA, leading to covalent cross-linking of DNA molecules (**Part B**). Alkylating agents in use in veterinary medicine include the nitrogen mustards and nitrosoureas. The major dose-limiting toxicity of these agents is bone marrow suppression. These agents are carcinogenic and mutagenic.

Nitrogen Mustards

Cyclophosphamide is used to treat lymphoreticular neoplasms and some carcinomas. This drug can be given orally or IV and is well distributed to most tissues, but not the CNS. It undergoes activation in the liver by the CP450 system to aldophosphamide, which is further metabolized in target tissue to phosphoramide mustard and acrolein. Besides myelosuppression, cyclophosphamide may cause vomiting, diarrhea, or a sterile hemorrhagic cystitis. Cystitis is due to chemical irritation from urinary excretion of acrolein. The incidence of this side effect can be minimized by encouraging diuresis during treatment or by concurrent administration of mesna (sodium 2-mercaptoethanesulfonate), which binds to and detoxifies the irritant metabolites. Ifosfamide is a structural analogue of cyclophosphamide that is also activated in the liver. Due to its greater propensity to cause cystitis, coadministration of mesna is recommended.

Chlorambucil is a slow-acting nitrogen mustard. It is generally well tolerated orally and is indicated in the treatment of chronic lymphocytic leukemia and multiple myeloma. Chlorambucil's myelosuppressive action is usually moderate, gradual in onset, and rapidly reversible. Melphalan is another slow-acting nitrogen mustard that is used to treat multiple myeloma in dogs and cats.

Nitrosoureas

The nitrosoureas include carmustine and lomustine. The agents are considered bifunctional because they alkylate DNA and carbamylate lysine residues in proteins. Carmustine and lomustine are highly lipophilic and thus cross the blood-brain barrier. This has made them useful in the treatment of brain tumors. They can be administered orally or IV. Although bone marrow depression is observed and may be severe, it is often cumulative and delayed. Cytotoxic effects on the liver, kidneys, and CNS have been reported.

Antimetabolites

Folic Acid Analogues

Methotrexate, a folic acid analogue, is actively transported into cells where it inhibits dihydrofolate reductase and thymidylate synthase, enzymes necessary for purine and pyrimidine synthesis (*see* **Part B**). It is a cell cycle–specific agent, being most effective against cells in the S phase. Methotrexate is readily absorbed from the gastrointestinal tract and is well distributed to tissues, except the CNS. Most of the drug is excreted unchanged in the urine by a combination of glomerular filtration and active tubular secretion. Severe myelosuppression and gastrointestinal toxicity are dose limiting. In addition, hepatotoxicity has been reported. In high-dose protocols, methotrexate toxicity to normal tissue can be limited by leucovorin (folinic acid) rescue. Leucovorin is a fully reduced folate coenzyme that is more rapidly transported into normal cells than tumor cells. It does not require reduction by dihydrofolate reductase. Methotrexate is primarily used to treat lymphoreticular neoplasms.

Pyrimidine Analogues

The pyrimidine analogue 5-fluorouracil is phosphorylated in cells to 5-fluoro-2-deoxyuridine-5-monophosphate (5FdUMP), which blocks thymidylate synthase activity, thereby inhibiting DNA synthesis. Fluorouracil is also metabolized to 5-fluorodeoxyuridine triphosphate and 5-fluorouridine triphosphate, which are incorporated into and alter the function of DNA and RNA (*see* **Part B**). Since gastrointestinal absorption is variable, fluorouracil is most often given IV. It is distributed widely including to the CNS. It undergoes extensive hepatic metabolism and excretion in the bile and urine. Although it is more effective against logarithmically growing cells, no cell cycle specificity has been assigned to this drug. Fluorouracil is most toxic to the bone marrow and gastrointestinal tract. In cats, the drug shows such severe CNS toxicity that it should not be used. It is most often used to treat carcinomas of the gastrointestinal tract.

Cytosine arabinoside is a pyrimidine analogue that is phosphorylated intracellularly to ara-cytidine triphosphate, an inhibitor of DNA synthesis (*see* **Part B**). It is considered S phase specific. It is capable of penetrating the CNS and has been incorporated into combination chemotherapy protocols for canine and feline lymphoma. The dose-limiting toxicity is myelosuppression and gastrointestinal upset.

Mitotic Inhibitors

Vinca alkaloids are natural substances obtained from the periwinkle plant. The vinca alkaloids bind to tubulin, a microtubular protein, preventing polymerization of microtubules (*see* **Part B**). This disrupts cellular mitosis and thus these drugs are cell cycle specific at the M phase. The vinca alkaloids must be given IV and are vesicants. They are rapidly distributed to many tissues, but not the CNS. They are metabolized in the liver and excreted into bile. The 2 most commonly used drugs, vincristine and vinblastine, although similar in structure and activity, differ in tumor efficacy and side effects. Vincristine is the more commonly used agent and is a frequent component in combination protocols for canine and feline lymphoma. It is also used to treat transmissible venereal tumors. Gastrointestinal toxicity is most often dose limiting, with anorexia, vomiting, and constipation being the most common side effects. The drug is only mildly myelosuppressive. A peripheral neurotoxicity may occur with long-term use. Vinblastine, which is incorporated into some protocols for lymphoma and sarcomas, is less well tolerated in small animals and is considered a potent myelosuppressive agent.

49 Antineoplastic Agents II: Antibiotics, Enzymes, Platinum Compounds, and Miscellaneous

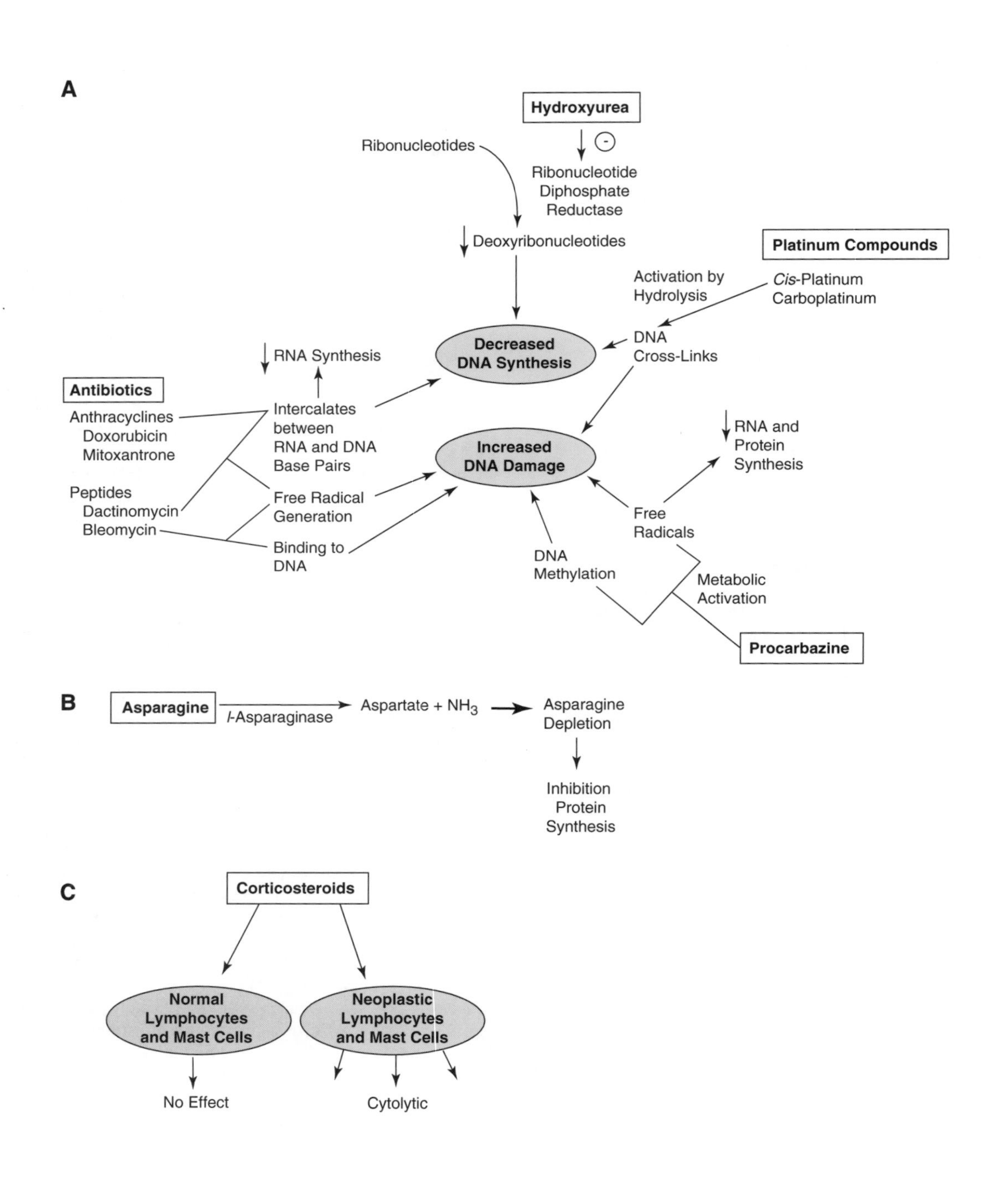

Antibiotics

The antineoplastic antibiotics in use in veterinary medicine include the anthracyclines (doxorubicin, mitoxantrone), dactinomycin, and bleomycin. These drugs primarily exert their cytotoxic action by interfering with DNA or RNA synthesis (**Part A**). Although they are most active on cells in the S phase, they are not considered cell cycle dependent. All of these agents must be given IV and their major dose-limiting toxicity is myelosuppression.

Doxorubicin is an anthracycline derivative that has wide activity against both sarcomas and carcinomas. It is incorporated into most combination protocols for canine and feline lymphosarcomas and has also been used as a single agent. Doxorubicin intercalates between DNA base pairs and thus interferes with DNA synthesis. It also leads to the generation of free radicals, which cause membrane damage and DNA strand breaks. Doxorubicin is metabolized in the liver and excreted in the bile and urine. Although the major acute dose-limiting toxicity is myelosuppression, doxorubicin can cause severe gastrointestinal upset (vomiting, hemorrhagic diarrhea). Rapid IV administration may result in the release of histamine, which is manifested as severe pruritus and facial swelling. Pretreatment with antihistamines and slow administration can prevent this reaction. Severe tissue damage can result from extravascular administration. Myocardial toxicity manifested by CHF secondary to cardiomyopathy limits the cumulative dose of doxorubicin that can be administered. The risk of cardiomyopathy increases above a cumulative dose of 250 mg/m^2 in dogs. Mitoxantrone was developed as a doxorubicin analogue and shares many of the same actions. However, it is not a vesicant and is somewhat less myelosuppressive and cardiotoxic.

Dactinomycin is a chromopeptide antibiotic with a similar mechanism of cytotoxicity as the anthracycline antibiotics. The drug is given IV and is excreted unchanged in the urine and bile. Toxicity includes myelosuppression and gastrointestinal signs. It does not penetrate the CNS. Although it has not received widespread use in veterinary medicine, it is generally considered effective against lymphoreticular neoplasms and germ cell tumors in humans.

Bleomycin is actually a mixture of glycopeptides with a terminal amino tripeptide that binds to DNA and a heavy metal-binding component that binds to Cu^{2+} and Fe^{2+}. The net result is inhibition of DNA synthesis and free radical generation. Bleomycin is most active in the G2 phase. The drug is widely distributed, but not in the CNS. It is metabolized by aminopeptidase present in many tissues. Since the skin and lungs are deficient in this enzyme, bleomycin accumulates in these organs and they are the major sites of toxicosis. Pulmonary fibrosis and changes in skin pigmentation occur. In addition, hypersensitivity reactions marked by fever, chills, and myalgia occur in many patients. Since bleomycin is not toxic to the bone marrow or gastrointestinal tract, it may be incorporated into combination protocols without fear of overlapping toxicity.

Enzymes

L-Asparaginase is used to treat canine and feline lymphomas. It hydrolyzes asparagine to aspartic acid and ammonia (**Part B**). Since malignant lymphocytes have low levels of L-asparaginase synthase, they rely on uptake of asparaginase from the extracellular environment to survive. L-Asparaginase is G1 phase cell cycle dependent. It is administered IM. Toxicity includes induction of an anaphylactic reaction, pancreatitis, and hepatotoxicity.

Platinum Coordination Complexes

Cis-platinum, diaminedichloroplatinum, is an inorganic platinum complex. It enters cells by passive diffusion, and following hydrolysis of the chloride moieties is activated to a reactive intermediate that cross-links DNA, leading to DNA damage and decreased DNA synthesis (*see* **Part A**). After IV administration it has a biphasic elimination curve with an initial $t^1/_2$ of about 20 min followed by a terminal $t^1/_2$ of 5 days. The latter may represent binding of the drug to tissue macromolecules. Highest tissue concentrations are found in the liver, ovaries, testes, and uterus. Nephrotoxicity is the major dose-limiting toxicity. It is dose related and cumulative. Diuresis during administration reduces nephrotoxicity. The drug may also cause ototoxicity and neurotoxicity, although rarely. Its use in cats is contraindicated because cats develop a fatal pulmonary vasculitis. Transient nausea and vomiting are common after administration and may be severe. Some veterinary patients show prolonged anorexia (1–2 weeks) after administration. Myelosuppression occurs but is less severe than with other drugs. It is used to treat a variety of carcinomas and osteosarcoma.

Carboplatin is an analogue of *cis*-platinum that is less nephrotoxic and ototoxic and less likely to induce emesis. It can be given without concurrent diuresis. It does not accumulate in the body, being eliminated in the urine within 3–6 hours. Its dose-limiting toxicity is myelosuppression, which typically occurs with a nadir of 2 weeks.

Corticosteroids

Corticosteroids are cytotoxic for neoplastic lymphocytes and have been incorporated into most combination chemotherapy protocols for canine and feline lymphosarcoma (**Part C**). They are also cytotoxic to mast cells and used to treat mastocytosis. Glucocorticoids are not cell cycle specific. They are readily absorbed from the gastrointestinal tract and well distributed to tissues including the CNS. The drugs are metabolized in the liver and excreted by the kidneys. Side effects include polyuria, polyphagia, immunosuppression, gastrointestinal disturbances, pancreatitis, and glucose intolerance. Long-term use leads to iatrogenic hyperadrenocorticism (*see* Chapter 59).

Miscellaneous Agents

Hydroxyurea inhibits ribonucleotide diphosphate reductase, thus limiting the availability of deoxyribonucleotides for DNA synthesis (*see* **Part A**). The drug is S phase cell cycle dependent. It is absorbed well from the gastrointestinal tract and is distributed to most tissue including the CNS. Most of the drug is excreted unchanged in the urine. The primary use of the drug in veterinary patients is to treat polycythemia vera and essential thrombocytosis. The dose-limiting toxicity is bone marrow depression. Other side effects include gastrointestinal disturbances.

Procarbazine is a non–cell cycle–specific drug used in the treatment of lymphoreticular neoplasms. The drug is well absorbed after oral administration and must undergo extensive metabolic activation by the hepatic CP450 system. The active metabolites result in the methylation of DNA and the generation of reactive free radicals that damage DNA (*see* **Part A**). Active metabolites are excreted in the urine. Side effects include vomiting and diarrhea, but the major dose-limiting toxicity is myelosuppression, which may be delayed for weeks after starting therapy. Since the drug is also a monoamone oxidase inhibitor and readily penetrates the blood-brain barrier, psychic disturbances may occur. It should not be given concurrently with tricyclic antidepressants, sympathomimetic agents, and foods with high tyramine content such as liver and cheese. It is a potent carcinogenic and teratogen.

50 Gastrointestinal Drugs: Drugs Controlling Emesis

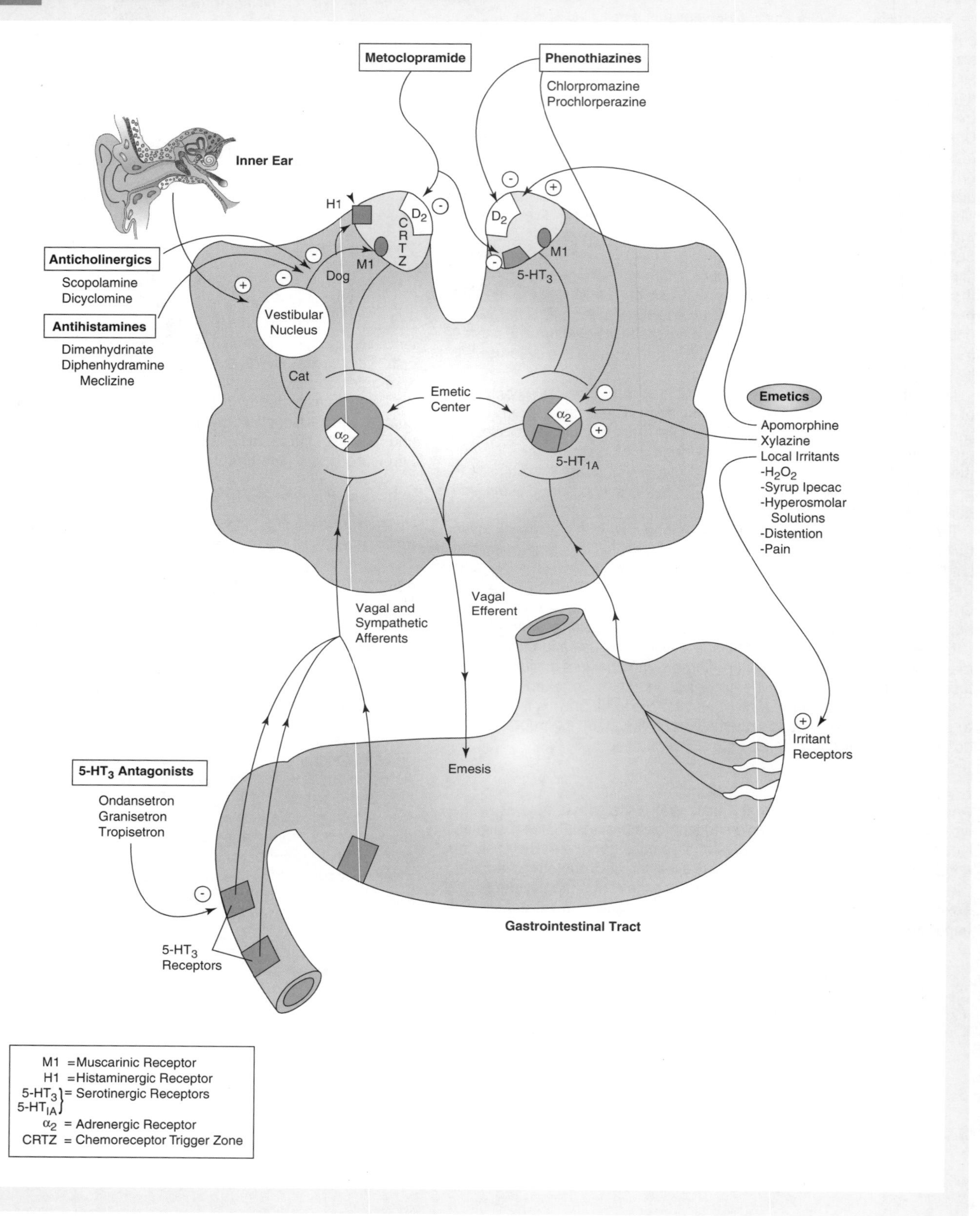

Physiology of Vomiting

Vomiting is a highly coordinated protective reflex. It is initiated by stimuli that travel via afferent neuronal pathways to the emetic center in the lateral reticular formation of the medulla oblongata (**Figure**). The emetic center integrates this afferent information and relays efferent signals to the gastrointestinal tract to initiate vomiting. Afferent input into the emetic center arrives from 4 sources: 1) the chemoreceptor trigger zone (CRTZ), 2) the vestibular system, 3) peripheral sensory receptors, and 4) higher central nervous system centers.

The CRTZ is located in the area postrema in the lateral walls of the fourth ventricle. It is functionally devoid of the blood-brain barrier and, therefore, sensitive to blood-borne toxins. Dopaminergic neurotransmission is important in the CRTZ, although serotonin ($5\text{-}HT_3$), cholinergic (M1), histamine (H1), and α_2 adrenergic receptors are also present. Afferent input to the emetic center from the vestibular system originates in the semicircular canals of the inner ear, travels via cranial nerve VIII to the vestibular nucleus, and arrives at the emetic center after passing through the CRTZ. Motion sickness and labyrinthitis cause vomiting through this pathway. Neurotransmission is via muscarinic (M1) and histaminergic (H1) receptors. Peripheral sensory receptors are located throughout the body, but are concentrated within the gastrointestinal tract. Stimulation of these peripheral receptors, which are $5\text{-}HT_3$ serotoninergic, occurs in response to pain, inflammation, distention, or rapid changes in osmolality. Impulses are transmitted via sympathetic and vagal nerves to the emetic center. Neurotransmission from higher CNS centers is mediated by adrenergic and histaminergic pathways.

Antiemetics

Phenothiazines

The phenothiazines, chlorpromazine and prochlorperazine, are α_2 adrenergic receptor antagonists with antidopaminergic and weak antihistamine and anticholinergic properties. Phenothiazines are capable of blocking the vomiting reflex at both the CRTZ and the emetic center. These drugs are metabolized by the liver to glucuronide and sulfate conjugates and excreted in the urine. They are well absorbed orally, but have significant first-pass hepatic metabolism. Side effects include mild sedation and systemic hypotension (due to peripheral vasodilation secondary to α_1 adrenergic blockage). The phenothiazines are relatively contraindicated in animals with seizure disorders, as these agents may lower the seizure threshold. In general, phenothiazines are more effective antiemetics in dogs than cats.

Metoclopramide

The antiemetic effect of metoclopramide, a dopamine and $5\text{-}HT_3$ serotonin receptor antagonist, is mediated by blockage of neurotransmission in the CRTZ. It also has peripheral anticholinergic effects which are largely responsible for the drug's prokinetic effects (*see* Chapter 53). Metoclopramide is well absorbed from the gastrointestinal tract, but has a bioavailability of only 50% due to significant hepatic first-pass metabolism. Since metoclopramide has a short $t^1/_2$ (60–90 min), it must be administered frequently by intermittent bolus injection or given as a continuous-rate infusion. Metoclopramide is excreted unchanged in the urine and in the bile following hepatic metabolism. Side effects of metoclopramide include hyperactivity, restlessness, tremors, and constipation. These side effects are more frequent in horses and cats.

Peripheral 5-HT₃ Serotonin Antagonists

The $5\text{-}HT_3$ serotonin receptor antagonists block neurotransmission in the CRTZ and, more importantly, interfere with the action of serotonin on peripheral receptors in the gastrointestinal tract. These drugs are particularly effective in preventing emesis induced by cytotoxic chemotherapy agents. Ondansetron is the prototypical drug. Side effects in humans include headache, diarrhea, and mild increases in liver enzymes. Other $5\text{-}HT_3$ serotonin receptor antagonists include granisetron and tropisetron.

Opioids

Butorphanol is a synthetic opioid agonist-antagonist. The drug's antiemetic effect is most likely mediated by blockade of opioid receptors in the emetic center, as well as modulation of input from higher CNS centers. Sedation is a common side effect. It has been shown to be effective in preventing chemotherapy-induced emesis in dogs.

Antihistamines

In dogs, antihistamines are effective against vomiting associated with motion sickness. Antihistamines differ in their duration of action and the degree of sedation they cause. Dimenhydrinate, diphenhydramine, and meclizine are the most used antihistamines to control motion sickness. Antihistamines work poorly to control motion sickness in cats. This may be related to the fact that motion sickness in cats is not mediated by the CRTZ, but instead impulses pass directly from the vestibular system to the emetic center.

Anticholinergics

Antimuscarinic agents that penetrate the blood-brain barrier such as scopolamine and dicyclomine can be used to control motion sickness in dogs. Side effects of these agents include drowsiness, xerostomia, and constipation. Anticholinergics that do not cross the blood-brain barrier such as atropine, glycopyrrolate, propantheline, and isopropamide have no central antiemetic action. Any ability they possess to suppress emesis is related to inhibition of vagal afferent impulses, relief of gastrointestinal muscle spasms, and inhibition of gastrointestinal secretions. Since these drugs may also cause gastrointestinal ileus, they may actually promote vomiting and are relatively contraindicated in vomiting patients.

Emetics: Central Acting

Apomorphine

Apomorphine is an opioid dopaminergic agonist that induces emesis by activating the CRTZ. In general, it is an effective emetic only in dogs. It can be given parenterally or tablets can be placed in the conjunctival sac. Emesis occurs within 2–10 min. The latter is the preferred method because the tablet can be removed once the animal vomits, and complications of excessive administration such as protracted vomiting or CNS depression can be avoided.

Xylazine

Xylazine is an α_2 adrenergic receptor agonist that works as a centrally acting emetic in cats. The dose required to induce emesis in cats is far below that needed for tranquilization. Potential side effects include bradycardia.

Emetics: Peripheral Acting

Local Irritants

A number of agents can induce emesis by causing irritation to the gastrointestinal mucosa. Oral ingestion of a saturated solution of table salt (NaCl) or 3% hydrogen peroxide often causes vomiting in dogs and cats. Ipecec syrup is a local gastric irritant with some additional central emetic actions. It should never be used in cats, and repeated administration to dogs without induction of emesis should be avoided, owing to potential heart, hepatic, and renal toxicity.

51 Gastrointestinal Drugs: Drugs that Inhibit Gastric Acid Secretion

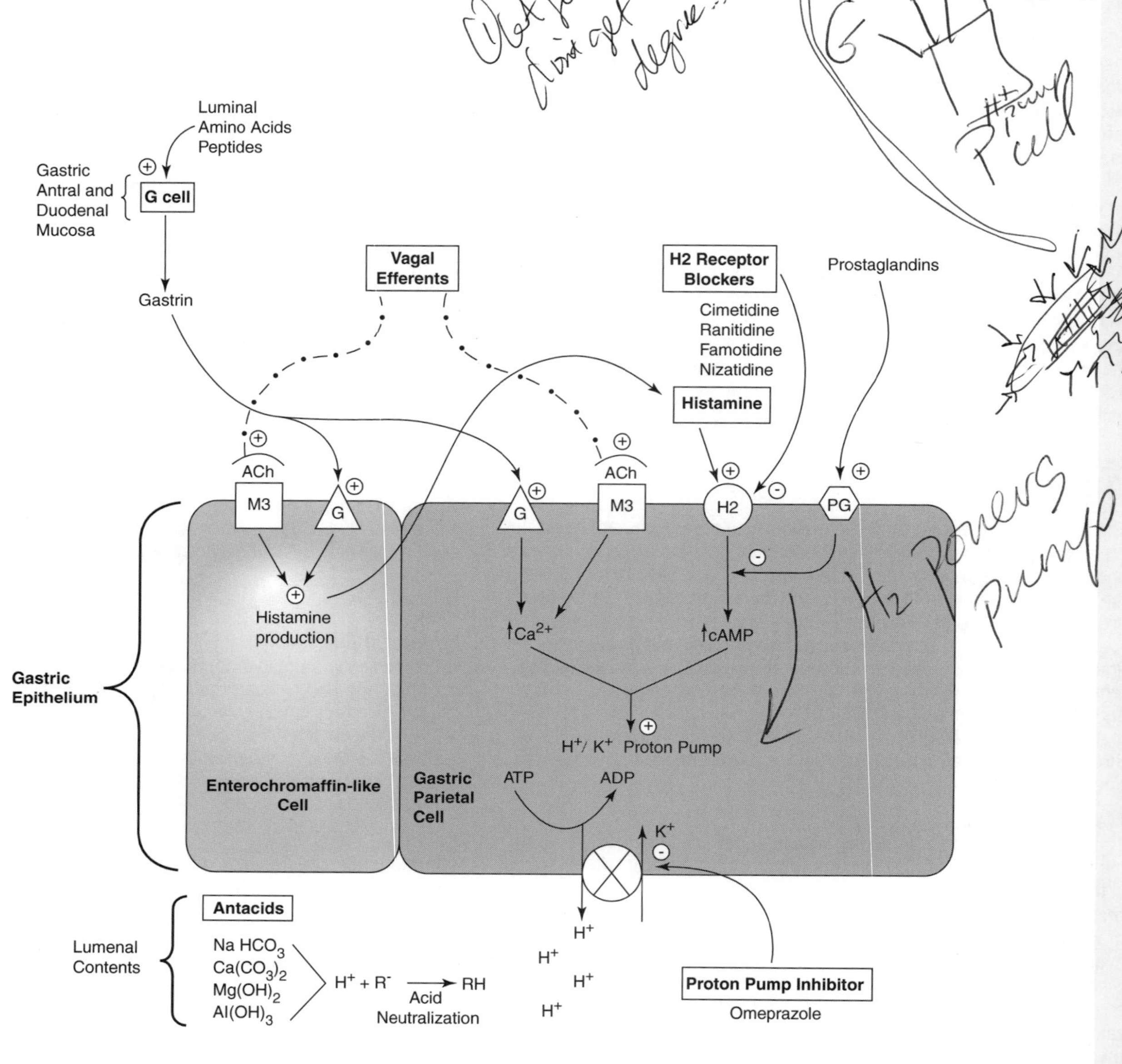

Parietal

Maintenance of Gastric Mucosal Integrity

The integrity of the gastric mucosa is maintained by a balance between gastric acid secretion and gastric mucosal defense mechanisms. Acid secretion by the gastric parietal cells is modulated by cholinergic neurotransmission, endocrine secretion of the peptide hormone gastrin, and paracrine secretion of histamine and prostaglandins. Simultaneous occupation of the gastrin, histamine (H2), and acetylcholine (ACh) (muscarinic, M3) receptors is necessary for optimum stimulation of acid secretion. When engaged, these receptors modulate signal transduction events that increase cAMP (histamine) or calcium (gastrin, ACh), which in turn activates the luminal H^+/K^+ ATPase proton pump on the parietal cell luminal surface, leading to H^+ secretion (**Figure**). Drugs can modulate gastric acid secretion either by blocking the interaction of secretagogues with their receptors, inhibiting intracellular events (decrease cAMP), or by blocking the ATPase pump.

The gastric mucosa has several protective mechanisms whereby it creates a barrier to protect itself from its harsh luminal environment. Gastric mucosal cytoprotection is mediated by a hydrophobic mucous layer, mucosal bicarbonate secretion, and maintenance of a high rate of mucosal blood flow and epithelial cell turnover. The mucous layer is a water-insoluble gel that adheres to the mucosal surface and promotes the formation of a stable unstirred layer. This layer traps bicarbonate secreted by the gastric mucosa, and together the mucous layer and the bicarbonate protect the gastric epithelium from acid back diffusion. Gastric mucosal blood flow is supplied by a dense capillary network beneath the epithelial surface and is maintained in large part by the vasodilatory effects of prostaglandins. A high rate of mucosal blood flow provides the constituents and energy necessary for the production of gastric secretory products and is vitally important for the maintenance of a high rate of gastric epithelial cell turnover.

Drugs that Inhibit Acid Secretion

H2 Receptor Antagonists

H2 antagonists reversibly inhibit the interaction of histamine with parietal cell H2 receptors (*see* **Figure**). The H2 antagonists in clinical use today, cimetidine, ranitidine, famotidine, and nizatidine, are analogues of histamine. The drugs differ in their potency, duration of action, disposition, and drug interactions. The order of potency as gastric acid inhibitors is as follows: famotidine > ranitidine = nizatidine > cimetidine. Despite differences in potency, all are equally effective in lowering gastric pH and promoting ulcer healing in humans. Cimetidine is the prototypic drug. It is readily absorbed from the gastrointestinal tract although food will delay absorption. It is given 3 or 4 times a day. Cimetidine inhibits the hepatic CP450 system and can interfere with the clearance of drugs metabolized by this system. Thus concurrent administration of cimetidine will prolong the $t^{1/2}$ of a number of drugs including theophylline, phenobarbital, coumadin, quinidine, and many benzodiazepines. Cimetidine also decreases hepatic blood flow, which may decrease the clearance of flow-limited drugs such as lidocaine and propranolol. Cimetidine undergoes some hepatic metabolism, but is largely excreted unchanged in the urine.

Ranitidine also inhibits hepatic CP450 enzymes, but not to the same extent as cimetidine. It undergoes partial hepatic metabolism with excretion into the bile (30% of an IV dose, 70% of an oral dose), the rest excreted unchanged in the urine. It is given 2 or 3 times a day. Famotidine has poor oral availability (37%), but is considerably more potent than cimetidine or ranitidine. It is given 1 or 2 times a day. Nizatidine has excellent bioavailability (97%) and is given once a day. Neither famotidine nor nizatidine inhibit hepatic CP450 enzymes and both are primarily eliminated unchanged in the urine. Ranitidine and nizatidine have promotility effects on the stomach (*see* Chapter 53). Since H2 antagonists increase gastric pH, they can interfere with the absorption of drugs that require an acidic environment for absorption, such as ketoconazole.

All of the H2 receptor antagonists are safe drugs. Cimetidine can cause CNS (mental confusion) side effects in humans and cutaneous reactions have been reported in cats. Cimetidine has antiandrogenic effects in humans (gynecomastia and impotence). Rapid IV infusion of H2 antagonists can cause bradycardia and hypotension due to the release of histamine.

Proton Pump Inhibitors

Omeprazole is a substituted benzimidazole that irreversibly inhibits the proton pump on the gastric parietal cell (*see* **Figure**). Since omeprazole can prevent gastric acid secretion by any secretagogue, it is the most potent gastric acid inhibitor. A single daily dose results in virtual anacidity. As the drug is a weak base and unstable at acid pH, it is supplied as an enteric coated capsule to prevent inactivation in the stomach. Omeprazole is absorbed in the alkaline environment of the duodenum and selectively partitions in the intracellular acidic environment of the gastric parietal cell. In this acidic environment, the drug is protonated and converted to its active form, which irreversibly inhibits the proton pump. In humans, the drug is highly protein bound and undergoes hepatic metabolism to inactive metabolites, which are excreted in the urine. Adverse effects are rare. A consequence of the profound inhibition of gastric acid secretion is increased gastrin levels. Omeprazole-induced hypergastrinemia causes parietal cell hyperplasia in laboratory animals and has led to the development of gastric carcinoids in rats. The clinical significance of these observations in veterinary patients is unknown. Omeprazole inhibits hepatic CP450 enzymes. Prolonged administration and the maintenance of low gastric pH may predispose to bacterial overgrowth in the gastrointestinal tract. The primary clinical indication for proton pump inhibitors is for the treatment of severe erosive esophagitis or ulcers associated with gastrinomas or mastocytosis. An equine product formulated as a paste is available to treat gastric ulcers in performance horses and foals.

Antacids

Antacids are basic compounds that neutralize acid in the gastric lumen. Antacids are available as suspensions of aluminum hydroxide, magnesium hydroxide, combinations of aluminum and magnesium hydroxide, and tablets with calcium carbonate or calcium carbonate–magnesium hydroxide combinations (*see* **Figure**). Aluminum or magnesium hydroxide–containing compounds are the most effective antacids. Aluminum-containing antacids have the added benefit of inactivating pepsin, binding bile acids, and inducing local prostaglandin synthesis. Antacids have a quick onset of action (within 30 min), but their duration of action is short (2–3 hours). Although antacids are as effective as H2 blockers in healing ulcers and are over-the-counter medications, their short duration of action limits their clinical usefulness. Failure to use antacids frequently results in acid rebound (i.e., induction of an increase in total acid secretion secondary to blockage of the inhibitory influence of gastric acid pH on gastrin release). This problem is circumvented if gastric acid is neutralized continuously as it is formed. The side effects of antacid therapy are cation dependent. They include constipation (aluminum and calcium), diarrhea (magnesium), hypercalcemia (calcium), and hypophosphatemia (aluminum). Since antacids alter gastric pH, they may alter the rates of dissolution, absorption, and bioavailability of a number of drugs. In general, it is best to avoid concurrent administration of antacids and other drugs. The use of sodium bicarbonate as an antacid should be avoided. As sodium bicarbonate is water soluble and thus absorbed systemically, it can promote the development of alkalosis. In addition, overuse may result in rapid generation of carbon dioxide and lead to gastric distention.

52 Gastrointestinal Drugs: Cytoprotective and Anti-inflammatory Agents

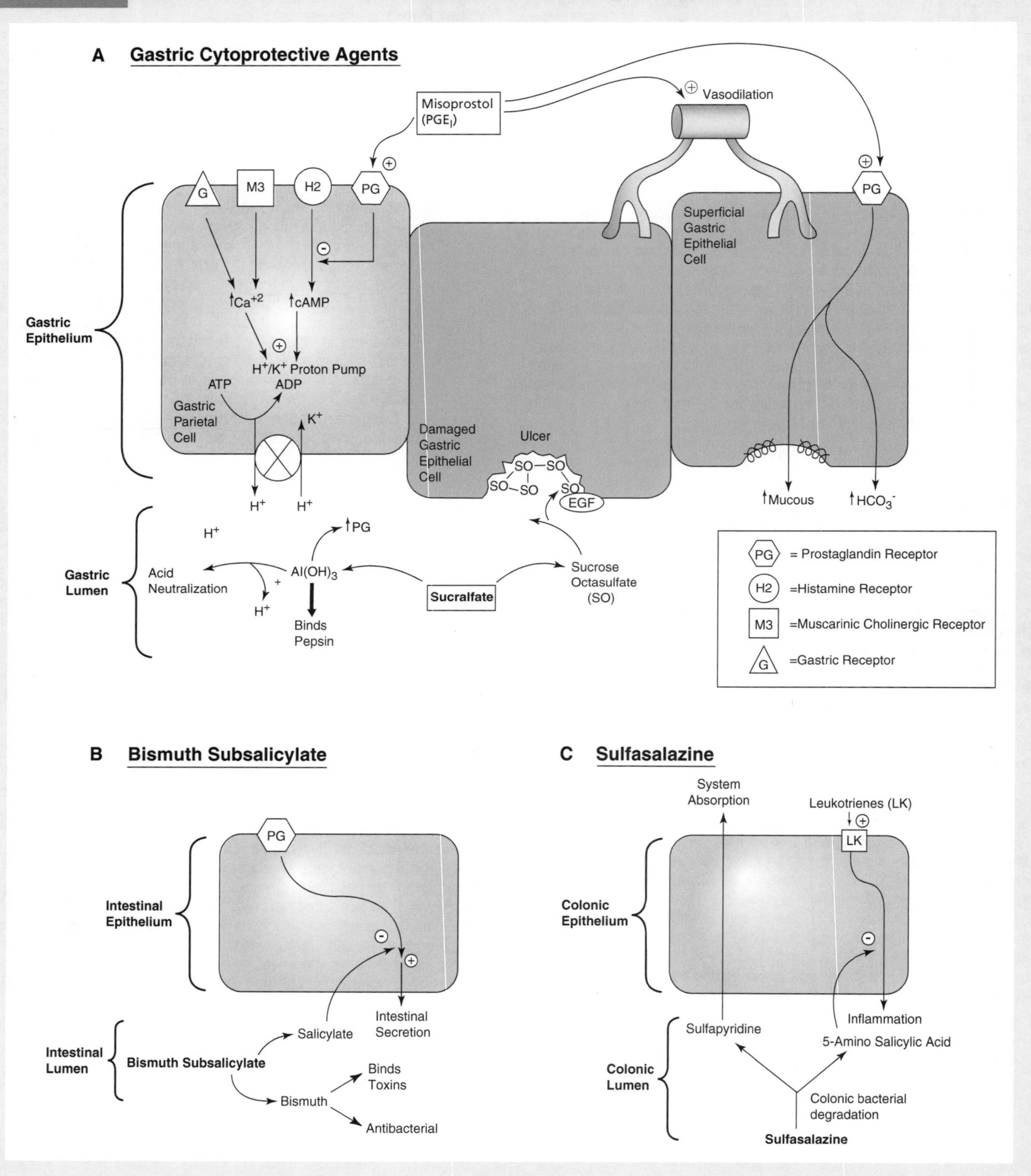

Gastric Cytoprotective Agents

Prostaglandins

Prostaglandins in the E series decrease gastric acid secretion and stimulate gastric mucosal protection (**Part A**). Misoprostol, a methyl ester analogue of PGE_1, is a gastric cytoprotective agent. It's cytoprotective effect is mediated by enhancement of mucosal blood flow, which results in increased bicarbonate secretion, mucus production, and epithelial cell turnover. It also binds to the prostaglandin receptor on the gastric parietal cell and decreases intracellular cAMP levels leading to decreased activity of the luminal H^+/K^+ ATPASE pump. This results in a decrease in gastric acid secretion. Misoprostol is well absorbed after oral administration. It undergoes significant hepatic first pass metabolism to misoprostol acid that is active. The free acid undergoes further hepatic biotransformation prior to excretion in the urine. Misoprostol may stimulate intestinal motility and secretion, leading to diarrhea. This side effect, however, is most often self-limiting. Misoprostol stimulates uterine contraction and should not be administered to or handled by pregnant individuals. The primary indication for misoprostol is in the prevention of nonsteroidal anti-inflammatory drugs such as aspirin and ibuprofen.

Sucralfate

Sucralfate is a mixture of sucrose octasulfate and aluminum hydroxide (*see* **Part A**). After oral administration, sucralfate dissociates in the acid environment of the stomach into aluminum hydroxide and sucrose octasulfate. The latter undergoes polymerization to form a paste-like complex that binds to damaged gastric epithelial cells. The insoluble complex forms a barrier that protects the ulcer from further damage. Sucralfate also binds to and inactivates bile acids and pepsin. The aluminum hydroxide is not of sufficient quantity to alter gastric pH, but does stimulate the formation of local mediators that protect the gastric epithelium such as prostaglandins and growth factors. Sucralfate binds epidermal growth factor, resulting in its accumulation in ulcer beds. Sucralfate enhances mucosal blood flow by increasing prostaglandin synthesis and stimulating the production of the local vasodilatory compound, nitric oxide.

Sucralfate is minimally absorbed after oral administration and thus quite safe. Local effects include constipation (due to the aluminum hydroxide) and interference with the absorption of other concurrently administered oral drugs. It is recommended that sucralfate be given at least 2 hours apart from other drugs. Sucralfate works best in the presence of an acid pH, but does not require it. The question as to whether it should be administered at the same time as H2 blockers is a mute one, since there is no additional therapeutic effect of coadministration. Sucralfate is indicated in the treatment of gastric and duodenal ulcers and erosive esophagitis.

Bismuth Subsalicylate

Bismuth subsalicylate has several beneficial actions on the gastrointestinal tract (**Part B**). The compound is cleaved in the intestine to bismuth and salicylate. Bismuth absorbs toxins (including endotoxin), coats ulcerated mucosal surfaces and has mild anti-bacterial action against gastric (*Heliobacter* sp.) and intestinal (*E. coli, Salmonella* sp., *Camphylobacter* sp) pathogens. The salicylate has an antiprostaglandin effect on the intestine that decreases intestinal secretions in many diarrheal diseases. Bismuth is minimally absorbed whereas salicylate is almost completely absorbed. The latter undergoes hepatic metabolism followed by renal excretion. Bismuth subsalicylate should be used with caution in cats since they are predisposed to salicylate toxicosis (due to poor hepatic glucuronidation of the compound). Bismuth subsalicylate may change stool color to resemble the presence of digested blood (melena). Bismuth is radiopaque and may confuse radiographic evaluation of the gastrointestinal tract.

Sulfasalazine

Sulfasalazine is used to treat inflammatory conditions of the large intestine (**Part C**). It is composed of sulfapyridine linked by a diazo bond to the diazonium salt of salicylic acid. Sulfasalazine escapes absorption in the small intestine and is delivered to the colon where the azo bond is split by resident bacteria, releasing sulfapyridine and 5-aminosalicylic acid (mesalamine). Its exact mechanism of action is not known, but the 5-aminosalicylic acid may decrease proinflammatory leukotriene production. Only a small portion of the sulfapyridine and 5-aminosalicylic acid is absorbed systemically. The amount that is absorbed is metabolized by the liver and excreted in the urine. Since cats have a decreased ability to metabolize salicylates, the drug should be used cautiously in them. The most common side effect is dry eye syndrome (rare), which is associated with the sulfapyridine portion. Olsalazine and balsalazide are similar to sulfasalazine but in olsalazine the azo bond links two 5-aminosalicylic acid molecules and in balsalazide 5-aminosalicylic acid is linked to an inert vehicle. There is little experience with these more expensive preparations in veterinary patients.

Metronidazole

The antibiotic metronidazole is capable of inhibiting cell-mediated immunity and has proved useful in the management of inflammatory gastrointestinal conditions. Metronidazole may also inhibit leukocyte–endothelial cell adhesion.

53 Gastrointestinal Drugs: Drugs Affecting Motility

Gastrointestinal Promotility Agents

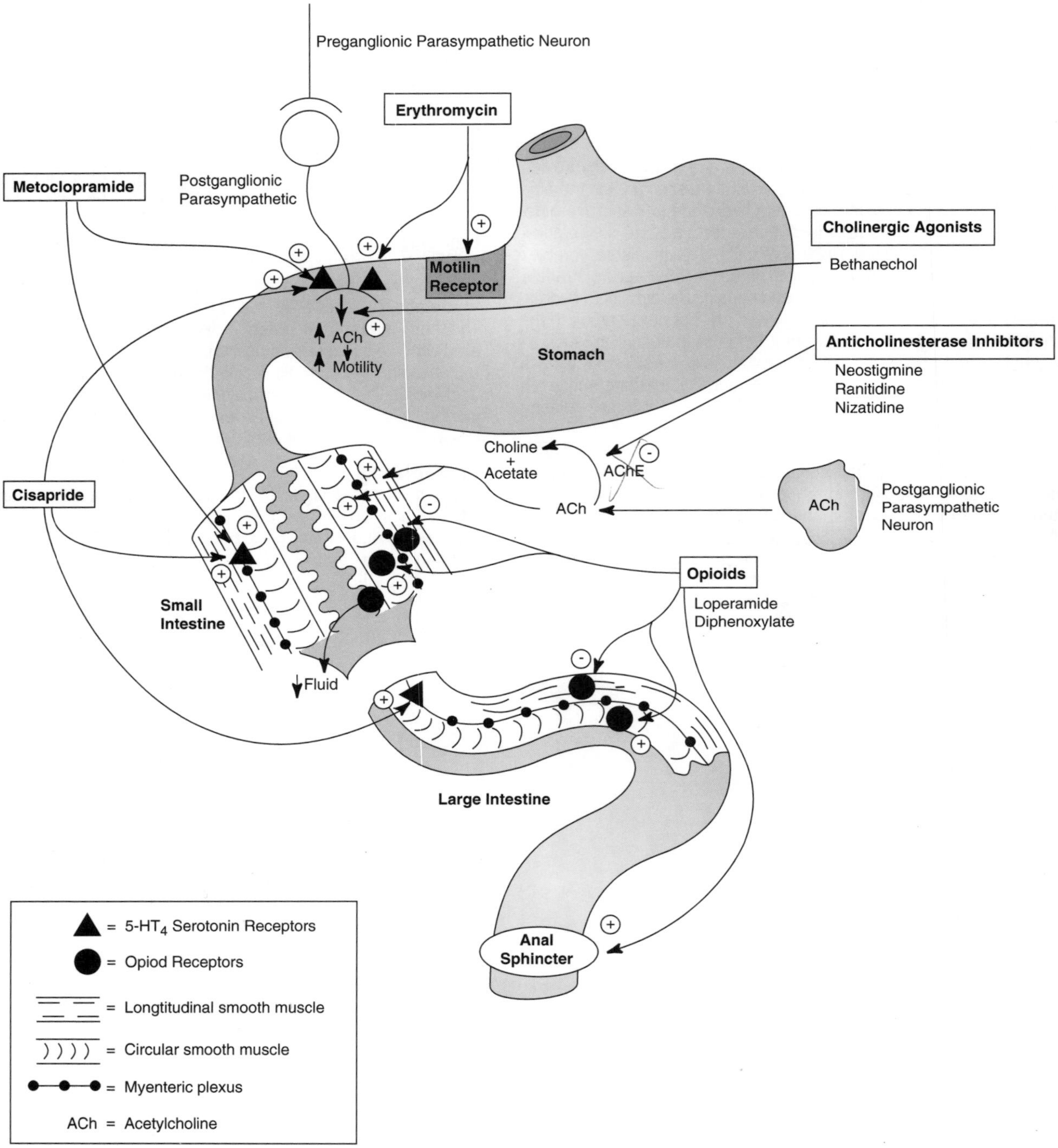

Preganglionic Parasympathetic Neuron

Erythromycin

Metoclopramide

Postganglionic Parasympathetic

Motilin Receptor

Cholinergic Agonists

Bethanechol

↑ ACh

↑ Motility

Stomach

Anticholinesterase Inhibitors

Neostigmine
Ranitidine
Nizatidine

Choline + Acetate

AChE

ACh

ACh

Postganglionic Parasympathetic Neuron

Cisapride

Opioids

Loperamide
Diphenoxylate

Small Intestine

↓Fluid

Large Intestine

Anal Sphincter

▲ = 5-HT$_4$ Serotonin Receptors

● = Opiod Receptors

= Longtitudinal smooth muscle

= Circular smooth muscle

= Myenteric plexus

ACh = Acetylcholine

Physiology of Gastric and Intestinal Motility

The contractile activities of the gastric smooth muscle are regulated to perform 3 functions: 1) receptive relaxation following ingestion of a meal, 2) mixing and breakdown of ingested material, and 3) coordination of gastric contraction with pyloric and duodenal contraction so that the gastric contents are propelled into the small intestine. The movement of contents through the intestinal tract is the net effect of 2 types of motility. Rhythmic segmentation is characterized by contractions of the circular muscle layer and increases resistance to intestinal flow. This allows digestive and absorptive processes time for completion. Peristalsis results from constriction of the longitudinal muscle layer and functions to move intestinal contents aborally. In the fasted state, large "housekeeping" contractions referred to as the migrating motility complex sweep through the gastrointestinal tract to propel indigestible solids aborally.

Regulation of gastrointestinal smooth muscle activity occurs at 3 levels: 1) extrinsic autonomic nervous system, 2) intrinsic (myenteric plexus) nervous system, and 3) receptors mediated by neuropeptides. Acetylcholine (ACh) is the primary excitatory neurotransmitter of the intrinsic and extrinsic nervous systems. Adrenergic stimulation of the intrinsic nervous system inhibits gastrointestinal motility. Nonadrenergic, noncholinergic neural inhibition of gastrointestinal smooth muscle motility is mediated by nitric oxide. A number of neuroactive peptides including vasoactive intestinal peptide, substance P, serotonin, prostaglandins, and gastrin can modulate gastrointestinal motility.

Metoclopramide

Metoclopramide (2-methoxy-5-chloro-procainamide) has antidopaminergic and cholinergic properties. Its antiemetic action is mediated centrally by antagonism of dopamine neurotransmission in the chemoreceptor trigger zone (*see* Chapter 50). Metoclopramide's peripheral promotility effect is mediated by an enhancement of ACh release from postganglionic cholinergic neurons (**Figure**). This effect may be mediated by presynaptic 5-HT$_4$ serotonin receptors. Metoclopramide increases lower esophageal sphincter pressure and is thus useful in preventing gastroesophageal reflux. Metoclopramide promotes gastric emptying by increasing the amplitude and frequency of gastric antral contractions, inhibiting fundic receptive relaxation, and coordinating gastric pyloric and duodenal motility. Metoclopramide has no effect on motility in the distal part of the small intestine or the colon. It is well absorbed after oral administration, with a $t_{1/2}$ of 60–90 min in dogs. In dogs, metoclopramide undergoes hepatic metabolism to glucuronide and sulfate conjugates, which are excreted in bile and urine. Extrapyramidal side effects are occasionally noted, particularly in cats and horses.

Cisapride

Cisapride is a substituted benzamide that stimulates motility in the proximal and distal parts of the intestinal tract including the colon (*see* **Figure**). It increases lower esophageal sphincter pressure, promotes gastric emptying, and stimulates small and large intestinal contractions. Its prokinetic effects are due to enhancement of cholinergic neurotransmission in the myenteric plexus. It is about 8 times more potent than metoclopramide as a prokinetic agent. Since cisapride is not absorbed across the blood-brain barrier, it lacks metoclopramide's central antiemetic action and is not associated with extrapyramidal side effects.

Cisapride activates 5-HT$_4$ serotonin receptors on myenteric neurons, resulting in the release of ACh. In the feline colon, cisapride also stimulates contraction by a noncholinergic mechanism, which may involve direct stimulation of serotoninergic receptors on smooth muscle cells. Cisapride is well absorbed after oral administration, with a $t_{1/2}$ of 5 hours in dogs and cats. It is metabolized in the liver and excreted in the bile (75%) and urine (25%) and appears to be well tolerated. It has recently been withdrawn from the market due to fatal cardiac toxicity in humans.

Erythromycin

Erythromycin is a macrolide antibiotic. At low doses this antibiotic has effects on gastric and intestinal motility (*see* **Figure**). In cats, rabbits, and humans, erythromycin's promotility action is mediated by its effect as a motilin agonist. As a motilin agonist, it stimulates contractions similar to those of the phase III migrating motility complex. In cats, erythromycin also increases lower esophageal sphincter pressure. In dogs, erythromycin's gastric prokinetic action is not completely understood, but may involve stimulation of 5-HT$_3$ serotonin receptors on postganglionic cholinergic neurons, resulting in the release of ACh and/or substance P.

Erythromycin-induced contractions begin in the stomach and migrate to the duodenum, jejunum, and ileum. It has no effect on colonic motility. Erythromycin is metabolized in the liver and excreted into the bile (66%) and urine (33%). The prokinetic dose is one-tenth of the antimicrobial dose. Side effects at this dose are minimal.

Cholinergic Agents

In addition to their effects on gastric secretion, the H2 receptor antagonists ranitidine and nizatidine stimulate gastrointestinal motility by inhibiting AChE activity (*see* **Figure**). Both agents are better at stimulating gastric motility than intestinal motility. These promotility effects occur at the doses recommended for antisecretory activity.

Neostigmine is an AChE inhibitor that has been used to stimulate intestinal motility in postoperative ileus in horses. It is only active parenterally. Side effects include skeletal muscle weakness, colic, diarrhea, and nausea.

Bethanechol is a synthetic muscarinic cholinergic agonist. It can increase the amplitude of gastric contraction and has been used in horses to promote gastric emptying. Side effects include salivation, diarrhea, and abdominal pain.

Opioids

Opioids bind to μ and δ receptors on small and large intestinal smooth muscle (*see* **Figure**). They increase segmental circular smooth muscle contractions and decrease peristaltic longitudinal contractions. The net effect is to prolong intestinal transit time. Opioids modulate intestinal smooth muscle activity by inhibiting ACh release. Opioids also increase rectal sphincter tone, and decrease intestinal, pancreatic, and biliary secretion. These drugs are particularly effective in the symptomatic treatment of nonspecific diarrhea. Diphenoxylate hydrochloride is a meperidine derivative marketed in combination with atropine sulfate. The atropine is added so that its unpleasant side effects will limit the drug's abuse potential. Loperamide hydrochloride, a synthetic opioid marketed over the counter, has no systemic opioid agonist activity and does not cross the blood-brain barrier. Loperamide is slowly and incompletely absorbed after oral administration and therefore has low abuse potential. The opioids also have antisecretory activity in the small intestine, which enhances their efficacy as antidiarrheal drugs. Side effects of loperamide are rare, but include constipation, abdominal cramping, vomiting, hypersalivation, and bradycardia.

Anticholinergics

Anticholinergics reduce both peristaltic and rhythmic segmentation. The net effect is to create a functional dynamic ileus. Since even mild peristaltic contractions are capable of propelling contents through the intestine in the absence of rhythmic segmentation, intestinal transit may actually increase with the use of anticholinergics. Anticholinergics have no effect on intestinal secretion.

54 Gastrointestinal Drugs: Laxatives and Cathartics

A Stimulants

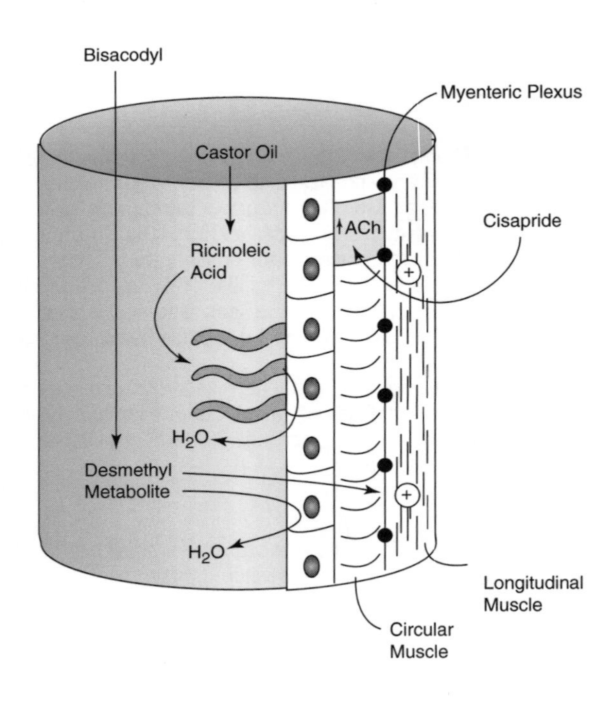

B Osmotic Agents

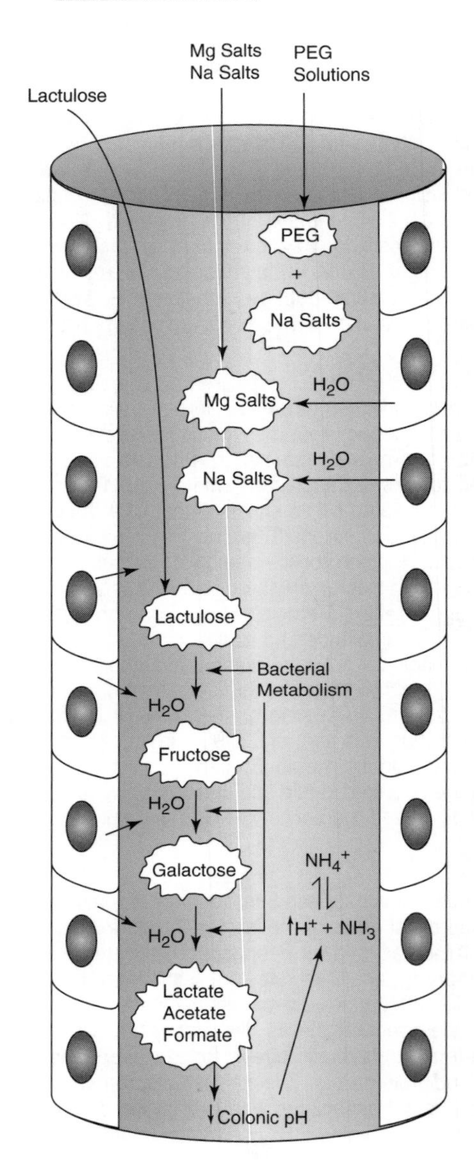

C Bulk-Forming/Lubricants/Surface-acting Agents

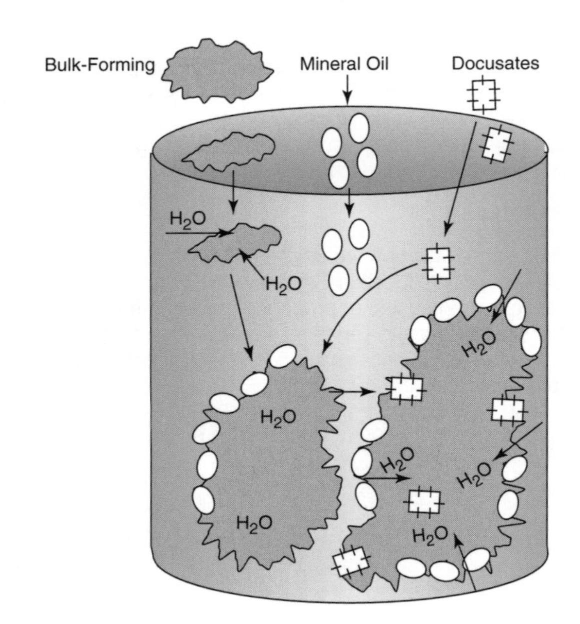

Cathartics and laxatives are a group of drugs that promote defecation. The terms *laxative* and *cathartic* imply different intensities of drug effect. Laxatives promote elimination of a soft formed stool, whereas the action of cathartic results in a more liquid stool. Cathartic and laxatives are used for: 1) cleansing of the bowel prior to radiographic procedures, elective surgery, or colonoscopy; 2) elimination of unabsorbed toxins in cases of acute poisoning; 3) reduction of fecal impactions; and 4) facilitation of defecation in patients with painful anorectal conditions. These agents may be classified into 5 groups: stimulants, osmotic agents, bulk-forming agents, lubricants, and surface-active agents.

Stimulants

These agents work on intestinal mucosa to reduce the net absorption of electrolytes and water, or by increasing colonic motility (**Part A**). Castor oil is a bland, nonirritating oil that following ingestion is hydrolyzed in the intestine to release ricinoleic acid, which stimulates colonic motility and secretion. The response to castor oil is prompt (4–6 hours in small animals and 12–18 hours in large animals) and results in thorough evacuation of the bowel.

Bisacodyl is a diphenylmethane derivative. After oral ingestion, its main action is in the colon. It is rapidly converted by intestinal and bacterial enzymes to its active desacetyl metabolite. Only a small amount is absorbed. It is subsequently glucuronidated and excreted in the urine and bile. Long-term use of the diphenylmethanes can damage the enteric myenteric plexus and cause inflammatory colitis.

Cisapride stimulates colonic motility by enhancing the release of acetylcholine from the myenteric plexus (*see* Chapter 33). Until its recent withdrawal from the market, cisapride was often combined with stool softeners in the chronic management of constipation due to feline idiopathic megacolon.

Osmotic Agents

Agents that act as osmotic laxatives include Mg^{2+} and Na^{+} salts, lactulose, and polyethylene glycol (PEG)–electrolyte solutions. Various salts of Mg^{2+} and the sulfate, phosphate, and tartrate salts of Na^{+} are poorly absorbed from the intestinal tract and act as osmotic particles in the lumen, causing the osmotic movement of water into the gut lumen (**Part B**). The increased luminal volume stretches the mucosa, stimulating mechanoreceptors that reflexly cause an increase in peristaltic activity. Sodium sulfate is the most effective and least expensive Na^{+} salt, but it is bitter tasting. The phosphate salts are more pleasant tasting. Sodium phosphate enemas are employed for rectal administration. Magnesium sulfate has an intensely bitter taste, while magnesium oxide and magnesium hydroxide are more pleasant tasting. When given at small doses, the Na^{+} and Mg^{2+} salts result in a laxative effect within 6–8 hours, whereas larger cathartic doses are capable of evacuating the bowel in <3 hours. Systemic toxicity can be associated with absorption of salt. The use of sodium phosphate enemas in cats is contraindicated, owing to the development of hyperphosphatemia and hypocalcemia. The sodium salts should not be used in patients with renal, hepatic, or heart failure. To avoid dehydration with these agents, their administration should be accompanied by an adequate amount of water to ensure no net loss of water from the body.

Lactulose is a disaccharide that escapes absorption in the small intestine. In the small and large intestine, lactulose exerts an osmotic effect to draw water into the lumen. This effect is enhanced by bacterial metabolism of the disaccharide into fructose and galactose and then to lactate, acetate, and formate. These metabolites are not well absorbed and contribute to the osmotic pressure. They also reduce colonic pH, which may enhance colonic motility and secretion. Lactulose is commonly employed in the management of feline idiopathic megacolon.

Lactulose is also used in the treatment of hepatic encephalopathy. Its beneficial actions in this condition include: 1) ion trapping of NH_3 (a decrease in colonic pH promotes protonation of NH_3), with subsequent excretion of NH^{4+} in the feces; 2) alteration in bacterial metabolism so that less NH_3 is produced; and 3) reduction in NH_3 absorption due to enhanced colonic transit time.

PEG-electrolyte solutions are mixtures of sodium sulfate, sodium bicarbonate, sodium chloride, and potassium chloride in an isotonic solution that contains PEG. When large volumes of this nonabsorbable mixture are ingested, copious diarrhea ensues. These products are used solely for the cleansing of the bowel for colonoscopy and radiographic examinations.

Bulk-Forming Laxatives

Dietary fiber is indigestible plant cell wall material. These plant wall materials include fibrillar polysaccharides (cellulose) and matrix polysaccharides (pectins, hemicellulose). High-fiber foods include whole grains, bran, vegetables, and fruits. Dietary fiber acts as a laxative by binding water in the colonic lumen, thereby softening stool and supporting the growth of bacteria, which increases fecal mass. Bacterial metabolism of fiber generates osmotically active metabolites that contribute to the laxative action by drawing water into the lumen. Palatable sources of dietary fiber for veterinary patients include bran or linseed mashes in horses and unprocessed wheat bran or canned pumpkin for small animals. Several dietary supplements containing psyllium husk or semisynthetic celluloses are available as bulk laxatives. Bulk-forming laxatives are usually effective within 24 hours of ingestion and reach a maximum effect after several days of repeated administration. Adverse effects of bulk laxatives are rare, although flatulence and borborygmi may occur. All of the agents should be given with adequate amounts of water to augment their action and prevent dehydration.

Lubricants

The lubricant laxatives soften and lubricate the fecal mass, which in turn facilitates expulsion (**Part C**). Mineral oil is a mixture of aliphatic hydrocarbons obtained from petroleum. The oil is indigestible and not appreciably absorbed. After administration for several days, it softens and lubricates stool. It is commonly given to equine patients by stomach tube to treat large bowel impactions. Its use in small animals is limited by its objectionable taste and the toxic consequences that might result from accidental inhalation (lipid pneumonia). In feline patients, white petroleum jelly or flavored petroleum-based products are used to treat constipation associated with the accumulation of trichobezoars.

Surface-Acting Agents

The docusates (docusate sodium sulfosuccinate, docusate calcium sulfosuccinate, and docusate potassium sulfosuccinate) are anionic surfactants (**Part C**). They have a detergent-like action that promotes the mixing of lipids, water, and other fecal material in an overall effect of softening the stool. They may also stimulate intestinal water and electrolyte secretion. They should not be given concurrently with mineral oil or other laxatives or drugs as they can enhance the intestinal absorption of many compounds.

55 Endocrine: Pituitary and Reproductive Hormones

A Reproductive Hormones

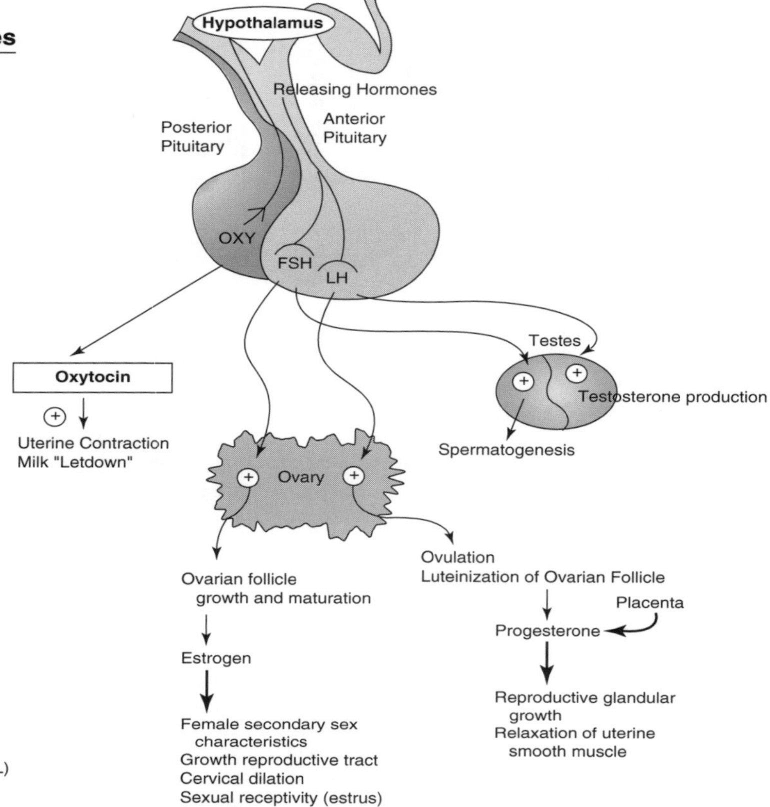

B Estrous Cycle

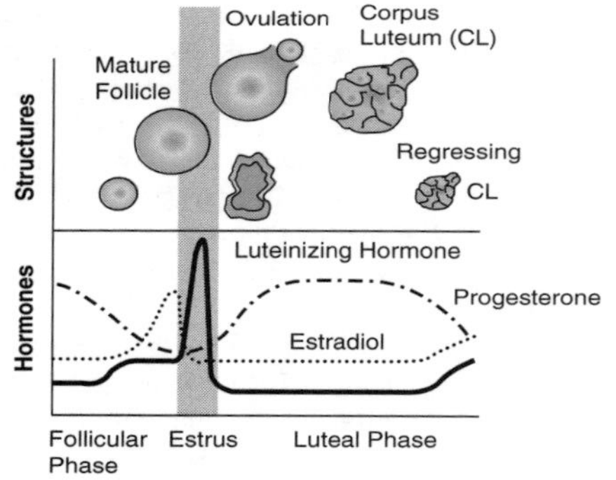

LH = Luteinizing Hormone
FSH = Follicle Stimulating Hormone
OXY = Oxytocin
CL = Corpus Luteum

C Pharmacological Use of Reproductive Hormones

Drug/Hormone	Use
Oxytocin	Uterine contraction in retained placenta, uterine inertia, and uterine prolapse Milk "letdown"
Gonadotropins HCG PMSF Gonadorelin Fertirelin Buserelin	Luteinization of follicular cysts Stimulate ovarian follicle development (superovulation) Induce ovulation
Progesterone Medroxyprogesterone Megestrol Melengestrol Altrenogest	Estrous synchronization Estrous postponement
Estrogen Estradiol Diethylstilbestrol	Growth promotion Canine urethral incompetence Prevention of ovum implantation
Testosterone Testosterone Mibolerone	Androgenize cows Estrous postponement Growth promotion
Prostaglandins $PGF_{2\alpha}$ Cloprostenol Fluprostenol Fenprostalene	Estrous synchronization Induce parturition Arbortifactant Treatment of canine open pyometra

The hypothalamus and pituitary gland regulate the secretion of reproductive hormones. The hypothalamus secretes releasing hormones, which mediate the release of prolactin, follicle-stimulating hormone (FSH), and luteinizing hormone (LH) from the anterior pituitary. The hypothalamus also directs the secretion of oxytocin from the posterior pituitary (**Part A**).

Reproductive Hormones

Gonadotropins

The gonadotropins LH and FSH are secreted from the pituitary gland under the control of gonadotropin-releasing hormone (GnRH) (*see* **Part A**). In females, FSH stimulates follicle development and estrogen secretion, while LH evokes ovulation and promotes luteinization, which increases progesterone secretion (*see* **Parts A** and **B**). Human chorionic gonadotropin (hCG) and pregnant mare serum gonadotropin (PMSG) are glycoproteins secreted from the placenta during pregnancy. These nonpituitary gonadotropins mediate a long-lasting effect. PMSG has an FSH-like effect while HCG has an LH-like effect (**Part C**). Gonadorelin is a synthetic human gonadotropin. Buserelin, leuprolide, and histrelin are synthetic GnRH preparations. These drugs are given parenterally to induce LH release and are used to achieve luteinization of follicular cysts in cows. Buserelin has also been used to stimulate ovarian activity in mares. FSH preparations are used to induce superovulation in cows. PMSG is used to stimulate graafian follicle development and thus estrus in pigs. HCG is used to induce ovulation in horses. Since it may be immunogenic, no mare should receive more than 2 injections in the same breeding season.

Prostaglandins

$PGF_{2\alpha}$ and its analogues (cloprostenol, fluprostenol, and fenprostalene) are luteolytic (*see* **Part C**). In cows, horses, and goats, $PGF_{2\alpha}$ is used for estrus induction and synchronization. $PGF_{2\alpha}$ is also used to induce abortion or parturition (via a decrease in progesterone level and induction of uterine contraction) in cows and pigs. It is somewhat less effective in inducing abortion in dogs. It is generally recommended that natural $PGF_{2\alpha}$ be used in dogs, rather than the more potent analogues, which cause an increased incidence of adverse effects. $PGF_{2\alpha}$, along with concurrent antibiotics, is used in the bitch and queen to treat open pyometras, as prostaglandins will stimulate uterine contraction and facilitate cervical dilation. $PGF_{2\alpha}$ therapy is continued for 3–5 days or until vaginal discharge is gone and uterine size is significantly reduced. Side effects seen in dogs undergoing treatment with prostaglandins include vomiting, ataxia, anxiety, abdominal cramping, diarrhea, and hypersalivation. Pretreatment with anticholinergics and a regular program of walking for 30–40 min after administration lessen side effects. In the cow, $PGF_{2\alpha}$ is used to treat retained placenta.

Progestins

Progesterone is secreted by the corpus luteum of cycling animals and the placenta of pregnant animals. Progesterone increases glandular growth after priming with estrogens, desensitizes uterine smooth muscle, and has anabolic actions (*see* **Parts A**, **B**, and **C**). Synthetic progesterones, known as progestins, include medroxyprogesterone, megestrol, melengestrol, and altrenogest. The progestins are used in many species for estrous synchronization. Administration of progestins mimics the effects of the corpus luteum and thus inhibits estrus. Following drug withdrawal, there is shortened interval to estrus, which results in estrous synchronization. Altrenogest is an orally administered progestin approved for use in controlling the estrous cycle of the mare. Although not approved for this use, it has been used to promote maintenance of pregnancy. Melengestrol is an orally active progestin supplied as a premix for cattle and sheep to synchronize estrus. Controlled internal drug release devices (CIDRs) containing progesterone for intravaginal use have been marketed to synchronize or control estrus in cattle, goats, and sheep. A progesterone-releasing intravaginal device (PRID) is a stainless-steel coil coated with silicone rubber that contains 6.75% progesterone. The PRID is used to synchronize estrus in cows.

Megestrol acetate prevents estrus in the bitch when treatment is started in anestrus. The agent is given once a day orally. It is not approved for use in cats but is effective in postponing estrus in this species. The drug is conjugated by the liver and excreted in the urine. Side effects in dogs include endometrial hyperplasia, endometritis, and pyometra. Since progestins inhibit uterine smooth muscle activity and relax the cervix, they predispose to uterine infection. Side effects in cats are adrenal suppression, induction of diabetes mellitus, hepatotoxicity, and the development of mammary hyperplasia or neoplasia. Due to its anti-inflammatory activity, megestrol is also used in cats to treat inflammatory disorders.

Depot injections of medroxyprogesterone can be used to postpone estrus in dogs and cats. Since these injections have been associated with a very high incidence of cystic endometrial hyperplasia and pyometra, their use is not recommended. Medroxyprogesterone is antigonadotropic, antiestrogenic, and antiandrogenic. It can also suppress ACTH and cortisol release. Due to these actions, it has also been used to treat aggressive behavioral disorders in dogs and cats and as an anti-inflammatory agent in cats.

Gonadal Hormones

Testosterone is secreted by the testis and is responsible for masculinization (*see* **Parts A** and **C**). It also has anabolic effects to increase protein synthesis, and to promote the growth of bone, cartilage, and other tissues. It enhances erythropoiesis by promoting the secretion of erythropoietin. Preparations include natural testosterone esters and mibolerone. The esters of testosterone have been used to androgenize cows to produce a "teaser" animal that facilitates the identification of cows in estrus and to treat urinary incontinence and dermatitis in male dogs. Mibolerone given orally prevents estrus in the bitch when given at least 30 days prior to the onset of proestrus. Adverse effects of testosterone include infertility or oligospermia, development of perianal adenomas, perineal hernias or prostatic disorders, masculinization of females, and hepatotoxicity.

Estrogens are secreted from the ovaries and placenta. They stimulate and maintain the reproductive tract, cause cervical dilation, stimulate mammary gland growth, increase sexual receptivity, and have protein anabolic effects. The most commonly used steroidal estrogen is estradiol, which is available as the cypionate, benzoate, and valerate ester. Zeranol and DES are nonsteroidal estrogens. Estradiol and zeranol are used for their growth-promoting effects in cattle and sheep. Estradiol cypionate used to be used as a "mismating" shot in dogs, but because of a high incidence of complications that accompany its use including pyometra and bone marrow hypoplasia, it is now contraindicated. Likewise, the treatment of prostatic hyperplasia or perianal tumors with estrogens is considered unsafe, owing to the possibility of fatal bone marrow toxicity as well as induction of prostatic metaplasia or prostatitis. DES is used in small animals to treat incontinence associated with decreased urethral sphincter tone.

Oxytocin

Oxytocin is synthesized in the hypothalamus and stored in the posterior pituitary (*see* **Parts A** and **C**). It induces contraction of uterine smooth muscle and results in milk letdown. It is used to induce parturition, reverse uterine inertia, induce milk letdown, treat retained placenta, and promote uterine involution following uterine prolapse repair. After parenteral administration, its onset of action is quick (1–5 min), with a duration of action up to 20 min. Oxytocin is metabolized to an inactive intermediate by the liver and excreted in the urine. Oxytocin is contraindicated in dystocia or when the cervix is not dilated. Side effects are rare. Overzealous administration can result in uterine rupture or fetal injury.

56 Endocrine: Thyroid Drugs

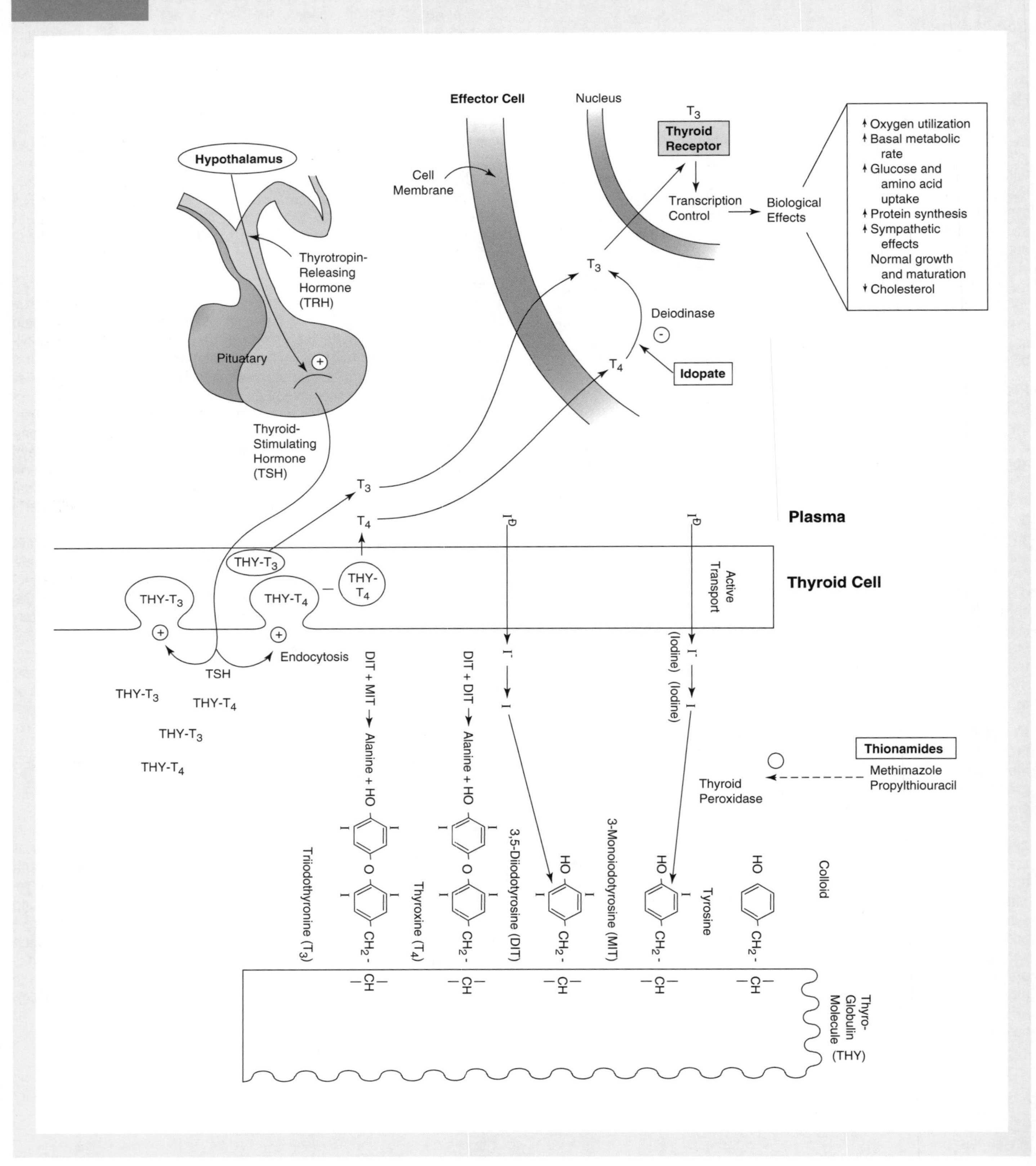

Thyroid

The functions of the thyroid hormone are to 1) maintain a basal level of metabolism in tissues that is optimal for their function, 2) stimulate oxygen consumption in most cells in the body, 3) regulate lipid and carbohydrate metabolism, and 4) promote normal growth and maturation. The hormones secreted by the thyroid gland are thyroxine (T_4) and triiodothyronine (T_3) (**Figure**). Iodide is the raw material for thyroid hormone synthesis. The thyroid gland actively concentrates iodide by transporting it into colloid. In the gland, iodide is rapidly oxidized to iodine and bound to tyrosine, a reaction catalyzed by thyroid peroxidase. Thyroglobin is synthesized within the thyroid gland and secreted into colloid. T_3 and T_4 are synthesized by iodination and condensation of tyrosine molecules, which are bound in peptide linkage to thyroglobin. T_3 and T_4 remain bound to thyroglobin until they are secreted. In serum, both T_3 and T_4 are highly protein bound (99.9%), with only a small percentage of each present in the free form. Free T_3 is the active hormone. T_3 can be formed from T_4 by deiodination of T_4 in the peripheral tissues. Thyroid hormone binds to a nuclear receptor and modulates the transcription of several genes. Thyroid hormone secretion is stimulated by the secretion of thyroid stimulation hormone (TSH) by the pituitary. The release of TSH is under the control of thyroid-releasing hormone (TRH) released from the hypothalamus. The secretion of both TSH and TRH is inhibited by free T_3 and free T_4. Clinical illness can occur when there is either a deficiency or an excess of thyroid hormone.

Thyroid Supplements

Thyroid hormone supplements can be divided into 3 classes: 1) crude hormones prepared from thyroid glands, 2) synthetic L-thyroxine, and 3) synthetic L-triiodothyronine. Crude thyroid hormone prepared from slaughterhouse thyroid glands is available in the form of desiccated thyroglobin. These preparations are useful to treat large animals with thyroid disorders as they are relatively inexpensive. These preparations have the disadvantage of variable biological activity and a short $t^1/_2$. Synthetic L-thyroxine is recommended as first-line therapy for thyroid hormone replacement in small animals. Therapy with levothyroxine results in normalization of both T_3 and T_4 levels. Response to therapy with levothyroxine usually occurs in 4–6 weeks. Replacement therapy with L-triiodothyronine is less desirable. This drug only normalizes T_3 levels. Normalization of T_3 levels may not meet the varied demands for thyroid hormone among different organs in the body. Organs such as the pituitary and the brain that have high requirements for T_3 maintain these levels by an increased capacity to 5′-deiodinate T_4 to T_3. Without normalization of T_4, administration of T_3 may shortchange these organs. Synthetic T_3 is used only when documented malabsorption of T_4 exists or when a peripheral deiodination defect is present. When thyroid hormone supplements are used at clinically appropriate doses, there is little concern for overdosage since mild overreplacement is asymptomatic in most animals. This resistance to iatrogenic thyrotoxicosis is related to an efficient capacity to clear thyroid hormone via biliary and fecal excretion.

Antithyroid Drugs

The thioureylenes are actively accumulated by the thyroid gland where they block thyroid synthesis by inhibiting reactions catalyzed by thyroid peroxidase (*see* **Figure**). These reactions include oxidation of iodine, the coupling of iodotyrosyl groups to form the ether linkage of T_3 and T_4, and iodination of tyrosyl residues in thyroglobin. They have no effect on the thyroid's ability to trap iodine nor do they inhibit the release of thyroid hormone. Methimazole is now the antithyroid drug of choice in treating thyrotoxicosis in small animals. The drug is available for oral administration, but has variable bioavailability. It is excreted in the urine as metabolites and unchanged drug. As intrathyroidal hormone stores must be depleted before a response to therapy occurs, there is a lag of 1–3 weeks before clinical improvement is noted and a significant reduction in serum T_4 occurs. The use of methimazole in cats has been associated with a number of side effects, including 1) hematological abnormalities (leukopenia, thrombocytopenia, hemolytic anemia, agranulocytosis, lymphocytosis, and eosinophilia), 2) gastrointestinal disturbances (anorexia and vomiting, which may be self-limiting), 3) dermatological reactions (facial pruritus and excoriations), and 4) hepatotoxicity. Many cats (52%) will become antinuclear antibody positive during chronic methimazole therapy, but a clinical syndrome resembling systemic lupus is rare. A reduction in dose results in normalization of the titer. Propylthiouracil has fallen from favor in the treatment of hyperthyroidism because of the high incidence of side effects associated with its use. These include gastrointestinal disturbances, anorexia, and the development of immune-mediated disorders.

Ipodate is a radiopaque organic iodine contrast agent (*see* **Figure**). The compound inhibits the 5′-deiodination of T_4 to T_3. In cats it decreases T_3, but not T_4. It is well tolerated by cats, but response to therapy may be short-lived (3–6 months). It does not work well in severe hyperthyroidism.

57 Endocrine: Drugs Affecting Glucose Metabolism

A Regulation Insulin Secretion

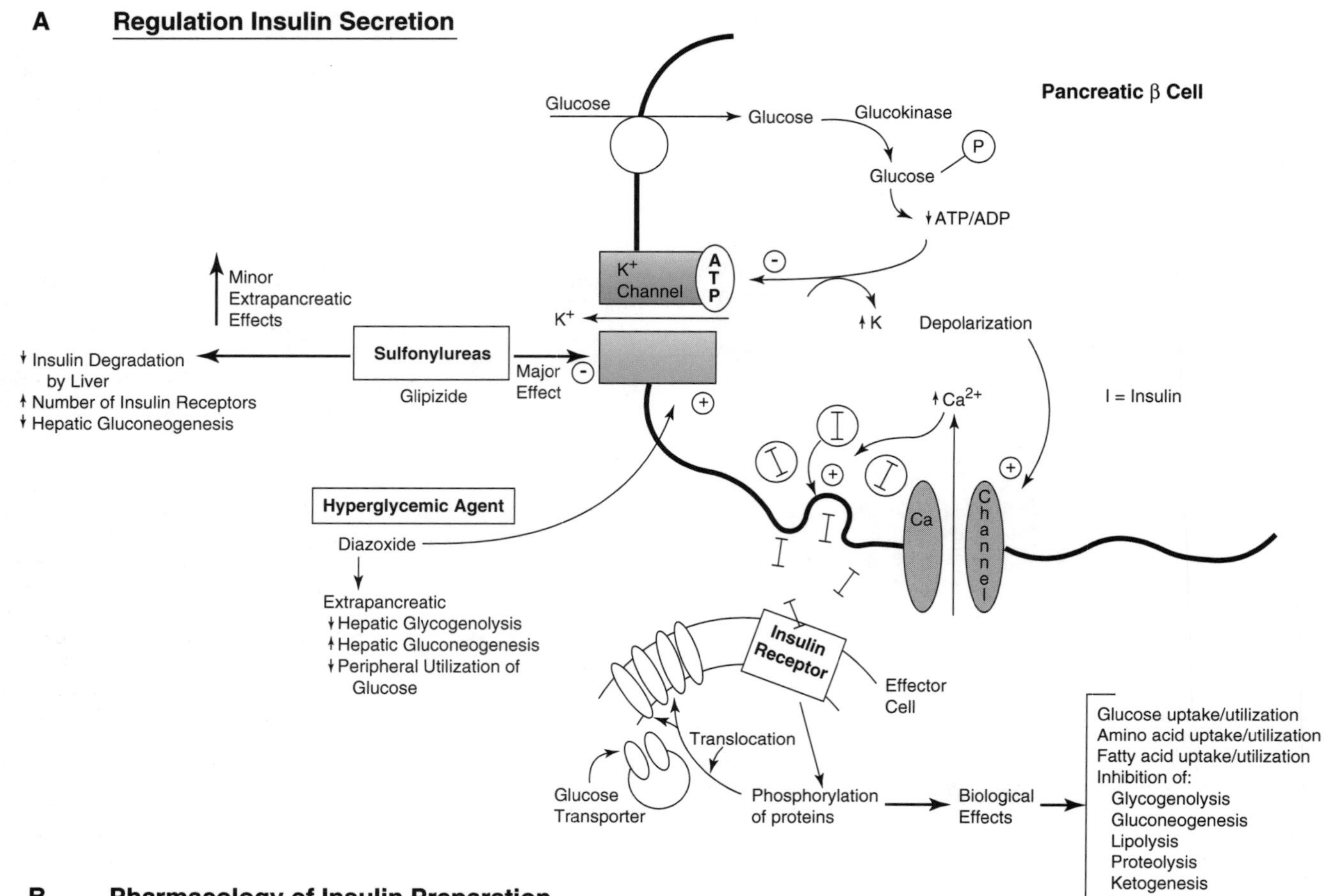

B Pharmacology of Insulin Preparation

Type of Insulin	Route of Administration	Action (hr) Cat Peak	Cat Duration	Dog Peak	Dog Duration
Rapid acting					
Regular (Crystalline)	IV IM SC	1 1–4 1–4	1–4 3–8 4–10	1 1–4 1–4	1–4 3–8 4–10
Semilente	SC			4–8	12–16
Intermediate acting					
NPH	SC	2–8	4–12	8–12	8–16
Lente	SC	2–8	6–14	2–10	8–24
Long acting					
PZI*	SC	4–10	12–30	5–20	30
Ultralente	SC	4–10	8–24	4–16	8–28

*Limited available. Specialty pharmacies will compound

Insulin

Insulin is a small peptide hormone produced by the β cells of the pancreas. Insulin secretion is primarily regulated by blood glucose concentration. Insulin is anabolic and facilitates cellular uptake and metabolism of glucose in all tissues except the liver, brain, and red blood cells. It inhibits gluconeogenesis, glycogenolysis, lipolysis, ketogenesis, and proteolysis (**Part A**). The biological activity of insulin preparations is determined using a bioassay that is based on the capacity of insulin to lower blood glucose concentration. Most preparations are at 100 U/mL. Syringes to administer insulin are standardized to match the concentration of insulin in the vial. Insulin preparations are divided into fast, intermediate, and long acting (**Part B**). Although there is considerable species variation, short-acting preparations (regular, semilente) take effect within 30 min and last 2–5 hours. Intermediate-acting insulins [lente, neutral protamine Hagedorn (NPH)] have an onset of action within 2–3 hours that last 4–12 hours. The long-acting insulins [ultralente, protamine zinc (PZI)] have an onset of action of about 4 hours and last 6–24 hours. It is important to remember that the time course of action of any insulin preparation varies among species, among individuals of the same species, and from day to day in any individual. All of the insulin preparations are given SC except for regular insulin, which can also be given IV or IM. The insulin suspensions are prepared in special buffers to promote stability. The acetate buffered insulin (lente insulin) is somewhat less soluble and more slowly absorbed.

The amino acid sequence of insulin shows some species variation. Canine insulin is identical to porcine insulin, and differs from human insulin by 1 amino acid and from bovine by several amino acids. Feline insulin differs from bovine insulin by 1 amino acid and from porcine and human insulin by 3 amino acids. The major insulin on the market is human recombinant insulin, although beef, pork, and beef-pork combination are still available. The human products tend to have a more rapid onset of action and shorter duration of action than the beef or pork products.

Oral Hypoglycemic Agents

Sulfonylureas

Sulfonylureas increase insulin secretion and improve insulin resistance. Sulfonylureas inhibit ATP-dependent potassium channels in the pancreatic β cells, and by doing so, result in depolarization of the membrane and release of insulin (*see* **Part A**). Sulfonylurea therapy improves tissue sensitivity to circulating insulin by increasing insulin receptor binding or improving postreceptor signal transduction. They also inhibit hepatic glycogenolysis and decrease the hepatic extraction of insulin. The second-generation sulfonylurea glipizide has been used to treat type II diabetes mellitus in cats. Glipizide is well absorbed after oral administration and is metabolized in the liver and excreted in the urine. The most common side effect is hypoglycemia. Other side effects include gastrointestinal upset and hepatotoxicity. Sulfonylureas may promote the progression of pancreatic amyloidosis.

Metformin

Metformin, a biguanide hypoglycemic agent, works by inhibiting hepatic glucose release and restoring peripheral tissue sensitivity to insulin. It has no effect on insulin release and therefore can not cause hypoglycemia as a side effect. Side effects in humans include vomiting, and diarrhea. The drug is excreted unchanged in urine.

Acarbose

Acarbose B, an α-glucosidase inhibitor, interferes with intestinal glucose absorption by decreasing fiber digestion and thus glucose liberation. It has only mild glucose lowering ability. It should not be used in animals that are not overweight. Acarbose is not systemically absorbed and can be combined with other oral hypoglycemic agents. Side effects include flatulence and diarrhea.

Thiazolidinediones

The thiazolidinediones work by increasing the sensitivity of skeletal muscle and adipose tissue to insulin. The first generation drug, troglitazone, has been withdrawn from the market due to severe hepatotoxic reactions in humans. The second generation products, rosiglitazone and pioglitazone, are currently on the market but have not been used extensively in veterinary patients.

Hyperglycemic Agents

Diazoxide is a nondiuretic benzothiadiazine that inhibits the release of insulin from the β cells by increasing the β cells' permeability to potassium, resulting in membrane (*see* **Part A**) hyperpolarization. It also has extrapancreatic effects to promote hepatic glycogenolysis and decrease glucose uptake by the liver. It is used to treat hypoglycemia associated with insulin-secreting tumors in dogs. The drug is partially metabolized by the liver and excreted in the active and metabolized form by the kidneys. Side effects include gastrointestinal disturbances, sodium and water retention, induction of diabetes mellitus, and hematological abnormalities. Thiazide diuretics may potentiate the hypoglycemic action of diazoxide.

Somatostatin inhibits insulin and glucagon release from the pancreatic β cells, growth hormone release from the pituitary, and the release of several gastrointestinal hormones. The only preparation available for parenteral administration is octreotide. It has been used in veterinary medicine with limited success in dogs to treat insulinomas and gastrinomas and in cats to treat acromegaly.

58 Endocrine: Mineralocorticoids and Adrenolytic Drugs

A Steroid Synthesis

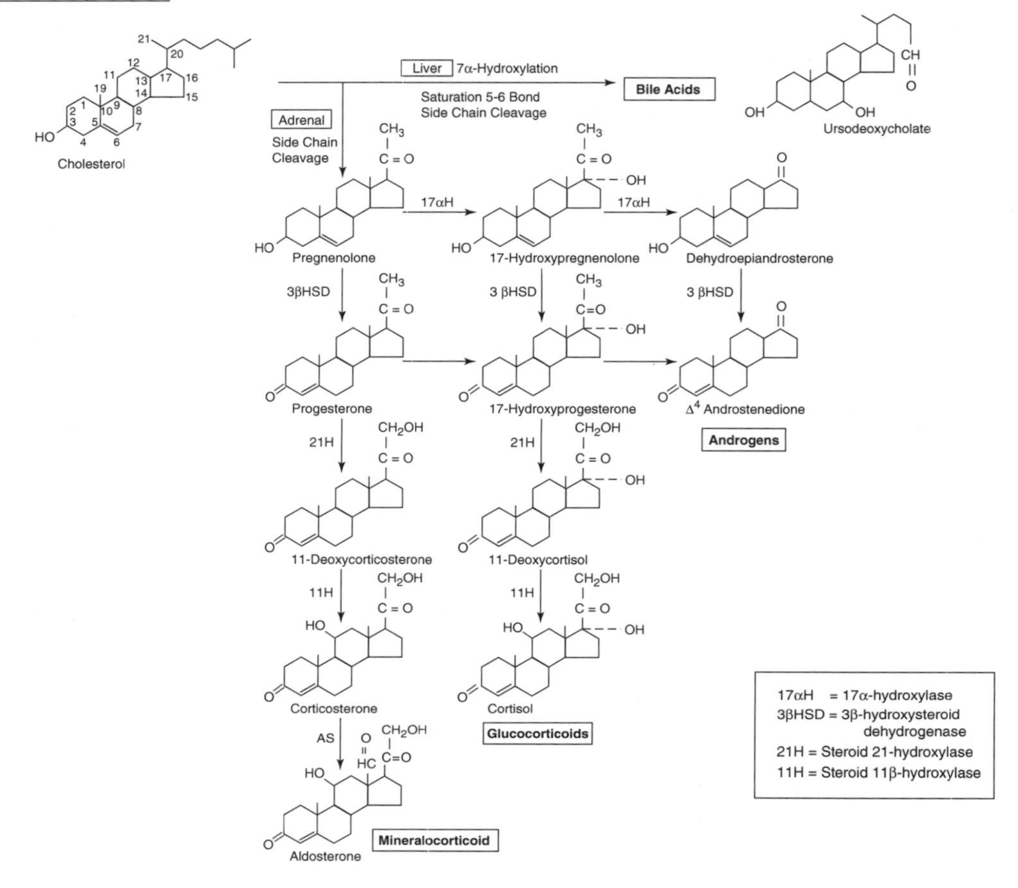

B Modulation of Adrenal Steroidogenesis

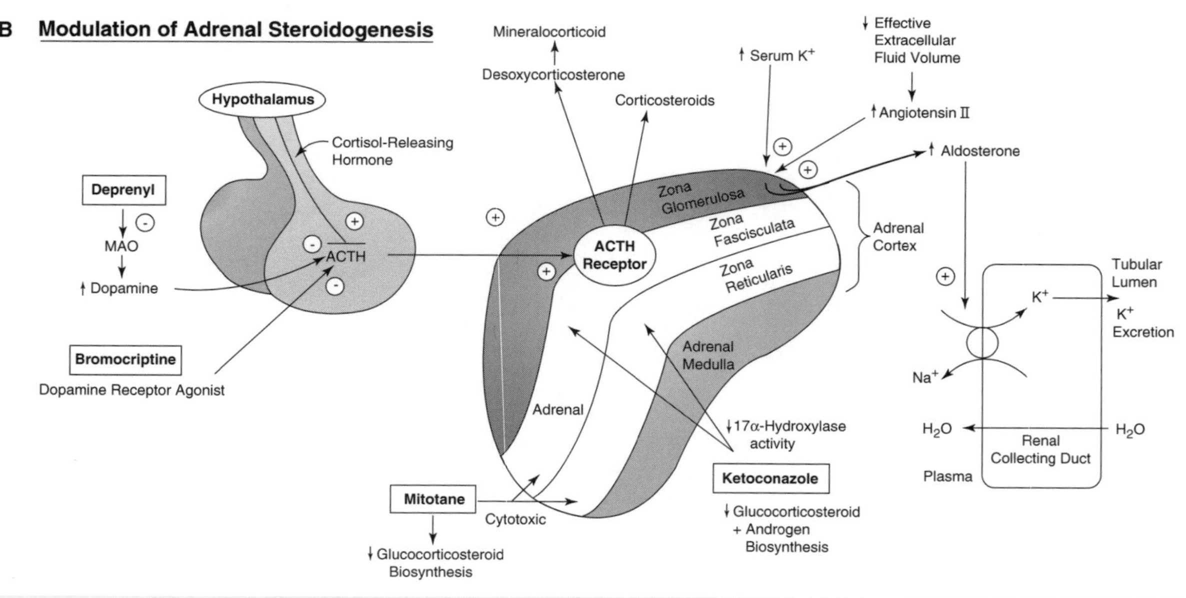

The three main classes of steroids synthesized in the body are: 1) adrenocortical hormones including corticosteroids and mineralocorticoids, 2) reproductive hormones such as estrogen and androgens, and 3) bile acids. All of these steroids are synthesized from cholesterol (**Part A**). The adrenocortical hormones are synthesized in the adrenal cortex; the reproductive hormones, primarily in the gonads although small amounts are produced by the adrenals; and the bile acids, exclusively in the liver. Corticosteroids, the reproductive hormones, and bile acids are discussed in Chapters 55, 59, and 60, respectively.

Cortisol releasing hormone, produced by the hypothalamus, stimulates the pituitary to release adrenocorticotropin (ACTH). ACTH binds to receptors on the adrenal gland to stimulate the production of corticosteroids and mineralocorticoids (**Part B**). Mineralocorticoid secretion is also stimulated by increases in serum K^+ or serum angiotensin II levels.

Mineralocorticoids

The pharmacological use of mineralocorticoids is limited to physiological replacement in the setting of hypoadrenocorticism. Mineralocorticoids are produced in the zona fasciculata and zona glomerulosa of the adrenal cortex (**Part B**). The major mineralocorticoid, aldosterone, is produced by the zona glomerulosa under the control of circulating angiotensin II levels. The zona fasciculata produces desoxycorticosterone under the control of ACTH (**Part B**). The major stimuli for the secretion of aldosterone are decreases in whole-body volume status and increases in serum potassium concentrations. Aldosterone works on the distal nephron and collecting duct of the kidneys to stimulate sodium (and thus water) retention and potassium excretion. Aldosterone is not available as a pharmacological agent due to its short duration of action. Desoxycorticosterone pivalate (DOCP) is available for IM injection to treat hypoadrenocorticism. DOCP is a sustained-release formulation whose action lasts up to 3 weeks. An empirical starting dose of 2.2 mg/kg given IM every 25–30 days is initiated and dose and dose interval are adjusted based on evaluation of serum electrolyte concentrations. Side effects of therapy are rare, but with overdosage include polyuria/polydipsia, hypertension, hypernatremia, and hypokalemia. DOCP has very little glucocorticoid activity and therefore concurrent supplementation with prednisone or cortisone is recommended.

Fludrocortisone acetate is an orally available mineralocorticoid used in treating hypoadrenocorticism. It is generally started at an empirical dose of 0.1 mg/10 lb once a day and the dose is adjusted based on evaluation of serum electrolytes. Fludrocortisone has substantial glucocorticoid activity, so replacement with prednisone may not be necessary. Side effects of therapy are rare except with overdosage, which can cause hypertension, hypokalemia, and edema. Some animals will show clinical signs of corticosteroid excess.

Adrenolytic Drugs and Corticosteroid Synthesis Inhibitors

The primary indication for therapy with corticosteroid synthesis inhibitors or adrenolytic drugs is the treatment of spontaneous hyperadrenocorticism. The majority of cases of hyperadrenocorticism in dogs, cats, and horses are due to the presence of a functional adrenocorticotropin-secreting tumor in the pituitary gland (PDH). Rarely, dogs may also get functional adrenocortical tumors. In PDH, excess secretion of ACTH results in bilateral adrenal hyperplasia and hypersecretion of corticosteroids. Therapy is aimed at decreasing ACTH release by the tumor or suppressing the synthesis of corticosteroids from the adrenal glands.

Drugs that Decrease ACTH Secretion

Several drugs have been used to control pituitary ACTH secretion. These include the antiserotoninergic agent cyproheptadine, and the dopaminergic drugs bromocriptine and deprenyl (*see* **Part B**). Dopamine can inhibit ACTH release from pituitary tumors. Bromocriptine mesylate is a dopamine agonist and a prolactin inhibitor. It is used as an inexpensive treatment for horses with pituitary adenomas. It is administered parenterally and is metabolized by the liver. Side effects include gastrointestinal distress, sedation, lethargy, and hypotension. It works poorly in canine patients. Deprenyl is a monomine oxidase enzyme inhibitor that inhibits the enzymatic destruction of dopamine and thus raises brain dopamine levels. The drug is well tolerated by canine patients. Unfortunately, dopaminergic drugs are only successful in lowering cortisol levels in a small percentage of dogs. Cyproheptadine is an antihistamine that also has antiserotoninergic effects in the CNS. It is infrequently used to control hyperadrenocorticism due to poor efficacy.

Adrenolytic Drugs

Mitotane[1-(*o*-chlorophenyl)-1-(*p*-chlorophenyl)-2,2-dichloroethane,*o,p*-DDD] is an adrenolytic drug. This drug causes a relatively selective destruction of the zona fasciculata and zona reticularis of the adrenal cortex and thus destroys the corticosteroid-producing tissue (*see* **Part B**). Very little information about the pharmacokinetics of mitotane in domestic animals is known. It is a very lipophilic drug and in humans only 40% of an oral dose is biologically available. In humans it undergoes some hepatic metabolism followed by urinary excretion, but the majority is excreted unchanged into the bile.

Mitotane is commonly used to treat both pituitary-dependent hyperadrenocorticism and benign and malignant adrenocorticol neoplasms in dogs. It is typically given in a daily loading dose to a point where adrenal corticosteroid secretion is minimal (usually <5 µg/dL) and there is no response to ACTH stimulation. Then a maintenance dose equal to the daily loading dose given weekly is instituted. The exact loading dose necessary to treat hyperadrenocorticism is highly variable between patients so each animal needs to be monitored closely during loading therapy for signs of iatrogenic hypoadrenocorticism. This is typically manifested by the appearance of lethargy, anorexia, and gastrointestinal upset from 4 to 7 days after starting the loading therapy. If these signs occur, mitotane should be discontinued, an ACTH stimulation test performed, and the animal started on physiological replacement with corticosteroids. Rarely animals develop destruction of the zona glomerulosa and concurrent mineralocorticoid deficiency.

The side effects of mitotane therapy are generally mild, with the most common being gastrointestinal upset. This can usually be avoided by dividing the daily dose and giving it with a meal. Other side effects include central nervous system depression, ataxia, hepatotoxicity.

Drugs that Inhibit Corticosteroid Secretion

Ketoconazole inhibits cortisol production in the adrenals and has been used successfully to treat canine hyperadrenocorticism (*see* **Parts A** and **B**). Ketoconazole inhibits 17α-hydrolase activity. Since the inhibition of corticosteroid synthesis is reversible, the drug must be given on a daily basis. Side effects include vomiting, anorexia, and hepatotoxicity. Ketoconazole also inhibits the adrenal production of estradiol and testosterone and has been used to treat disorders associated with excess production of these sex hormones.

59 Endocrine: Corticosteroids

A Secretion and Action of Corticosteroids

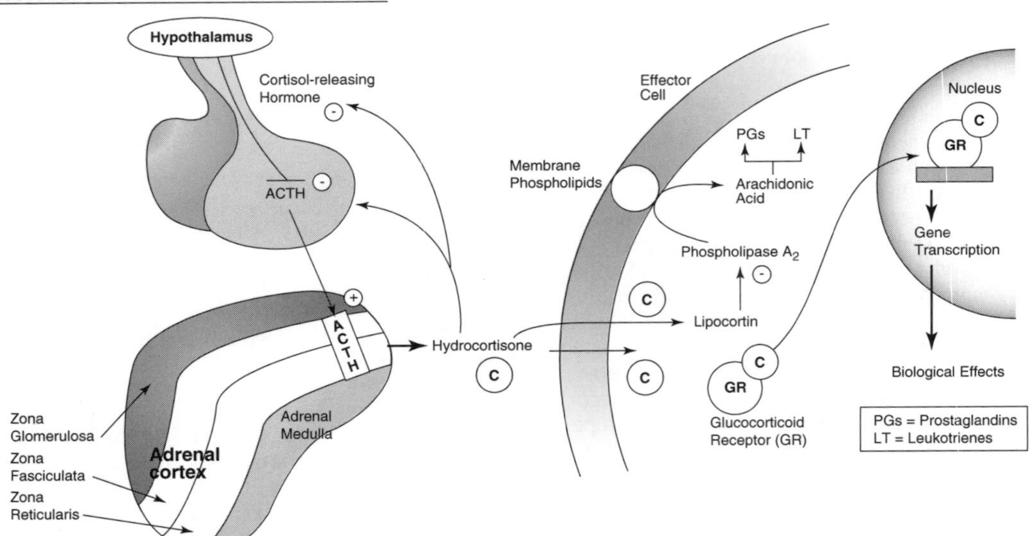

B Physiological Effects of Corticosterioids

- ↑ Hepatic gluconeogenesis, glycogenesis and protein synthesis
- ↑ Peripheral glucose utilization
- ↑ Proteolysis (mobilization of amino acids from skeletal muscle to liver)
- ↓ Lipolysis in adipose tissue
- ↑ Mineralocorticoid activity (mild) which promotes Na^+/H_2O retention

Pharmacological Effects of Corticosteroids

Anti-inflammatory/immune modulation

- ↓ Activation phospholipase A2
- ↓ Macrophage cytokine production/release
- ↓ Endothelial leukocyte adhesion

C Structure Activity Relationship

Hydrocortisone

Drug	Glucocorticoid Potency	Substitution	Mineralocorticoid Potency
Hydrocortisone	1	-	1
Prednisone ← Liver	4	Double Bond C1-2	0.8
Prednisolone	4	Double Bond C1-2	
Methylprednisone	5	6α Methylation	0.5
Triamcinolone	5	Double Bond C1-2 9 Fluorination 16α Hydroxylation	0
Dexamethasone	25	Double Bond C1-2 9 Fluorination 16α Methylation	0

D Physiological Effect/Pharmacological Manifestations of Corticosteroid Use

Effect	Due to:
Polyuria/polydipsia	Mineralocorticoid action
Increased susceptibility to infection	Anti-inflammatory/immune modulation
Gastrointestinal ulceration	Inhibition of GI prostaglandin production
Pancreatitis	?Change in viscosity pancreatic secretion
Laminitis (horses)	?Potentiation of catecholamine-mediated vasoconstriction
Steroid hepatopathy (dogs>>>>cats)	↑ Hepatic glycogen production
Increased serum Alkaline phosphatase (dog)	Induction of corticosteroid isoenzyme
Mild polycythemia	↑ Erythropoietin
Mature neutrophilia	↑ Release from bone marrow ↓ Marginization
Lymphopenia/eosinopenia	Redistribution away from periphery
Polyphagia, mood alteration	? CNS effects of corticosteroids
Hyperglycemia/glucose intolerance/diabetes mellitus	Acute: ↑hepatic gluconeogenesis with ↓peripheral utilization of glucose Chronic: pancreatic β cell exhaustion
Iatrogenic Hyperadrenocorticism	Chronic suppression of hypothalamic-pituitary-adrenal axis with adrenal atrophy

Corticosteroids are synthesized on the adrenal gland under the control of the pituitary hormone ACTH (**Part A**). The major circulating corticosteroid in most species is hydrocortisone (cortisol). The pituitary release of ACTH is controlled by cortisol-releasing hormone from the hypothalamus. Cortisol is normally secreted at a low maintenance rate, with increased amounts made available during stress. Negative feedback of cortisol on ACTH secretion occurs at both the pituitary and the hypothalamic level. The varied physiologic and pharmacologic effects of corticosteroids are summarized in **Part B**.

Anti-inflammatory Actions

At high concentrations seen during exogenous administration or during chronic stress, corticosteroids suppress the inflammatory response. Corticosteroids inhibit phospholipase A_2 (*see* **Part A**), and therefore prevent the breakdown of arachidonic acid into the potent proinflammatory mediators prostaglandins and leukotrienes. Corticosteroids also inhibit the production or release of inflammatory cytokines from leukocytes. They also inhibit leukocyte migration to inflammatory sites, decrease the phagocytic activity of the reticuloendothelial system, and inhibit fibroblast proliferation and collagen deposition.

Immune-Modulating Actions

At even higher doses, corticosteroids are immunosuppressive. The anti-inflammatory and immune actions of corticosteroids are intimately linked, as they both largely result from inhibition of leukocyte and macrophage functions. Corticosteroids do not prevent humoral or cell-mediated immune responses but rather inhibit the manifestations of these reactions. Their inhibitory effects are mediated by the inhibition of cytokine release and response, and include: 1) prevention of interleukin-1 and tumor necrosis factor release by activated macrophages, 2) inhibition of interleukin-2 release by T lymphocytes, 3) prevention of the action of interleukin-2 on T lymphocytes, and 4) inhibition of interferon's effects on macrophages.

Chemical Formulations

Corticosteroids share the same basic steroid structure (**Part C**). The keto groups at C3 and C20, the double bonds at C4-5, and the hydroxyl group at C11 are essential for corticosteroid activity. The potency of corticosteroids are compared to hydrocortisone, which is given a value of 1. The addition of a double bond at the C1-2 position (prednisone) increases potency by 4 times. Methylation of prednisone at the 6α position increases corticosteroid activity slightly but decreases mineralocorticoid activity. Fluorination at the 9α position (dexamethasone, triamcinolone) increases the corticosteroid potency and decreases mineralocorticoid activity.

The addition of an ester group at C21 alters the water solubility and influences the duration of action. Phosphate and hemisuccinate esters, which are readily water soluble, have a rapid onset and short duration of action when given parenterally. Acetate and diacetate esters are poorly water soluble and can be given as repository IM or SC injections. Acetonide esters have a rapid onset and an intermediate duration of action.

Uses of Corticosteroids

Physiological Replacement

Physiological corticosteroid replacement therapy is required with hypoadrenocorticism. Since animals produce approximately 1 mg of cortisol/kg/day, an equipotent dose of prednisone (0.1–0.2 mg/kg) is used for physiological daily replacement. During stressful episodes, the replacement dose is increased 2–5 times.

Shock Therapy

The value of corticosteroid use in all forms of shock is controversial. Septic or endotoxic shock is the most responsive to corticosteroid administration. Methylprednisolone sodium succinate and hydrocortisone sodium succinate (30–50 mg/kg IV) have shown some benefit. The best therapy for shock is aggressive fluid therapy.

Anti-inflammatory

This is the most common use for corticosteroids in veterinary medicine. Anti-inflammatory doses of prednisone (0.5–1.0 mg/kg/day) are used to treat a range of inflammatory or allergic disorders, particularly those involving the skin and respiratory tract. Some rules should be considered. Before using anti-inflammatory corticosteroid therapy one should attempt to identify and remove the cause of inflammation. Since corticosteroids work nonspecifically, they will decrease inflammation due to any stimuli including an infectious agent. One should establish a goal for therapy and then attempt to use the smallest dose of corticosteroids for the shortest period of time possible. Short-acting oral corticosteroids (prednisone) should be used preferentially with a goal of eventually establishing alternate-day therapy (ADT). The side effects of long-term corticosteroid therapy, which are primarily related to suppression of the hypothalamic-pituitary-adrenal axis, can be dramatically decreased by using ADT.

Immunosuppressive Therapy

Immunosuppressive doses of prednisone (2 mg/kg/day) are commonly employed as first-line therapy for immune disorders such as immune-mediated hemolytic anemia, thrombocytopenia, polyarthritis, and systemic lupus erythematosus. High doses are maintained until the disease is in remission and then slowly tapered to the lowest dose that keeps the disease in remission, preferably given as ADT. If high doses of corticosteroids are necessary to maintain remission, alternate immunosuppressive drugs (*see* Chapter 41) should be added to the therapy to permit lowering of the corticosteroid dose.

Side Effects

The most important side effects of short-term (<2 weeks) corticosteroid therapy are (**Part D**): 1) increased susceptibility to infection, 2) polyuria with secondary polydipsia, 3) polyphagia, 4) behavioral and mood changes (depression, panting, lethargy), 5) diarrhea, and 6) the development of pancreatitis. The use of corticosteroids in dogs with spinal cord trauma to decrease cord swelling has been associated with an increased risk of colonic perforation. In dogs, corticosteroids induce production of the steroid-specific isoenzyme of alkaline phosphatase, increase serum levels of alanine aminotransferase and γ-glutamyl transpeptidase, and result in glycogen deposition in the liver, causing hepatomegaly (steroid hepatopathy). Although these hepatic changes themselves are seldom associated with functional hepatic failure, they frequently complicate the biochemical evaluation of liver function. Corticosteroids can induce parturition in late pregnancy and are teratogenic in early pregnancy. In horses, corticosteroid use can induce laminitis. Corticosteroids alter circulating blood cell counts. Animals develop a mature neutrophilic leukocytosis with a monocytosis, eosinopenia, and lymphopenia. The red cell volume is often high normal due to corticosteroid enhancement of erythropoietin production.

The most serious consequence of long-term corticosteroid use is suppression of the hypothalamic-pituitary adrenal axis, resulting in adrenal atropy (secondary adrenocortisol deficiency) and iatrogenic Cushing's syndrome. Affected animals develop clinical signs similar to dogs with spontaneous hyperadrenocorticism including loss of hair, thinning of skin, muscle wasting and weakness, abdominal redistribution of fat stores (potbellied appearance), recurrent infections, and reproductive disorders. The gluconeogenic effects of corticosteroids can lead to pancreatic β cell exhaustion and the development of diabetes mellitus.

Withdrawal from long-term corticosteroid use should be gradual to allow time to re-establish a normal hypothalamic-pituitary-adrenal axis. Signs of too rapid corticosteroid withdrawal include mental dullness, weakness, anorexia, vomiting, and behavioral changes.

60 Endocrine: Antidiuretic Hormone, Growth Hormone, and Miscellaneous Steroids

A Antidiuretic Hormone and Growth Hormone

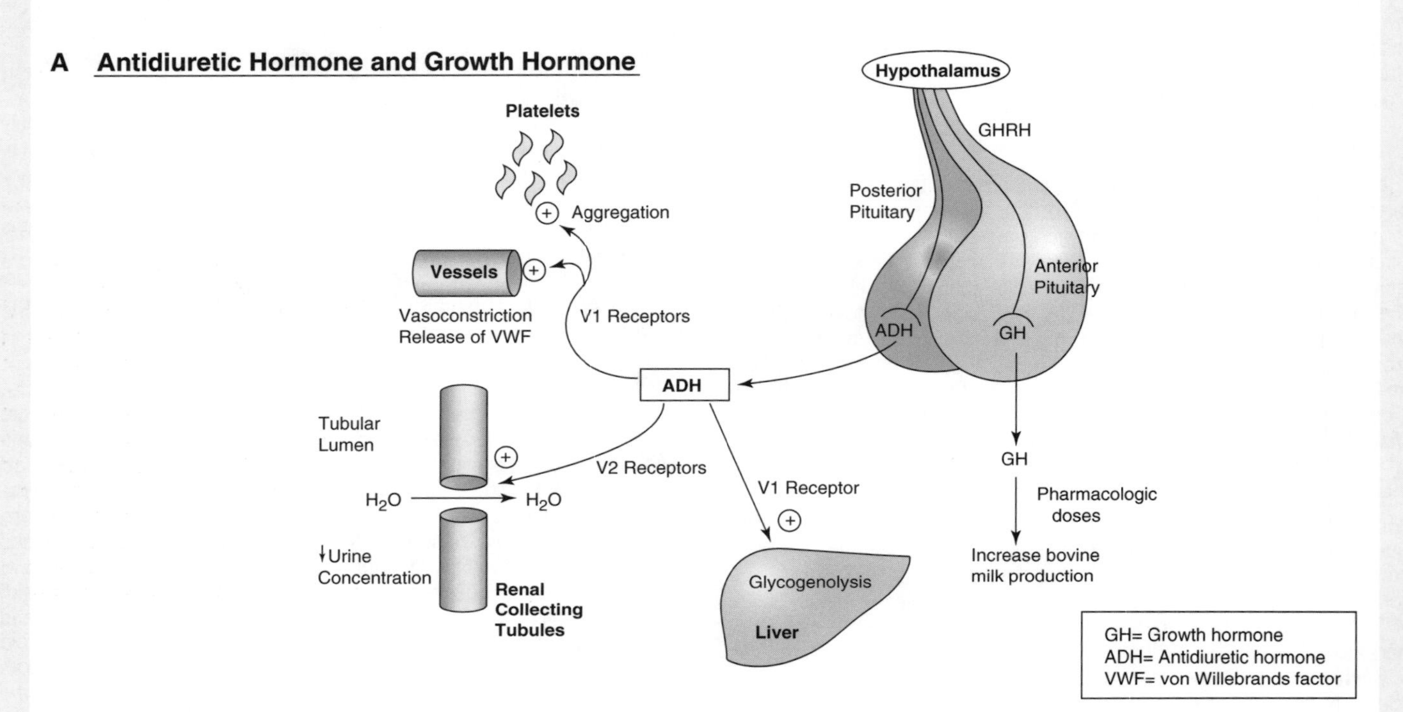

B Anabolic Steroids

Drugs	Actions	Side Effects
Stanozolol Boldenone Undecylenate Nandrolone	Increase erythropoiesis Promote muscle development Stimulate appetite	Hepatotoxicity Electrolyte and water retention Decreased spermatogenesis Estrus suppression

C Ursodeoxycholate

CH=O
OH
OH

Biological Effects →

- H Chloresis
- Hepatic cytoprotection
- Immune modulation

Antidiuretic Hormone (ADH)

ADH is secreted by the neurohypophysial system and binds to 2 receptors (**Part A**). The VI receptors are on vascular smooth muscle, hepatocytes, and platelets and mediate vasoconstriction, glycogenolysis, and platelet aggregation, respectively. The V2 receptors are on the renal collecting ducts. Ligand binding to V2 receptors makes the tubules permeable to water, enabling the production of a concentrated urine, and stimulates the release of von Willebrand factor. Desmopressin (DDAVP; 1-desamino-8-D-arginine vasopressin) is a long-acting synthetic analogue of ADH that works primarily at V2 receptors. Its main use is for the treatment of diabetes insipidus. DDAVP can be given parenterally, but is also active when given intranasally or into the conjunctival sac. The minimum dose required by these latter routes is 5–10 times the parenteral dose. The administration of 1–4 drops once or twice a day controls polyuria in most animals with diabetes insipidus. DDAVP is used to treat canine patients with von Willebrand disease, as it causes a dose-dependent increase in the release of von Willebrand factor and significantly shortens the bleeding time in these dogs. The parenteral route is indicated in the treatment of von Willebrand disease because hemostatic effects require 5–10 times the dose needed to control polyuria. Side effects of therapy are rare. Conjunctival irritation can occur. Since high levels of ADH can prevent excretion of a free water load, excessive doses of DDAVP can lead to fluid retention and hyponatremia or if severe overhydration develops, cerebral edema.

Growth Hormone

Growth hormone (GH) is normally synthesized and secreted by the pituitary gland under the influence of GH releasing hormone produced by the hypothalamus (*see* **Part A**). GH is used in cattle to enhance growth or milk production. Bovine GH is a prolonged-release injectable formulation of a recombinant bovine growth hormone. It is given SC every 14 days. Adverse effects include mild hyperthermia, clinical and subclinical mastitis, reduced feed intake, swelling at the injection site, mild anemia, decreased fertility, and increased gestational problems (early birth, twinning, retained placenta).

Anabolic Steroids

Anabolic steroids are used in veterinary medicine 1) to treat nonregenerative anemias, 2) to enhance growth and performance, and 3) to stimulate appetite (**Part B**). The anabolic steroids are synthetic androgenic compounds that have been developed to have enhanced anabolic effects and decreased virilizing effects. Anabolic steroids are useful in anemia because they stimulate erythropoietin production and may accelerate heme synthesis and red cell proliferation. In the setting of adequate protein and calories, the anabolic steroids promote the development of muscle mass and can reverse catabolism. They increase nitrogen retention, lean body mass, and body weight. Due to their substantial abuse potential in humans, anabolic steroids are now controlled drugs.

In small animals, stanozolol and nandrolone decanoate are used and in the horse, stanozolol and boldenone undecylenate are used. Stanozolol is available for oral or parenteral administration. The anabolic effects of stanozolol persist for up to 1 week. Nandrolone is given parenterally as a weekly injection and is considered somewhat more effective in stimulating erythropoiesis. Boldenone is given by IM injection every 8 weeks to horses. Little information on the pharmacokinetics of these drugs in veterinary patients is known. The side effects of anabolic steroids are hepatotoxicity (particularly in cats); sodium, calcium, phosphate, potassium, and water retention; potentiation of prostatic disease, decreased spermatogenesis; and estrus suppression. Anabolic agents should not be used in pregnant or breeding animals.

Bile Acids

Bile acids are synthesized exclusively in the liver from cholesterol. The only bile acid used pharmacologically in small animals is ursodeoxycholic acid (**Part C**). Originally isolated from black bear bile, ursodeoxycholate is now synthesized commercially to treat cholesterol gallstones and as a hepatoprotective agent in humans. Ursodeoxycholate lowers plasma and bile cholesterol levels by inhibiting cholesterol secretion into bile and reducing intestinal reabsorption of cholesterol and hepatic synthesis of cholesterol. Ursodeoxycholate has the following potential benefits in treating hepatic disease. First, it stimulates bicarbonate-rich bile flow (choleresis). Second, chronic administration results in the replacement of the normal relatively hydrophobic pool of bile acids with the relatively hydrophilic ursodeoxycholate. Since it is well known that hydrophobic bile acids are cytotoxic agents at high concentrations (such as those seen in hepatic diseases), their replacement by ursodeoxycholate may be hepatoprotective. Third, ursodeoxycholate may work by protecting mitochondrial membrane integrity. Last, ursodeoxycholate appears to have immune-modulating effects that may be related to interactions with the corticosteroid receptor.

61 Immunosuppressive Drugs

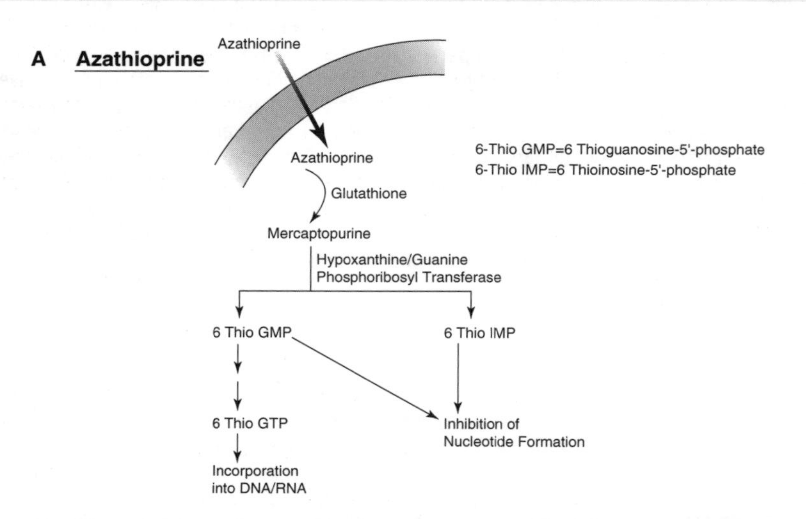

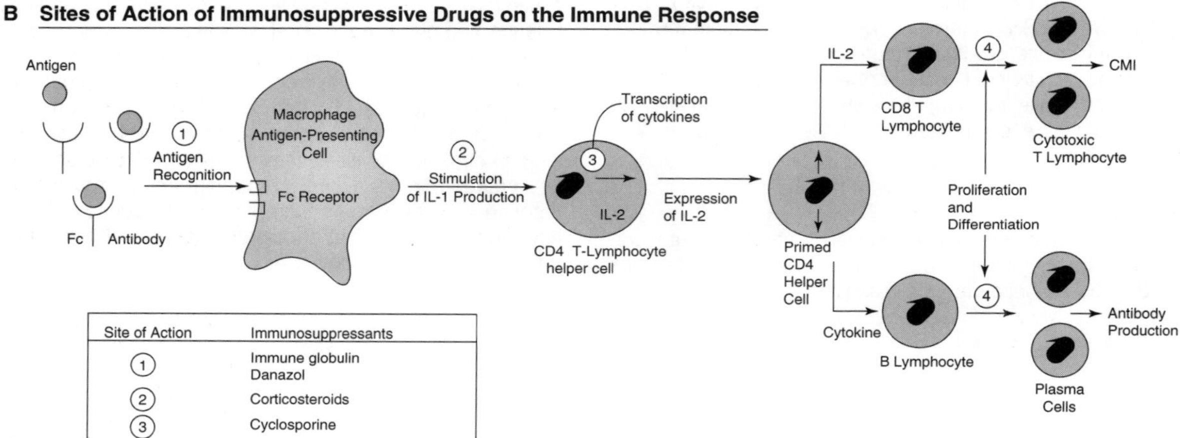

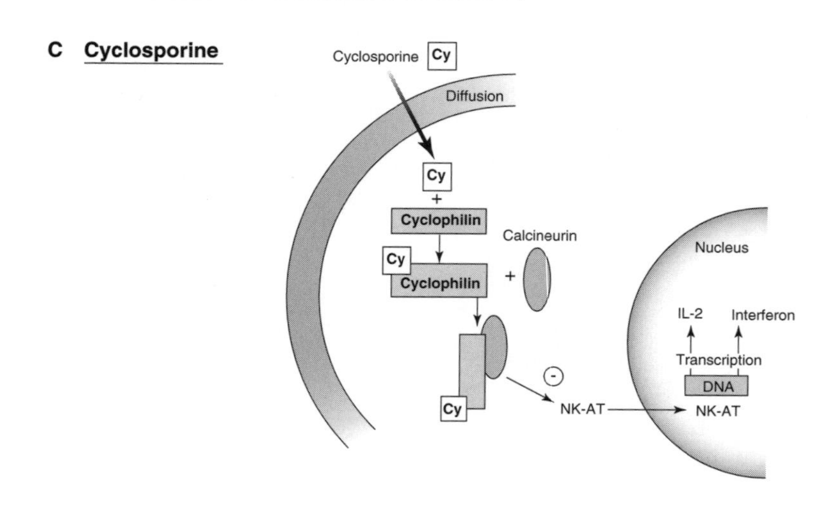

Agents that Inhibit the Proliferation and Expansion of Immune Cells

Azathioprine

Azathioprine is a prodrug that undergoes slow activation by nucleophiles, such as glutathione, present in several tissues to form the active drug mercaptopurine (**Part A**). Mercaptopurine is a hypoxanthine and guanine purine analogue that is converted to 6-mercaptopurine nucleotides, which interfere with DNA and RNA synthesis. Mercaptopurine is metabolized by enzymatic S-methylation in several tissues and oxidation by xanthine oxidase in the liver. The metabolites are excreted in the urine. Azathioprine has a greater inhibitory effect against T-lymphocyte–mediated immune responses than B-lymphocyte antibody production. It also decreases killer and natural killer cell populations. Clinical response to azathioprine may be delayed for several weeks. In veterinary medicine it is used as a second-line immunosuppressive agent for a variety of immune-mediated diseases. It is often combined with corticosteroids to permit lowering of the corticosteroid dose. Side effects of therapy include reversible bone marrow suppression (leukopenia anemia, thrombocytopenia). Cats are sensitive to azathioprine-induced bone marrow suppression, so much so that the drug is seldom used in this species. This species sensitivity may be related to low activity of the enzyme thiopurinemethyltransferase in cats. This enzyme is important in inactivation of the active metabolites of azathioprine. Azathioprine-induced pancreatitis and hepatotoxicity have been reported in dogs.

Alkylating Agents

Cyclophosphamide is an alkylating agent that interferes with DNA function (*see* Chapter 48). Cyclophosphamide is converted to its active form, aldophosphamide, in the liver. Aldophosphamide enters cells, where it is further metabolized to phosphoramide and acrolein. Cyclophosphamide is a rapidly acting immunosuppressive agent that inhibits both cell-mediated and humoral immune responses (*see* **Part B**). Antibody production by B lymphocytes is decreased and B- and T-lymphocyte numbers decrease. Cyclophosphamide also interferes with neutrophil and macrophage function. Cyclophosphamide is frequently used as second-line therapy in corticosteroid refractory cases of severe life-threatening immune disorders such as immune-mediated thrombocytopenia or immune-mediated hemolytic anemia. Side effects include myelosuppression, gastrointestinal upset, and sterile hemorrhagic cystitis (associated with urinary excretion of the acrolein).

Chlorambucil is an alkylating agent that has a much slower onset of action than cyclophosphamide. It is also a prodrug that is metabolized in the liver to its active form, phenylacetic acid. Phenylacetic acid is further metabolized and the metabolites excreted in the urine. The most common side effect is myelosuppression.

Agents that Prevent Activation of the Immune System

Corticosteroids

Corticosteroids are considered first-line therapy for most immune-mediated diseases in veterinary patients. They are discussed in Chapter 59.

Cyclosporine

Cyclosporine is a hydrophobic peptide that is produced commercially by fungal fermentation. Cyclosporine enters T lymphocytes and binds to a cytoplasmic receptor protein, cyclophilin (**Part C**). The cyclophilin-cyclosporine complex binds to and inhibits calcineurin, a calcium-dependent phosphatase. Inhibition of calcineurin prevents dephosphorylation of regulatory cytoplasmic proteins that normally mediate the movement of the transcription factor NF-AT to the nucleus. This blocks the transcription of genes regulated by NF-AT. One of these gene products is interleukin-2. Blockage of interleukin-2 production leads to impaired cytotoxic and helper T-lymphocyte proliferation (*see* **Part B**). Cyclosporine also inhibits transcription of interferon-α, an important signal for macrophage activation.

Absorption of cyclosporine after oral administration is quite variable. Bioavailability, which can be enhanced by administration with a fatty meal, is low, 15%–60% depending on the individual. The drug undergoes extensive hepatic biotransformation and the metabolites are eliminated in bile. Cyclosporine is available for oral and parenteral administration and as a topical ointment to treat immune-mediated ocular disease in the dog (*see* Chapter 19).

Therapeutic drug monitoring is necessary to maintain cyclosporine blood concentrations in the proper range. Trough concentrations should be kept in the 250–400-ng/mL range. Side effects of therapy are increased risk of infection and lymphoreticular malignancies. Cats frequently object to the unpalatable taste, but otherwise tolerate the drug well. Side effects in dogs have included gastrointestinal disturbances, weight loss, gingival hyperplasia, shaking, and hirsutism. In humans, hepatotoxicity and hypertension have been reported.

Cyclosporine is used in veterinary medicine to prevent kidney rejection after renal transplantation in cats, to treat perianal fistulas and keratoconjunctivitis sicca in dogs, and to treat various immune-mediated disorders.

Agents that Interfere with Antibody/Antigen Presentation

Human Intravenous Immunoglobulin Therapy

Human immunoglobulin is a preparation of polyspecific immunoglobulin (IgG) that is obtained from the plasma of healthy human blood donors. The product is pooled from at least 1,000 donors, contains 90% biologically intact IgG, and is free of infectious agents and other immune products. Although it was first developed for the treatment of primary immunodeficiencies, it was soon discovered to have efficacy in treating immune-mediated disorders. IV administration of 1 or 2 doses 3–7 days apart may lead to rapid induction of a prolonged remission (2–6 weeks). Although the exact mechanism of action is unknown, it may involve blockage of Fc receptors on macrophages, leading to a decrease in B-lymphocyte autoantibody production (*see* **Part B**). Additional mechanisms of action include modulation of idiotypic antibody responses and enhancement of IgG catabolism. Immunoglobulin has been used with some success in treating immune-mediated thrombocytopenia and hemolytic anemia in dogs. The fear of allergic reactions to this human product has limited treatment in dogs to a single dose. This may explain, in part, the decreased efficacy in veterinary patients. Side effects in humans are occasional fever and vomiting.

Danazol

Danazol is a synthetic androgen that has been used as ancillary treatment in the management of chronic refractory cases of immune-mediated thrombocytopenia and immune-mediated hemolytic anemia. Although the exact mechanism of action is not known, danazol may reduce the binding of immunoglobulin and complement to red blood cells and platelets (*see* **Part B**).

Drugs Acting on Blood Elements

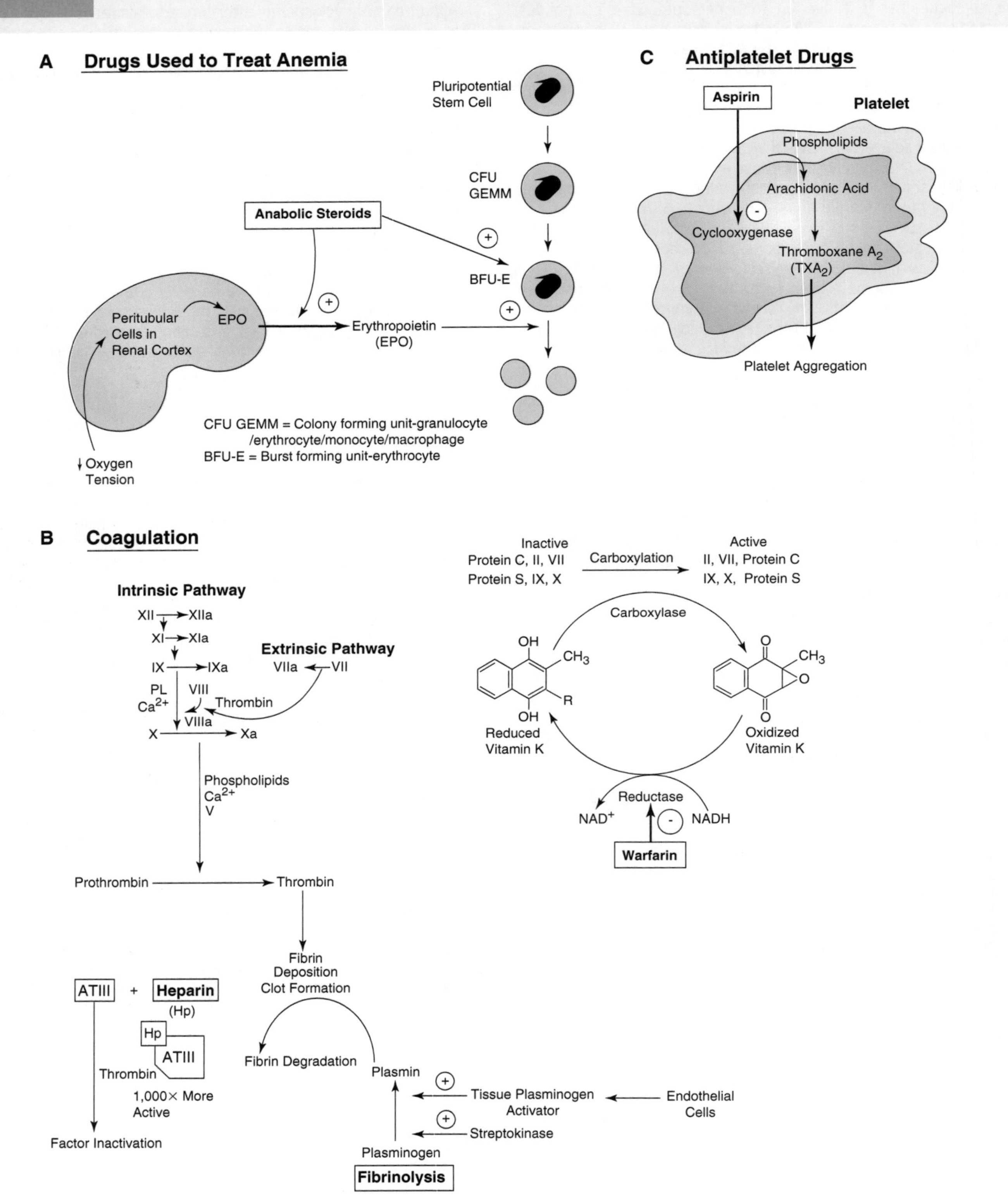

Drugs Used to Treat Anemia

Erythropoietin

Erythropoietin (EPO) is a hormone produced by the kidney that stimulates hematopoiesis. Binding of EPO to receptors on precursor erythrocytes stimulates their proliferation and differentiation into mature erythrocytes (**Part A**). Recombinant human EPO is available for the treatment of anemia associated with chronic renal failure. The drug is initially given SC (100 U/kg) 3 times a week until the target hematocrit is reached, and then the dosing interval is decreased to once or twice a week. In most patients with chronic renal failure, packed cell volume (PCV) normalizes in 3–4 weeks. Therapy with EPO may also be beneficial in other anemias associated with low EPO levels or EPO resistance.

Up to 30% of animals on chronic human recombinant EPO therapy develop antibodies that result in resistance to further treatment. In most cases, discontinuation of the drug results in disappearance of the antibodies and resolution of the anemia associated with their presence. In some animals, these antibodies will react with natural EPO, resulting in a refractory, transfusion-dependent aplastic anemia. Other adverse effects of EPO include allergic reactions at the site of injection, systemic hypertension, seizures, and iron depletion.

Anabolic Steroids

Anabolic steroids are synthetic compounds chemically related to testosterone. These compounds stimulate EPO production and directly induce proliferation of erythroid progenitor cells (**Part A**). Anabolic steroids are used in chronic nonregenerative anemia. Response to treatment with anabolic steroids is variable and the time to clinical improvement is long (up to 3 months). Nandrolone decanoate has the strongest hematological effect of the available preparations. The most serious side effect of anabolic steroid use is hepatotoxicity. They are controlled substances.

Hemoglobin Glutamer-200

Hemoglobin glutamer-200 (oxyglobin) is a solution of polymerized bovine hemoglobin approved for use as a blood substitute in dogs. It has a shelf life of 2 years at room temperature. When the reconstituted drug is given IV, it becomes oxygenated in the lungs and then releases oxygen to the tissues in the same way that endogenous hemoglobin does. The hemoglobin formulation also has colloidal properties similar to hetastarch. A dose of 15 mL/kg results in a 2.5 g/dL increase in hemoglobin, which persists for 24 hours. The $t^{1/2}$ is 30–40 hours, but some oxyglobin can be detected in the blood for 5–7 days after a single dose. Elimination of oxyglobin is the same as for endogenous hemoglobin. Adverse effects include discoloration of the mucous membranes and sclera, increased central venous pressure, and vomiting. Small amounts of unstabilized hemoglobin may be excreted in the urine, imparting a red color. Less commonly, ventricular arrhythmias, pulmonary edema, and fever have been seen.

Drugs Used to Treat Disorders of Coagulation

Anticoagulants

Pharmaceutical-grade heparin is a mixture of anionic sulfated mucopolysaccharides purified from bovine lung or pig intestinal mucosa, which have molecular weights ranging from 1,200 to 40,000. Heparin binds reversibly to the serine protease inhibitor antithrombin III (ATIII) (**Part B**). This binding induces a conformation change in ATIII that significantly (1,000 times) increases the ability of ATIII to inhibit activated coagulation factors, particularly thrombin and factor Xa. After factor inactivation, heparin dissociates from the ATIII-factor complex and is available for additional interactions. Heparin is also antithrombotic, as it binds to endothelial cell walls, imparting a negative charge that inhibits platelet adhesion and aggregation.

Heparin is only bioavailable when given parenterally. A high percentage of the administered dose binds avidly to plasma proteins, endothelial cells, and macrophages. It is metabolized by a heparinase in hepatic and reticuloendothelial cells, and the metabolites are excreted in urine. Individual responses to a given dose are highly variable due to variations in the degree of plasma protein binding and the amount of ATIII present. Response is best monitored by determination of prothrombin time (PT). The goal is prolongation of the PT by 1.5–2.0 times normal. Intravenous administration results in immediate anticoagulant activity of short duration (1–2 hours). SC administration results in a slower release of heparin, but comparable levels of anticoagulation. Heparin is indicated in the prevention and treatment of thromboembolic disorders and in disseminated intravascular coagulation.

Side effects of heparin therapy include hemorrhage and thrombocytopenia. In most cases, thrombocytopenia occurs after several days of therapy and resolves once the heparin is stopped. Other side effects include pain on injection, muscle necrosis at IM injection sites, mild transient increases in liver enzymes, suppression of aldosterone secretion and resultant hyperkalemia, and osteoporosis.

Vitamin K Antagonists

Warfarin is a coumadin derivative that interferes with vitamin K–dependent activation of coagulation factors II, VII, IX, and X and the anticoagulant proteins C and S. Warfarin blocks the reduction of vitamin K by vitamin K epoxide and thereby reduces the amount of available vitamin K (**Part B**). It takes several days of therapy with warfarin before coagulation factors are sufficiently depleted and the anticoagulation effect is seen. Blockage of the anticoagulant protein C, however, occurs more rapidly, leading to a short period of hypercoagulability. Heparin is routinely used for 2–5 days when warfarin therapy is started to avoid this.

Warfarin is available for oral administration and is well absorbed from the gastrointestinal tract. It is highly bound to plasma proteins and is metabolized by the hepatic CP450 system to inactive metabolites, which are excreted in the urine and bile. The $t^{1/2}$ between individuals and species is quite variable due to differences in genetic potential to metabolize the drug, variations in gastrointestinal absorption, interaction with other drugs and intestinal inactivation. Accurate dosing, therefore, must be based on drug response monitoring. The PT should be maintained at 1.3–2.0 times normal. The principal side effect of warfarin therapy is hemorrhage, which can be treated with high doses of vitamin K_1 and blood product support.

Thrombolytics

Streptokinase catalyzes the conversion of plasminogen to plasmin, which is the endogenous compound responsible for clot lysis (*see* **Part B**). The IV use of streptokinase has been reported in the treatment of feline thromboembolic disease. The major adverse reaction is bleeding. Allergic reactions and fever have been reported.

Antithrombotics

Aspirin irreversibly acetylates the active site of cyclooxygenase, preventing the formation of prostaglandins (**Part C**). The major prostaglandin in platelets is thromboxane A_2, which induces platelet aggregation. Since platelets cannot synthesize proteins, the action of aspirin on the platelet is permanent and new platelets must be made before normal platelet production of thromboxane A_2 is restored. The dose of aspirin required to decrease platelet thromboxane synthesis is well below that required for its anti-inflammatory action. Indications for the use of low-dose aspirin therapy include the treatment and prevention of arterial thrombosis in cats with cardiomyopathy and the prevention of pulmonary arterial thrombosis in dogs with heartworm adulticide disease.

QUESTIONS

Chapters 1 to 6

1. Diazepam, a benzodiazepine, is frequently used to treat status epilepticus, a life-threatening condition in which an animal has one seizure after another. Since diazepam is a highly lipid-soluble drug that when given orally has significant hepatic first-pass metabolism, what would be the best route of administration for immediate control of status epilepticus?

(A) Rectal

(B) Intramuscular

(C) Intravenous

(D) Oral

(E) Subcutaneous

2. Aminoglycosides are antibiotics that are organic bases. at physiological pH, they are highly charged but minimally bound to plasma proteins. Which of the following statements concerning the absorption and distribution of a drug with these properties is incorrect?

(A) It would be poorly absorbed from the gastrointestinal tract and likely are not bioavailable by oral administration

(B) Its volume of distribution should approximate that of the extracellular fluid

(C) It should be trapped in mastitic milk

(D) It should be readily distributed to the central nervous system

3. Assume that the pharmacological effect of drug A depends on activation by the hepatic cytochrome P450 (CP450) enzyme system and that the active drug is excreted exclusively by glomerular filtration in the kidneys. The serum concentration of active drug A following oral administration would be increased if drug A were given to an individual that:

(A) Had chronic renal disease

(B) Had chronic hepatic disease

(C) Was being treated concurrently with cimetidine

(D) Was being treated concurrently with phenobarbital

4. All of the following statements concerning phase II biotransformation reactions in the liver are true except:

(A) Cats have enhanced activity of the phase II enzyme glucuronyl transferase

(B) The products of phase II enzyme reactions are highly hydrophilic compounds

(C) Dogs have low levels of the phase II enzyme acetyltransferase

(D) Pigs have a relative deficiency in the enzymes necessary for sulfate conjugation

(E) The products of phase II enzyme reactions are always inactive metabolites

5. Consider a drug that is highly lipid soluble, has low plasma protein binding, and undergoes hepatic biotransformation to an active metabolite that is excreted in the urine. Which of the following statements concerning its predicted pharmacokinetics would be true?

(A) It will have a large volume of distribution

(B) It will have poor oral bioavailability

(C) Clearance of the drug will vary with the dose that is administered

(D) Its serum $t^1/_2$ will be very long

6. All of the following statements concerning the average steady-state concentration (Css) of a drug after multiple oral doses are true except:

(A) It is directly proportional to the drug's bioavailability

(B) It is directly proportional to the drug dose

(C) It is inversely related to the dosing interval

(D) It is inversely related to the drug's clearance

(E) It takes approximately 2 half-lives to obtain Css

7. All of the following statements concerning a drug's therapeutic index (TI) are true except:

(A) It is an indication of drug safety

(B) It is defined as the LD_{50} divided by ED_{50}

(C) The higher the TI, the safer the drug

(D) The greater the efficacy of a drug, the higher the TI

8. Drug A is likely to prolong or potentiate the pharmacological action of drug B if the following is true:

(A) Both drug A and drug B are more than 90% bound to albumin and drug A displaces drug B from its binding site on albumin

(B) Both drug A and drug B are excreted unchanged in the urine by glomerular filtration

(C) Drug B successfully competes with drug A for active secretion into renal proximal tubules

(D) Drug A binds to and inactivates drug B in the stomach

(E) Drug A is an antagonist at the same receptor that drug B is an agonist

Chapters 7 to 10

9. Which of the following statements concerning the parasympathetic nervous system is incorrect?

(A) The action of acetylcholine, which is the neurotransmitter for both preganglionic and postganglionic neurons in the parasympathetic system, is terminated by degradation by the enzyme acetylcholinesterase

(B) Parasympathetic preganglionic fibers originate in the thoracic and lumbar portions of the spinal cord and terminate on postganglionic fibers that are located close to the innervated organ

(C) The parasympathetic receptors on effectors organs are called muscarinic cholinergic receptors

(D) Receptor activation in the parasympathetic system results in miosis, hypersalivation, bronchoconstriction, decreased heart rate, urinary bladder contraction, and increased gastrointestinal motility

10. Which of the following statements concerning the action elicited by various adrenergic receptors is correct?

(A) β_1 receptors: decrease cardiac contractility

(B) β_2 receptors: bronchoconstriction

(C) α_1 receptors: vasodilation

(D) Dopamine receptors: renal and coronary vasodilation

(E) α_2 receptors: increase insulin secretion

11. Which of the following statements concerning acetylcholinesterase inhibitors is correct?

(A) Organophosphates work by carbamylating and irreversibly inactivating acetylcholinesterase

(B) Edrophonium is a short-acting acetylcholinesterase inhibitor that reversibly binds to the enzyme and undergoes rapid renal elimination

(C) Organophosphate toxicity can be treated with bethanechol

(D) Echothiophate is a topical acetylcholinesterase inhibitor used as a mydriatic agent in the treatment of glaucoma

(E) The preferred route for administration of neostigmine is oral

12. All of the following are potential clinical applications of atropine except:

(A) Bronchodilation

(B) Mydriasis

(C) Sinus bradycardia

(D) Decrease respiratory and salivary hypersecretion

(E) Vasoconstriction

13. You have prescribed phenylpropranolamine to increase urethral smooth muscle tone in a dog with urinary incontinence. You are concerned with all of the possible side effects except:

(A) Hypertension

(B) Dryness to the nasal mucosa

(C) Appetite suppression

(D) Bronchoconstriction

(E) CNS stimulation

14. All of the following adrenergic drugs can be used to treat hypertension except:

(A) Phenoxybenzamine

(B) Phentolamine

(C) Prazosin

(D) Norepinephrine

(E) Propranolol

Chapter 11

15. The major mechanism of action of topical anesthetic agents is:

(A) Inactivation of slow K^+ channels

(B) Blocking the opening of voltage-sensitive Na^+ channels

(C) Inhibition of the Na^+/K^+ ATPase

(D) Calcium channel antagonist

(E) Acetylcholine receptor antagonist

16. Which of the following statements concerning the pharmacological action of local anesthetics is correct?

(A) The lower the pK_a of the agent, the faster the onset of action

(B) The more hydrophobic (nonionized) the agent, the longer its duration of action

(C) Amide-linked local anesthetics undergo hydrolysis by plasma cholinesterases

(D) Signs of toxicosis include CNS depression and increased heart rate

(E) Addition of epinephrine to the local anesthetic shortens the duration of anesthesia

Chapter 12

17. Which of the following statements concerning the action of the depolarizing neuromuscular agent succinylcholine is correct?

(A) Succinylcholine is an acetylcholine receptor antagonist

(B) Succinylcholine is metabolized in the liver to inactive metabolites that are excreted in the urine

(C) Succinylcholine is a very-long-acting neuromuscular blocking agent in the horse

(D) Succinylcholine results in muscle paralysis and significant analgesia

(E) One of the side effects of succinylcholine therapy is an increase in bronchiolar and salivary secretions

18. The best nondepolarizing neuromuscular blocking agent to use in a patient with chronic renal failure would be:

(A) Atracurium

(B) Gallamine

(C) Pancuronium

(D) Tubocurarine

(E) Dantrolene

Chapters 13 to 17

19. Which of the following statements concerning the mechanism of action of digitalis is correct?

(A) Digitalis decreases cardiac contractility

(B) Digitalis inhibits the H^+/K^+ ATPase present on the apical membrane of the cardiac myocyte

(C) Digitalis increases the rate of discharge from the sinoatrial node

(D) Digitalis sensitizes baroreceptors and induces sustained inhibition of sympathetic activation

(E) Digitalis increases the rate of atrioventricular conduction

20. Which of the following statements concerning the use of dobutamine is correct?

(A) Dobutamine is best administered orally

(B) Dobutamine is primarily a β_1 adrenergic agonist

(C) Dobutamine should be given chronically on a daily basis for maximum pharmacological effects

(D) Dobutamine also binds to renal dopamine receptors and increases renal blood flow

(E) The net effect of administration of dobutamine is to decrease cardiac contractility

21. Which of the following statements concerning the pharmacological actions of enalapril is incorrect?

(A) Enalapril works by preventing the conversion of angiotensinogen to angiotensin I in the lung

(B) Enalapril is a balanced vasodilator that can be used to treat hypertension

(C) Enalapril is metabolized in the liver to the active drug enalaprilat

(D) Enalapril results in preferential renal glomerular efferent arteriolar vasodilation

(E) A potential side effect of enalapril is hyperkalemia due to inhibition of aldosterone secretion

22. Which of the following vasodilators is incorrectly matched with its mechanism of action?

(A) Nitroprusside: Increases guanylate cyclase activity leading to increased intracellular cGMP

(B) Hydralazine: Direct vascular smooth muscle relaxation

(C) Prazosin: α_1 adrenergic receptor antagonist

(D) Nitroglycerin: Results in generation of nitric oxide leading to increased cAMP

23. All of the following are pharmacological effects of quinidine except:

(A) Decrease in the rate of phase 0 depolarization in cardiac myocytes

(B) Prolongation of the atrial and ventricular refractory period

(C) Improved conduction through the atrioventricular node

(D) When coadministered with digoxin, decreases digoxin serum levels

(E) Modest negative inotropy

24. Cimetidine is a commonly prescribed H2 receptor antagonist used to treat gastric ulcers and gastritis. One of the side effects of cimetidine therapy is a decrease in hepatic CP450 enzyme activity and a decrease in hepatic blood flow. Cimetidine would affect the elimination of which class I antiarrhythmic agent?

(A) Lidocaine

(B) Procainamide

(C) Quinidine

(D) All of the above

25. Which of the following antiarrhythmic agents has no effect on β_1 adrenergic receptors on cardiac myocytes?

(A) Propranolol

(B) Amiodarone

(C) Sotalol

(D) Atenolol

(E) Diltiazem

26. A 6-year-old Doberman pinscher with known cardiomyopathy is in a state of collapse. A rapid physical examination reveals tachycardia with weak irregular pulses. An electrocardiogram shows supraventricular tachycardia with a rate of 240 beats/min. Which of the following is the most appropriate initial treatment?

(A) Diltiazem IV

(B) Verapamil IV

(C) Amiodarone IV

(D) Propranolol IV

(E) Amlodipine IV

Chapters 18 to 19

27. All of the following drugs can be used to treat primary glaucoma in the dog except:

(A) Atropine

(B) Timolol

(C) Dorzolamide

(D) Epinephrine

(E) Echothiophate

28. Scruffy is a 5-year-old English bulldog that you have just diagnosed with keratoconjunctivitis sicca (dry eye syndrome). The rest of the findings on ocular examination are normal. Which combination of drugs do you prescribe?

(A) Corticosteroids and antibiotics

(B) Cyclosporine and pilocarpine

(C) Topical nonsteroidal anti-inflammatory agents and atropine

(D) Topical antiviral agent and atropine

(E) Corticosteroids and phenylephrine

Chapter 20

29. Sissy is a 5-year-old spayed female poodle who you have diagnosed with urethral sphincter incompetence. She also has diabetes mellitus and systemic hypertension. The best drug to prescribe to increase urethral tone in Sissy is:

(A) Estrogen

(B) Bethanechol

(C) Ephedrine

(D) Phenylpropanolamine

(E) Diazepam

30. A cat with disease of the lower urinary tract, which you unblocked 2 days ago, now has evidence of an atonic bladder and a spastic urethra. Which of the following drug combinations would be of benefit?

(A) Phenoxybenzamine and atropine

(B) Phenylpropanolamine and bethanechol

(C) Estrogen and flavoxate

(D) Prazosin and bethanechol

(E) Imipramine and dantrolene

Chapters 21 and 22

31. All of the following are pharmacological actions of theophylline except:

(A) Bronchodilation

(B) Increase in mucociliary clearance

(C) Inhibition of mast cell degranulation

(D) Increase in strength of respiratory muscle contraction

(E) Suppression of ventricular arrhythmias

32. What is the major advantage of combining the use of a β_2 adrenergic receptor agonist like albuterol with a muscarinic cholinergic agonist like ipratropium in treating an equine patient with chronic obstructive pulmonary disease?

(A) They complement each other since both bind to the histamine receptor, resulting in bronchial smooth muscle relaxation.

(B) They complement each other by dilating bronchial smooth muscle in the large (muscarinic agonists) and small airways (β_2 agonists)

(C) The β_2 agonists suppress cough while muscarinic antagonists increase airway secretions

(D) Muscarinic antagonists increase the systemic absorption of β_2 agonists

(E) The combination works synergistically to relax bronchial smooth muscle by increasing intracellular cGMP

Chapters 23 and 24

33. All of the following are clinical indications for the use of furosemide except:

(A) Pulmonary edema secondary to congestive heart failure

(B) Serum hypercalcemia

(C) Serum hypokalemia

(D) Oliguric renal failure

(E) Exercise-induced pulmonary hemorrhage

34. Which of the following diuretics is incorrectly matched with its mechanism of action?

(A) Spironolactone: Blocks Na^+ absorption and K^+ excretion in the distal tubule

(B) Mannitol: Unabsorbed solute that decreases water resorption in the proximal tubule

(C) Acetazolamide: Inhibits proximal tubular resorption of HCO_3^- and Na^+

(D) Hydrochlorothiazide: Blocks Na^+ resorption in the distal tubule

(E) Triamterene: Blocks Na^+ and Cl^- resorption in the loop of Henle

Chapter 25

35. You have been charged with handling the anesthesia for a dog that requires an emergency splenectomy after being hit by a car. The dog has some degree of head trauma and has had several runs of ventricular tachycardia. You elect to use which of the following inhalational agents to maintain anesthesia?

(A) Metophane

(B) Halophane

(C) Isoflurane

(D) Nitrous oxide

36. The hepatic toxicity of inhalational anesthetic agents is most closely correlated with:

(A) The degree of hepatic metabolism of the agent

(B) The lipid solubility of the agent

(C) The blood gas partition coefficient of the agent

(D) The minimum alveolar concentration of the agent

Chapters 26 and 27

37. All of the following pharmacological effects are associated with dissociative anesthetics except:

(A) Maintenance of photic, pharyngeal, corneal, and pedal reflexes

(B) Mild increase in muscle tone

(C) Good somatic analgesia

(D) Decrease in cerebral blood flow and intracranial pressure

(E) Stimulation of the cardiovascular system

38. Propofol and the thiobarbiturates share all of the following properties except:

(A) They both have poor analgesic activity

(B) Anesthetic recovery with both agents depends in part on redistribution to other tissues

(C) They both can cause apnea at the time of induction if administered too rapidly

(D) They both directly depress the respiratory center in the brain and are used as euthanizing agents

(E) They both induce anesthesia rapidly (within 1–2 min)

Chapters 28 and 29

39. Which of the following statements regarding the mechanism of action or clinical indication for phenothiazines is incorrect?

(A) Phenothiazines are effective tranquilizers because of their antidopaminergic action in the brain stem

(B) Phenothiazines, when used as preanesthetic agents, decrease the quantity of potentially more dangerous drugs used to produce general anesthesia

(C) Phenothiazines are proarrhythmic and increase the incidence of catecholamine-associated arrhythmias seen with ultra-short-acting barbiturates and inhalational anesthetics

(D) Phenothiazines may cause hypotension due to α_1 adrenergic blockage and resultant vasodilation of peripheral vascular smooth muscle

(E) Phenothiazines decrease spontaneous muscle activity in response to stimuli but have no analgesic properties

40. Which of the following statements concerning the pharmacological actions of diazepam is incorrect?

(A) Diazepam is a lipid-soluble benzodiazepine that when given IV, quickly reaches its site of action, the γ-aminobutyric acid (GABA) receptor in the brain, to stop seizure activity

(B) Cats may suffer from an idiosyncratic hepatotoxic reaction from the administration of oral diazepam

(C) Diazepam relaxes skeletal muscle by interfering with interneuronal reflexes in the spinal cord and inhibiting the presynaptic release of acetylcholine

(D) Diazepam is excreted unchanged in the urine by a combination of tubular secretion and glomerular filtration

41. Which of the following statements concerning the cardiovascular and respiratory effects of xylazine is correct?

(A) Xylazine decreases myocardial sensitivity to catecholamine-induced arrhythmias

(B) Xylazine increases cardiac contractility

(C) After administration, an initial period of hypertension is followed by hypotension

(D) Xylazine improves arterial oxygenation in ruminants

(E) Xylazine decreases parasympathetic tone and may lead to sinus tachycardia

42. You have just administered xylazine to a horse with colic for its sedative properties. The horse will also benefit from what other properties of xylazine?

(A) Profound visceral analgesia

(B) Promotion of gastrointestinal motility

(C) Decreased sensitivity to auditory stimuli

(D) A reduction in blood glucose level

(E) Induction of emesis

Chapter 30

43. Which statement regarding opiate receptor subtypes is correct?

(A) Mu (μ) receptors are present in the brain stem and spinal cord and mediate analgesia

(B) Kappa (κ) receptors cause dose-dependent respiratory depression by a direct effect on the brain stem

(C) Delta (δ) and mu (μ) receptors result in inhibition of gastrointestinal circular smooth muscle contraction

(D) Mu (μ) receptors mediate dysphoric reactions to opiates

44. You have just finished a total hip replacement in a 6-year-old castrated male German shepherd and would like to give the dog an opiate for postoperative pain management. Which of the following opiates would be the most appropriate to use?

(A) Transdermal fentanyl patch

(B) Butorphanol IV

(C) Naloxone IV

(D) Morphine IM

Chapters 31 and 32

45. Which of the following statements concerning the pharmacological properties of NSAIDs is incorrect?

(A) They are weak acids that are highly protein bound

(B) They block the production of prostaglandins by inhibiting the enzyme cyclooxygenase

(C) Most undergo hepatic biotransformation and renal elimination

(D) They are prothrombotic

(E) They are antipyretic

46. Which of the following NSAIDs is incorrectly matched with its toxicity?

(A) Phenylbutazone: Renal papillary necrosis in foals

(B) Carprofen: Acute hepatic failure in dogs

(C) Acetaminophen: Methemoglobinemia in the cat

(D) Aspirin: Reversible inhibition of platelet aggregation

(E) Naproxen: Gastrointestinal ulceration in dogs

Chapter 33

47. Which of the following statements concerning the major mechanism of action of anticonvulsants is incorrect?

(A) Potassium bromide hyperpolarizes neuronal membranes by crossing Cl^- channels more readily than Cl^-

(B) Benzodiazepines act at the benzodiazepine receptor site, which regulates the GABA receptor and causes the Cl^- channel to open more frequently after GABA stimulation

(C) Phenobarbital prolongs GABA-mediated synaptic inhibition by causing the Cl^- channel to remain open for a longer period of time

(D) Phenytoin prolongs the duration of inhibitory neurotransmission by blocking K^+ channels

48. Rover is a 3-year-old golden retriever that has just had an episode of cluster seizures. He had 1–2 seizures an hour during this cluster and you would like to give him medication to prevent the emergence of another cluster. He is presently not on anticonvulsant therapy. The best choice would be:

(A) Phenobarbital IV

(B) Phenobarbital orally

(C) Potassium bromide orally

(D) Pentobarbital IV

(E) Diazepam IM

Chapters 34 and 35

49. Which of the following statements concerning the pharmacological actions of deprenyl is incorrect?

(A) It is an irreversible inhibitor of the mitochondrial enzyme monoamine oxidase

(B) It is metabolized in the liver to desmethylselegiline, which is the active drug

(C) It results in a decrease in dopamine and norepinephrine levels in the brain

(D) It can inhibit adrenocorticotropin release from the pituitary

(E) Therapeutic responses to deprenyl may be delayed up to 30 days

50. Which of the following behavior-modifying drugs would be safest to administer to a hyperthyroid cat with cardiomyopathy that is suffering from inappropriate elimination?

(A) Amitriptyline

(B) Buspirone

(C) Imipramine

(D) Megestrol acetate

Chapters 36 to 41

51. The potential risks associated with the use of any antibiotic include all of the following except:

(A) Promotion of bacterial drug resistance

(B) Direct host toxicity

(C) Interference with the protective function of normal bacterial flora

(D) Development of antibiotic residues in food animals

(E) Activation of latent viral infections

52. The development of antimicrobial drug resistance can be minimized by all of the following except:

(A) Using the antibiotic with the narrowest spectrum of action

(B) Administering the antibiotics for the shortest effective duration

(C) Basing antibiotic use on in vitro susceptibility testing

(D) Adhering to extralabel drug use policies in food animals

(E) Giving subtherapeutic doses of antibiotics to protect against recurrent skin infections

53. You have a dog with a severe life-threatening gram (−) meningitis. Which of the following β-lactam antibiotics would you use while awaiting the results of in vitro susceptibility testing?

(A) Penicillin G

(B) Ampicillin

(C) Third-generation cephalosporin

(D) Vancomycin

(E) First-generation cephalosporin

54. All of the following statements concerning the pharmacological actions of β-lactam antibiotics are correct except:

(A) They are weak acids that are generally excreted unchanged in the urine

(B) One of the most serious toxicities is the occurrence of an anaphylactic reaction

(C) They work by binding to penicillin-binding proteins involved in bacterial cell wall synthesis

(D) Clavulinic acid is an irreversible inhibitor of β-lactamases that alone has no intrinsic antibacterial action

(E) The β-lactam antibiotics are bacteriostatic

55. All of the following are potential toxic reactions to trimethoprim-sulfadiazine in the dog except:

(A) Inhibition of platelet function

(B) Keratoconjunctivitis sicca

(C) Immune-mediated hemolytic anemia

(D) Acute hepatic necrosis

(E) Folate deficiency

56. A fluoroquinolone would be a good choice for empiric antibiotic use in which of the following conditions?

(A) A 2-month-old foal with gram (−) pneumonia

(B) A cat with chronic renal failure from pyelonephritis due to gram (−) bacteria

(C) A dog with a cutaneous infection that is shown to contain gram (+) chains of cocci on Gram stain

(D) A cat with a cat bite abscess

(E) A lactating dairy cow with gram (−) mastitis

57. Which of the following statements concerning the renal toxicity of aminoglycosides is correct?

(A) The most accurate way to detect early renal damage during therapy with aminoglycosides is to monitor serum urea nitrogen and serum creatinine levels

(B) A 3-month-old, dehydrated, febrile puppy is at increased risk to develop renal toxicity

(C) Concurrent administration of NSAIDs decreases the incidence of renal toxicity

(D) The development of renal toxicity correlates with peak serum concentrations

58. Which of the following statements about the pharmacokinetics of aminoglycosides is incorrect?

(A) They are excreted unchanged in the urine

(B) They are polar organic bases

(C) High concentrations are found in the CNS

(D) They are inactivated in the presence of necrotic material

(E) They must be given parenterally

59. The drug of choice to treat a rickettsial infection in a dog with renal failure is:

(A) Doxycycline

(B) Erythromycin

(C) Clindamycin

(D) Tetracycline

(E) Chloramphenicol

60. Superficial pyodermas in the dog are associated most commonly with infection with β-lactamase–positive staphylococci. All of the following drugs would be effective therapy for the treatment of a superficial pyoderma in the dog except:

(A) Lincomycin

(B) Clindamycin

(C) Chloramphenicol

(D) Erythromycin

(E) Tetracycline

Chapter 42

61. Which of the following antifungal agents is incorrectly matched with its mechanism of action?

(A) Amphotericin B: Binds to ergosterol, resulting in the formation of membrane pores

(B) Griseofulvin: Binds to microtubules and disrupts mitosis

(C) Itraconazole: Inhibits ergosterol synthesis

(D) Flucytosine: Leads to the accumulation of 14-α-methyl sterols that interfere with energy metabolism

62. All of the following side effects may be seen with amphotericin B treatment except:

(A) Normocytic, normochromic anemia

(B) Hypokalemia

(C) Nephrotoxicity

(D) Fever

(E) Hepatotoxicity

63. Tina is a 5-year-old, domestic, short-hair cat that presents to you for evaluation of a 3-month history of chronic mucopurulent nasal discharge and a recent onset of anorexia and depression. Analysis of cytological smears of the discharge reveals *Cryptococcus* species. A sample of spinal fluid also demonstrates the organism. Which of the following antifungal agents would be best to treat this cat?

(A) Ketoconazole

(B) Amphotericin B

(C) Fluconazole

(D) Griseofulvin

(E) Flucytosine

Chapter 43

64. You have just diagnosed a dog with *Giardia* infection. All of the following drugs would be effective in treating the dog except:

(A) Metronidazole

(B) Furazolidone

(C) Fenbendazole

(D) Sulfonamides

(E) Albendazole

65. You have just diagnosed a horse with equine protozoal myelitis. Which of the following combinations of drugs might be effective in treating this disease?

(A) Trimethoprim-sulfadiazine and pyrimethamine

(B) Trimethoprim-sulfadiazine and monensin

(C) Metronidazole and imidocarb

(D) Amprolium and meglumine antimonate

(E) Fenbendazole and metronidazole

Chapters 44 to 47

66. Which of the following antihelminthic agents is incorrectly matched with its mechanism of action?

(A) Piperazine: Cholinergic receptor antagonist/GABA receptor agonist

(B) Fenbendazole: Binds to tubulin

(C) Levamisole: Cholinergic receptor agonist

(D) Pyrantel: GABA receptor antagonist

(E) Organophosphates: Indirect-acting cholinergic agonists

67. You have just diagnosed a dog with an infestation of whipworms and hookworms. Which medication would you prescribe?

(A) Fenbendazole

(B) Pyrantel

(C) Piperazine

(D) Organophosphates

68. Which of the following statements concerning the heartworm preventative formulation of ivermectin is correct?

(A) It is dangerous to administer to collies

(B) It directly kills adult heartworms

(C) Adverse reactions include mydriasis, hypersalivation, paresis, stupor, ataxia, and coma

(D) It decreases GABAnergic neurotransmission

(E) Mammals are not affected by ivermectin since they have no GABA channels

69. All of the following drugs can be used to treat infestation with the trematode *Fasciola hepatica* in cattle except:

(A) Albendazole

(B) Ivermectin

(C) Clorsulon

(D) Praziquantel

70. A client brings a stray kitten infested with fleas to your office. You estimate that the cat is about 8 weeks old. Which of the following ectoparasitic agents would you use to treat the cat?

(A) Organophosphates

(B) Permethrin

(C) Lufeneron

(D) Pyrethrin

(E) Imidacloprid

71. Which of the following ectoparasital drugs is paired incorrectly with its mechanism of action?

(A) Fipronil: Inhibition of GABAnergic neurotransmission in fleas and ticks

(B) Imidacloprid: Nicotinic cholinergic receptor agonist in fleas

(C) Carbamate: Inhibition of acetylcholinesterase in fleas and ticks

(D) Foramidines: Interference with monoamine oxidase activity in ticks

(E) Methoprene: Inhibition of energy production in fleas

Chapters 48 and 49

72. Myelosuppression is the dose-limiting toxicity for all of the following antineoplastic agents except:

(A) Cyclophosphamide

(B) Lomustine

(C) Vincristine

(D) Methotrexate

(E) Cytosine arabinoside

73. You have just diagnosed a cat with CNS lymphoma. Which of the following drugs would be most effective?

(A) Cyclophosphamide

(B) Methotrexate

(C) 5-Fluorouracil

(D) Cytosine arabinoside

(E) Vincristine

74. All of the following are potential side effects of doxorubicin therapy in the dog except:

(A) Myelosuppression

(B) Facial pruritus during administration

(C) Cardiotoxicity

(D) Hemorrhagic diarrhea

(E) Hemorrhagic cystitis

75. You have just diagnosed a 10-year-old cat with mammary carcinoma that spread to several regional lymph nodes. Following mastectomy with lymph node dissection, you chose which of the following chemotherapy agents to control residual disease?

(A) *Cis*-platinum

(B) L-Asparaginase

(C) Doxorubicin

(D) Corticosteroids

Chapters 50 to 54

76. Which statement concerning the neurotransmission of vomiting is incorrect?

(A) Peripheral sensory receptors are primarily dopaminergic

(B) Input from the vestibular system to the chemoreceptor trigger zone (CRTZ) is via histaminergic and muscarinic neurons

(C) Neurotransmission in the CRTZ is primarily dopaminergic

(D) Neurotransmission in the emetic center is mediated by α_2 adrenergic receptors

77. The most effective emetic agent in the cat is:

(A) Xylazine

(B) Apomorphine

(C) Metoclopramide

(D) Hydrogen peroxide

(E) Atropine

78. Which of the following statements concerning the pharmacology of metoclopramide is correct?

(A) Side effects include ataxia and depression

(B) It has a short half-life and for best control of vomiting should be administered by continuous-rate IV infusion

(C) Its peripheral prokinetic effects are due to stimulation of dopamine receptors in the stomach

(D) Oral doses of metoclopramide are smaller than parenteral doses due to a large hepatic first-pass effect

(E) It undergoes extensive hepatic metabolism and is excreted entirely in the bile

79. Which of the following statements concerning the pharmacological effects of sucralfate is incorrect?

(A) Sucralfate dissociates in an acidic environment to aluminum hydroxide and sucrose octosulfate

(B) Sucralfate binds to and inactivates pepsin and bile acids

(C) Sucralfate stimulates the local release of prostaglandins and growth factors

(D) Sucralfate is absorbed extensively and undergoes hepatic metabolism followed by urinary excretion

(E) Sucralfate results in the formation of a polymerized matrix over damaged gastric epithelium

80. Rover is an 8-year-old Labrador retriever you have just diagnosed with chronic inflammatory hepatic disease. Since dogs with chronic hepatic failure are at increased risk of developing gastric ulcers, you prescribe which of the following medications for lifelong management of this risk?

(A) Cimetidine

(B) Ranitidine

(C) Famotidine

(D) An antacid containing a combination of aluminum and magnesium

(E) Omeprazole

81. Holstein is a 14-year-old cat that you are treating for idiopathic megacolon, which is a disorder of colonic motility that results in recurrent bouts of constipation. Which of the following drugs should you prescribe to promote colonic motility?

(A) Metoclopramide

(B) Erythromycin

(C) Cisapride

(D) Loperamide

(E) Atropine

82. The beneficial action of loperamide in the symptomatic control of diarrhea is due to:

(A) An increase in segmental circular muscle contraction and a decrease in peristaltic longitudinal muscle contraction

(B) Binding to μ and κ receptors in the brain stem and spinal cord

(C) Stimulation of intestinal secretion

(D) A decrease in both segmental circular muscle contraction and peristaltic longitudinal muscle contraction

83. Otis is a 4-year-old yellow Labrador retriever that presents to you after ingestion of a chemical toxin. After inducing emesis with apomorphine, you administer activated charcoal to bind the residual toxin. Several minutes after giving the charcoal, you administer which of the following medications to promote defecation?

(A) Bisacodyl

(B) Canned pumpkin

(C) Magnesium sulfate

(D) Cisapride

(E) Docusate

84. Which of the following statements concerning lactulose is correct?

(A) It increases colonic pH

(B) It is useful in the treatment of hepatic encephalopathy because it enhances colonic ammonia absorption

(C) It is absorbed in the small intestine and metabolized in the liver, and the active metabolites are secreted into the colon

(D) It softens stool with a detergent-like action that promotes the mixing of stools with lipids and water

(E) It is metabolized in the colon to fatty acids

Chapters 55 to 60

85. Which of the following pituitary hormones is used to promote milk production in dairy cows?

(A) Antidiuretic hormone

(B) Somatotropin

(C) Adrenocorticotropin

(D) Testosterone

(E) Oxytocin

86. You want to treat a cow with luteinizing hormone (LH) to resolve follicular cysts but you find there is no LH in your truck. What can you use instead?

(A) Prostaglandin $F_{2\alpha}$

(B) Pregnant mare gonadotropin

(C) Synthetic gonadotropin-releasing hormone

(D) Adenocorticotropin

(E) Follicle-stimulating hormone

87. Desmopressin, a long-acting synthetic analogue of antidiuretic hormone, has all of the following actions except:

(A) Binds to V2 receptors on the renal collecting tubule, rendering them permeable to water

(B) Can be given intranasally or into the conjunctival sac to treat diabetes insipidus in the dog

(C) Can be given parenterally to increase the release of von Willebrand factor

(D) Stimulates peripheral vasoconstriction and hepatic glycogenolysis

88. Methimazole decreases thyroid hormone levels in cats with hyperthyroidism by:

(A) Inhibiting the uptake of iodide into the thyroid gland

(B) Blocking the release of thyroid hormone

(C) Preventing the peripheral conversion of T_4 to T_3

(D) Inhibiting thyroid peroxidase

(E) Direct cytotoxicity to the thyroid gland

89. The major disadvantage of using synthetic liothyronine (T_3) replacement to treat canine hypothyroidism is that:

(A) Dogs metabolize this form of the hormone so quickly, it is hard to maintain adequate serum concentrations

(B) Since T_4 is the active hormone, giving T_3 may not meet the hormonal demands of all organs

(C) It may not meet the demands of some organs that typically maintain high levels of active hormone by extracting T_4 and deiodinating it to T_3

(D) This form of the hormone is poorly absorbed from the gastrointestinal tract and undergoes significant hepatic first-pass metabolism

(E) Administration of this active form of the hormone increases the likelihood that dogs will exhibit signs of hyperthyroidism

90. Which of the following types of insulin would you administer to achieve the most rapid correction of hyperglycemia in a cat that presented with ketoacidotic diabetes mellitus?

(A) Regular

(B) Lente

(C) Ultralente

(D) NPH

91. All of the following are pharmacological actions of the sulfonylureas except:

(A) Inhibition of ATP-dependent K^+ channels on pancreatic β cells

(B) Decrease in hepatic gluconeogenesis

(C) Increased number of peripheral insulin receptors

(D) Decreased hepatic degradation of insulin

(E) Inhibition of insulin receptor phosphorylation

92. All of the following drugs can be used to treat canine hyperadrenocorticism except:

(A) Mitotane

(B) Ketoconazole

(C) Deprenyl

(D) Desoxycorticosterone pivalate

(E) Bromocriptine

93. Which of the following statements concerning anabolic steroids is incorrect?

(A) Anabolic steroids are drugs controlled by the Drug Enforcement Agency (DEA)

(B) Anabolic steroids stimulate erythropoietin production

(C) Anabolic steroids are potential hepatotoxins

(D) Anabolic steroids promote the development of muscle mass

(E) Use of anabolic steroids in the horse is illegal

94. Ursodeoxycholate has all of the following beneficial actions in hepatic disease except:

(A) Stimulation of hepatic DNA synthesis

(B) Replacement of the normal relatively hydrophobic pool of bile acids

(C) Stimulation of a bicarbonate-rich bile flow

(D) Immune modulating effect

(E) Maintenance of mitochondrial membrane integrity

95. All of the following clinicopathological abnormalities might accompany the use of corticosteroids in the dog except:

(A) An increase in serum alkaline phosphatase activity

(B) Mild polycythemia (increased red blood cell volume)

(C) Mature neutrophilic leukocytosis

(D) Lymphopenia

(E) Hyperkalemia

96. Trixie is a 5-year-old cocker spaniel who presents to your clinic for an acute onset of weakness, hematochezia, and epistaxis. On diagnostic workup you discover that she has immune-mediated thrombocytopenia. You start therapy with:

(A) Prednisone at 2 mg/kg/day

(B) Dexamethasone at 0.015 mg/kg day

(C) Prednisone at 0.125 mg/kg/day

(D) Prednisone at 0.5 mg/kg/day

(E) Dexamethasone at 2 mg/kg/day

97. The most serious side effect of long-term use of corticosteroids is suppression of the pituitary-adrenal axis. This suppression is best prevented by:

(A) Using short-acting preparations of corticosteroids gradually tapered to alternate-day therapy

(B) Using multiple single doses of long-acting repository preparations of corticosteroids

(C) Alternating the use of several corticosteroid preparations

(D) Using corticosteroid preparations that lack mineralocorticoid activity

Chapter 61

98. All of the following are side effects of azathioprine therapy except:

(A) Leukopenia

(B) Hepatotoxicity

(C) Pancreatitis

(D) Gastrointestinal ulceration

99. The mechanism of action of cyclosporine is to:

(A) Inhibit transcription of the gene encoding interleukin-2

(B) Inhibit lymphocyte DNA synthesis

(C) Block Fc receptors on macrophages

(D) Inhibit phospholipase A activation

Chapter 62

100. The most serious side effect of the use of human recombinant erythropoietin in cats with chronic renal failure is:

(A) Aplastic anemia due to the development of antibodies to the human product that cross-react with feline erythropoietin

(B) Systemic hypertension with the development of retinal detachments

(C) Depletion of iron stores

(D) Anaphylaxis

101. You are treating a 5-year-old intact male golden retriever for concurrent immune-mediated anemia and thrombocytopenia with high-dose corticosteroids. One of the most common side effects of IMHA in the dog is pulmonary thromboembolism. To prevent this from happening over the course of the next several days, you treat the dog concurrently with:

(A) Warfarin orally

(B) Aspirin orally

(C) Streptokinase IV

(D) Heparin SC

ANSWERS

1. The answer is C.
Intravenous (IV) administration is the best route because IV injection of any drug bypasses all barriers to absorption so that the drug is 100% bioavailable and essentially the entire dose of diazepam can be delivered to the brain. The hydrophobicity of diazepam permits the drug to distribute to the brain, since it will not be excluded by the blood-brain barrier. After oral administration diazepam is absorbed from the gastrointestinal tract, and undergoes significant hepatic first-pass metabolism to nordiazepam and oxazepam, both of which have only 25% of the anticonvulsant activity of the parent compound. The oral bioavailability of diazepam in the dog has been reported to be as low as 3%. The absorption of diazepam by intramuscular (IM) and subcutaneous (SC) routes is highly variable since the lipid-soluble drug does not diffuse well from the hydrophilic environment of the extracellular space into the vascular space. Therefore, these routes are of limited use in the emergency treatment of seizures. Intermittent IV administration of diazepam is frequently required to control cluster seizures in epileptic dogs. Dogs with cluster seizures have a pattern of multiple seizures occurring over a 1–2 day period. They are often refractory to chronic anticonvulsant therapy and present a source of frustration to owners who often must hospitalize their animal during clusters. Unfortunately most owners cannot be taught to give IV diazepam at home. An alternative to hospitalization is to have the owner give the injectable formulation of diazepam rectally. Following rectal administration, about 50% of the drug is absorbed directly into the systemic circulation and therefore bypasses uptake by the portal circulation and subsequent hepatic inactivation.

2. The answer is D.
Since the aminoglycosides are highly charged at physiological pH, they do not readily diffuse through the blood-brain barrier and are largely excluded from the CNS. Although plasma protein binding is low and this would not limit the drug's distribution, the fact that the drug exists in its ionized state limits distribution outside of the extracellular space. Normally aminoglycosides are not bioavailable after oral administration. Appreciable absorption of aminoglycosides may occur in the presence of gastrointestinal inflammation. Since mastitic milk is acidic, any unionized drug that penetrates into the udder rapidly becomes protonated and charged. This traps the ionized drug within the udder.

3. The answer is A.
The serum concentration of any drug that is excreted unchanged in the urine may be increased by the decrease in glomerular filtration rate that accompanies renal disease. An individual with chronic hepatic disease may have a decrease in the activity of the CP450 system, but this would only delay conversion of the prodrug to the active form and should not result in an increase in serum concentration of the active drug. Since cimetidine inhibits the hepatic CP450 system, it would result in decreased activation of the prodrug. Phenobarbital induces the hepatic CP450 system and could increase the rate at which the prodrug is converted to the active drug. In the presence of normal renal function, however, the drug would be eliminated at a faster rate as well, resulting in no net increase in serum concentration of active drug A.

4. The correct answer is A.
Cats actually have decreased levels of certain members of the glucuronyl transferase family. Since phase II reactions result in the formation of *inactive*, highly polar compounds that then readily can undergo biliary or renal excretion, cats are predisposed to the toxic effects of compounds that require phase II glucuronidation for elimination such as acetaminophen and aspirin. Dogs are known to have low levels of acetyltransferase activity, and this may predispose them to toxicity to compounds such as the sulfonamides which undergo elimination following acetylation. Pigs are known to be poor sulfate conjugaters. Cats are also somewhat deficient in their ability to conjugate xenobiotics to sulfate.

5. The correct answer is A.
Due to its high lipid solubility and low degree of protein binding, the drug should readily diffuse from the extracellular space into intracellular compartments. The volume of distribution (Vd) represents the theoretical volume in which the total amount of drug would have to be uniformly distributed to give the observed plasma concentration. It is defined mathematically as the ratio of the total amount of drug in the body to the serum concentration of the drug. Since highly lipophilic drugs readily pass from the vascular space to the interstitial space and then into intracellular compartments, the total amount of drug in the body will be larger than the serum concentration and the Vd will be quite large. Since the drug in the question is lipid soluble, it should be absorbed readily across the intestinal epithelium and oral bioavailability should be high. Highly lipid-soluble drugs may have low bioavailability if they have significant first-pass hepatic metabolism, but in this case the drug is converted to an active metabolite in the liver. Clearance (Cl) is the volume of blood cleared of drug per unit of time. Clearance for most drugs is constant over a wide range of drug doses. Only when drug uptake or elimination mechanisms are saturated would clearance become variable. Serum $t^{1/2}$ is defined as the time it takes for the total amount of drug in the body to decrease by 50%. It is a derived parameter that varies with both Cl and Vd and is equal to 0.693 Vd divided by Cl. A highly lipid-soluble drug with low tissue binding that has a large Vd does not necessarily have a long $t^{1/2}$ as this will also depend on the drug's clearance rate.

6. The answer is E.
It takes about 4 half-lives to obtain steady-state serum concentration (Css) of a drug. This is an important fact to know, especially when dealing with drugs with a long $t^{1/2}$ for which it may take several days or weeks to obtain Css and thus consistent pharmacological results. This frequently is encountered in establishing Css for phenobarbital and potassium bromide, the two most effective anticonvulsants for long-term control of seizures. In the case of phenobarbital, it takes 11–14 days to obtain Css while potassium bromide with a half-life of 25 days in dogs takes 3 months to obtain Css. The Css is defined mathematically as:

$$\text{Css} = \text{Bioavailability} \times \text{dose/dosing interval} \times \text{clearance}$$

Css will increase as the drug's dose and bioavailability increase and as the dosing interval is shortened or the clearance prolonged.

7. The answer is D.
The therapeutic index (TI) is defined as the ratio of the LD_{50} to ED_{50}. The ED_{50} is the dose of drug required to produce the desired effect in 50% of the population and is an indication of the drug's potency. The LD_{50} is the dose of the drug that is lethal to 50% of the population. Therefore, the larger the value for the TI, the safer a drug is. The efficacy of a drug is dependent on the affinity of the drug for its receptor and is defined mathematically by the Michaelis-Menton equation:

$$\text{Drug response} = E_{max}/Kd + \{\text{dose}\}.$$

where E_{max} indicates the efficacy of the drug. Drugs can have quite different potencies (i.e., ED_{50} values) but have the same efficacy (i.e., E_{max}).

8. The correct answer is A.
If drug A displaces drug B from albumin, the concentration of free drug B that is capable of distributing and interacting with its cellular receptor increases. Such an interaction occurs when nonsteroidal anti-inflammatory agents (NSAIDs) are given concurrently with warfarin. The result of NSAID displacement of warfarin from its albumin binding site is potentiation of warfarin's anticoagulation action, which can precipitate bleeding problems. If 2 drugs are excreted unchanged in the urine by glomerular excretion, their elimination is independent of one another. However, if

drug B competes with drug A for active secretion into the urine, drug A will be eliminated at a much slower rate while the elimination and pharmacological action of drug B will not be affected. If drug A binds to and inactivates drug B in the stomach, the bioavailability and thus pharmacological action of drug B will be diminished. Agonists are drugs that bind to and activate a receptor whereas antagonists are drugs that bind to but do *not* activate the receptor. Although it would depend partially on the binding affinity of each drug for the receptor, it is expected that the 2 drugs would cancel each other's effect.

9. The answer is B.
The parasympathetic preganglionic fibers originate in the midbrain, medulla oblongata, and sacral portion of the spinal cord. It is correct that the preganglionic fibers are quite long and synapse on postganglionic neurons that are close to the organ being innervated. Acetylcholine is the neurotransmitter released at both sympathetic and parasympathetic preganglionic neurons, but only in the parasympathetic system is acetylcholine released by postganglionic neurons. The receptors for acetylcholine on postganglionic neurons are nicotinic cholinergic receptors whereas acetylcholine receptors on effector organs are muscarinic cholinergic receptors. The major effects of parasympathetic stimulation on effector organs are presented in D.

10. The correct answer is D.
Dopamine receptors are in the renal, mesenteric, and coronary vasculature where they mediate vasodilation. Dopamine frequently is used clinically at low doses to increase renal blood flow to improve glomerular filtration rate in patients with oliguric or anuric renal failure. At higher doses, dopamine stimulates β_1 receptors and α_1 receptors, resulting in increased cardiac contractility and peripheral vasoconstriction, respectively. The β_2 receptors are present in smooth muscle of the vasculature and the bronchial, gastrointestinal, and genitourinary tracts, where they mediate smooth muscle relaxation. In the bronchial tree, the action of β_2 receptors is bronchodilation and this property is exploited clinically to reverse bronchoconstriction accompanying chronic bronchitis and asthma. The β_2 receptors are also present in the liver and in skeletal muscle where they promote glycogenolysis. The β_1 receptors are primarily present in the myocardium where they increase heart rate, cardiac contractility, and conduction. The α_1 receptors are present primarily on vascular and genitourinary tract smooth muscle. Stimulation of α_1 receptors results in peripheral vasoconstriction and closure of the urethral sphincter. This property allows exploitation of α_1 agonists as agents to maintain blood pressure and urinary continence. The α_2 receptors are present on vascular smooth muscle where they mediate vasodilation and on presynaptic adrenergic nerve terminals where they inhibit the release of norepinephrine. They are also present on pancreatic β cells where they decrease insulin release.

11. The correct answer is B.
Edrophonium reversibly binds to the active center of the acetylcholinesterase enzyme. Its duration of action is only 10–15 min since the drug undergoes rapid renal elimination. The drug is used clinically as an aid in diagnosing myasthenia gravis. Organophosphates work by irreversibly phosphorylating acetylcholinesterase. Organophosphate toxicity is marked by signs of overstimulation of nicotinic (muscle weakness, tremors) and muscarinic receptors (salivation, lacrimation, urination, and defecation—SLUD). Use of an anticholinergic agent, such as atropine, can reverse the signs of muscarinic toxicosis. Bethanechol is a direct cholinergic agonist and would exacerbate the signs of organophosphate toxicosis. Echothiophate is used as a miotic to facilitate aqueous drainage and thus lower the intraocular pressure in glaucoma patients. Neostigmine is poorly absorbed from the gastrointestinal tract and the preferred route of administration is parenteral (IM).

12. The answer is E.
Atropine is a direct-acting muscarinic cholinergic antagonist. Since muscarinic cholinergic receptors are not present on the peripheral vasculature, administration of atropine has little effect on vascular resistance. Atropine does cause large-airway bronchodilation; relaxation of the iris, leading to mydriasis; a decrease in vagal tone, leading to an increased heart rate; and inhibition of respiratory and salivary secretions.

13. The correct answer is D.
Phenylpropranolamine is an indirect-acting sympathomimetic drug that enhances the release of norepinephrine from sympathetic neurons. Its primary action is on α_1 receptors in the smooth muscle of the urethral sphincter where it works to increase tone. It is frequently prescribed to treat urethral muscle incompetence that develops following neutering. Since it is an α_1 agonist, it also will cause a degree of contraction of smooth muscle in the peripheral vasculature, leading to vasoconstriction and resulting in an elevation of blood pressure. Stimulation of α_1 receptors in the CNS may lead to hyperexcitability and suppression of appetite. Phenylpropranolamine is marketed as a nonprescription appetite suppressant in humans. Vasoconstriction of vessels in the nasal mucosa may lead to a sensation of dryness in the nose and mouth. For this reason, some dogs may actually drink more water while on the medication. Phenylpropranolamine is a frequent component of nonprescription nasal decongestants. As α adrenergic receptor stimulation has minimal effects on the bronchial smooth musculature, it would not be expected to cause bronchoconstriction. In fact, it may cause a mild degree of bronchodilation due to stimulation of β_2 receptors. Since phenylpropranolamine is also a β_1 agonist at higher doses, cardiac side effects such as tachycardia may be seen with the use of this drug.

14. The answer is D.
Norepinephrine is primarily an α_1 agonist causing vascular smooth muscle contraction, vasoconstriction, and increased blood pressure. Phenoxybenzamine, phentolamine, and prazosin are α_1 blockers that result in peripheral vasodilation and thus decreased blood pressure. Propranolol is a nonselective β adrenergic receptor antagonist. Although vascular β_2 receptors have only a limited role in the control of normal vascular tone, blockage of these receptors in hypertensive individuals lowers blood pressure.

15. The answer is B.
Local anesthetics block the opening of voltage-sensitive Na^+ channels, thus preventing the generation of an action potential in the nerve cell. Generation of an action potential occurs when large amounts of Na^+ are allowed to enter the cell through voltage-gated Na^+ channels. The increase in intracellular Na^+ causes transient depolarization of the cell (i.e., the membrane potential becomes positive), resulting in an action potential. Termination of the action potential depends on closure of the Na^+ channels coupled with opening of the slow K^+ channels that allow K^+ to escape from the cell. The Na^+/K^+ ATPase, which is present on the plasma membrane of all cells in the body, is necessary to maintain the negative resting cell membrane potential by keeping the intracellular Na^+ concentration low and intracellular K^+ concentration high. Engagement of the acetylcholine nicotinic receptor on the skeletal motor end plate opens a membrane channel that is permeable to both Na^+ and K^+. The movement of Na^+ and K^+ through this channel results in generation of the muscle action potential. Local anesthetics have no effect on the Na^+ channel coupled to the nicotonic cholinergic receptor.

16. The answer is A.
Since local anesthetics are weak bases, they exist in both ionized and nonionized forms depending on their pK_a and the surrounding pH. The lower the pK_a of the agent, the greater amount of drug present in the nonionized lipid-soluble form at physiological pH and the more rapid its penetration of the nerve cell membrane and the faster the rate of onset. The duration of action of local anesthetics is determined by the degree of protein binding once the agent is within the nerve cell. It is the cationic species (ionized form) that binds to an intracellular protein, which then interacts with and inactivates the Na^+ channel. Ester-linked agents undergo hydrolysis by plasma cholinesterases, while amide-type anesthetic agents are metabolized by the hepatic CP450 system. Since some degree of systemic absorption occurs even when the agents are applied topically, signs of toxicosis can develop. The most common side effects are seen in the CNS and cardiovascular system, but instead of tachycardia and depression, these agents are associated with CNS excitement

and seizures and decreases in cardiac conduction rate and myocardial contractility. Since inactivation of the local anesthetic agent depends on uptake into the systemic circulation, the greater the vascularity of the tissue, the faster the rate of elimination and the shorter the duration of action. Epinephrine is an α_1 adrenergic agonist that causes local vasoconstriction and therefore delays systemic uptake of the agent and prolongs the duration of action.

17. The answer is E.
Succinylcholine works as an agonist at the nicotinic cholinergic receptors at the neuromuscular junction. It may also stimulate muscarinic cholinergic receptors, which cause an increase in respiratory secretions. As an agonist at the acetylcholine receptor, succinylcholine results in transient muscle contraction. The drug prevents repolarization of the postsynaptic membrane, as it remains bound to the receptor and is not degraded by acetylcholinesterase. Further impulse generation in the muscle is prevented and a flaccid muscle paralysis occurs. Succinylcholine is metabolized by plasma cholinesterases. Succinylcholine has a rapid onset of action that is of very short duration in the horse, pig, and cat (2–8 min), but is more prolonged in the dog (25 min). None of the neuromuscular blocking agents have any effect on the neurotransmission of pain.

18. The answer is A.
Atracurium is inactivated by plasma esterases and by spontaneous degradation in the plasma (Hoffman reaction) and does not depend on renal excretion of active drug. Gallamine, pancuronium, and tubocurarine are excreted primarily unchanged in the urine. Dantrolene, a muscle relaxant, has a direct action on skeletal muscle but no effect at the neuromuscular junction.

19. The answer is D.
Digitalis increases cardiac contractility. This inotropic effect is due to inhibition of the plasma membrane Na^+/K^+ ATPase, which results in increased concentrations of intracellular Na^+, which in turn increase the transmembrane exchange of Na^+ for extracellular Ca^{2+}. The resultant increase in intracellular Ca^{2+} increases the amount of Ca^{2+} available for cardiac myocyte contraction. Digitalis also has a vagal-like effect on the heart, decreasing conduction through the sinoatrial and atrioventricular nodes, and is frequently used to slow the heart rate in the treatment of atrial fibrillation. This effect may lead to such adverse effects as partial or complete heart block.

20. The answer is B.
Dobutamine, a synthetic catecholamine, has its predominant action at β_1 receptors. Dobutamine is not available orally and has a very short $t^1/_2$ (2 min) so that it must be administered by continuous IV infusion. Since tolerance to the drug develops over a short period of time (48 hours) owing to β_1 receptor desensitization, the drug is used primarily for short-term ionotropic support in patients in myocardial failure. For reasons not completely understood, the benefits of 1 short-term infusion may extend over several weeks. Dobutamine does not activate dopamine receptors.

21. The answer is A.
Enalapril is an angiotensin I converting enzyme inhibitor and prevents the conversion of angiotensin I to angiotensin II in the lung and vascular endothelium. It is renin released by the juxtaglomerular apparatus in the kidney that converts angiotensinogen to angiotensin I. By preventing the production of angiotensin II, which is a potent vasoconstricting agent, enalapril causes both venous and arteriolar dilation (balanced vasodilation). Decreasing angiotensin II levels also inhibits aldosterone secretion from the adrenal. This prevents renal water and Na^+ retention and further decreases venous pressure. Enalapril is one of the most popular drugs used to control systemic hypertension in small animals. Because the action of aldosterone in the distal tubule of the kidney promotes Na^+ retention and K^+ excretion, inhibition of aldosterone release may result in decreased renal excretion of K^+ and hyperkalemia. Since angiotensin is important in maintaining renal perfusion via preferential vasoconstriction of the renal glomerular efferent arteriole, enalapril, by blocking angiotensin II production, causes efferent arteriolar dilation. The net effect is to decrease glomerular hypertension, which may contribute to ongoing renal pathology in some disease states. This effect may also lead to a decrease in glomerular filtration rate and the development of azotemia.

22. The answer is D.
Both nitroglycerin and nitroprusside lead to the formation of nitrous oxide, which stimulates increased guanylate cyclase activity leading to increased generation of cGMP. It is cGMP that mediates vascular smooth muscle relaxation. Hydralazine has a direct effect on vascular smooth muscle to cause relaxation independent of the sympathetic nervous system. In fact, administration of hydralazine can cause a decrease in blood pressure that is accompanied by sympathetic activation and reflex tachycardia. Prazosin is an α_1 adrenergic antagonist that acts as a balanced vasodilator.

23. The answer is D.
Concurrent administration of quinidine and digoxin results in an increase in digoxin serum levels. Quinidine, a class I antiarrhythmic agent, is a local anesthetic that decreases the rate of influx through fast Na^+ channels and thus decreases the maximal rate of depolarization in cardiac fibers (phase 0). Quinidine also inhibits the rate of spontaneous depolarization (phase 4) in automatic cells by blocking K^+ channels. This results in a prolongation of atrial and ventricular refractory periods. Quinidine has a vagolytic action, which also helps to prolong the atrial refractory period. This vagolytic action also leads to improved conduction through the atrioventricular (AV) node and may lead to an increase in ventricular rate. This is particularly true when the drug is given IV. Quinidine also has a mild negative ionotropic effect due to interference with transmembrane Ca^{2+} influx.

24. The answer is D.
Quinidine and lidocaine undergo hepatic metabolism by the CP450 system. Inhibition of these enzymes slows conversion of these drugs to their inactive metabolites and slows elimination. In addition, since lidocaine undergoes extensive hepatic first-pass metabolism, its clearance is blood flow limited. Therefore, any agent or physiological condition that slows hepatic blood flow will slow lidocaine elimination. Although procainamide undergoes excretion unchanged in the urine, cimetidine competes with procainamide for secretion into the renal tubules and thus could increase serum levels of procainamide.

25. The answer is E.
Diltiazem is a Ca^{2+} channel blocker. Propranolol is a nonselective β_1 and β_2 adrenergic receptor antagonist that also has some limited ability to block Na^+ channels at high concentrations. Amiodarone is primarily a K^+ channel blocker but also is a noncompetitive antagonist at β_1 and β_2 receptors. Amiodarone also blocks Na^+ and Ca^{2+} channels. Atenolol is a selective β_1 receptor antagonist.

26. The answer is A.
The key to answering this question is to realize that you want to give a drug that is effective against supraventricular arrhythmias but will not worsen the dog's underlying myocardial failure. Propranolol is a β_1 adrenergic antagonist and is effective against supraventricular arrhythmias but will substantially decrease myocardial contractility. Amiodarone is a class III antiarrhythmic that works by blocking K^+ channels. It is used primarily to treat life-threatening ventricular arrhythmias. Since the drug has a long half-life, it takes several weeks for the full pharmacological effect to become evident. Amlodipine, diltiazem, and verapamil are all Ca^{2+} channel blockers, but they differ in their selectivity for cardiac and vascular Ca^{2+} channels. Amlodipine's action is primarily on vascular smooth muscle, leading to vasodilation. It has minimal effects on cardiac myocytes and is not considered an antiarrhythmic agent. Both diltiazem and verapamil decrease conduction through the sinoatrial (SA) and AV nodes and are effective in the treatment of supraventricular arrhythmias. Verapamil also significantly decreases myocardial contractility and when given IV can lead to vasodilation and subsequent hypotension. Diltiazem, on the other hand, has negligible effects on cardiac contractility and causes only mild peripheral vasodilation.

27. The answer is A.
Atropine is an anticholinergic agent that causes mydriasis (which may compromise drainage of aqueous at the iridocorneal angle) and increases aqueous production. Timolol is a β_2 receptor agonist that decreases aqueous humor production. Dorzolamide is a topical carbonic anhydrase inhibitor that also decreases aqueous humor production. Epinephrine is an α and β adrenergic agonist that causes intense ocular vasoconstriction, which results in a reduction in aqueous humor production. Since epinephrine is also a mydriatic agent, it is often used in combination with a miotic agent such as pilocarpine (direct-acting parasympathomimetic) or echothiophate (indirect-acting parasympathomimetic).

28. The answer is B.
Cyclosporine is an immunosuppressant that inhibits T-lymphocyte activation. In dogs with keratoconjunctivitis sicca, cyclosporine restores normal tear production by decreasing lacrimal gland inflammation and by directly stimulating lacrimal gland secretion. Pilocarpine is a direct-acting cholinergic agonist that stimulates lacrimal gland secretion. Corticosteroids, antibiotics, topical nonsteroidal anti-inflammatory agents, antiviral medications, and mydriatics (atropine and phenylephrine) have no effect on the course of the disease but might be useful to manage secondary complications such as uveitis, bacterial conjunctivitis, or corneal ulceration.

29. The answer is A.
Estrogens have a permissive effect of a stimulation of urethral smooth muscle tone. Phenylpropanolamine and ephedrine are α adrenergic agonists that are highly effective in increasing urethral tone, but also cause peripheral vasoconstriction and can increase blood pressure. They also have effects at β receptors in the liver and adipose tissue to increase gluconeogenesis and lipolysis, which would be contraindicated in a diabetic. These agonists may also cause appetite suppression in some dogs, which could complicate diabetic regulation. Neither bethanechol, which is a cholinergic agonist, nor diazepam, which is a skeletal muscle relaxant, would have any effect on urethral smooth muscle tone.

30. The answer is D.
Prazosin is an α_1 adrenergic blocker that would relax urethral smooth muscle and bethanechol is a cholinergic agonist that would stimulate contraction of the detrusor muscle in the urinary bladder. Phenylpropanolamine, estrogen, and imipramine would increase urethral tone. Phenoxybenzamine is an α blocker that would relax the urethral sphincter, but atropine, an anticholinergic agent, would inhibit detrusor muscle contraction. Flavoxate also is a parasympatholytic drug that would decrease detrusor contraction. Dantrolene is a skeletal muscle relaxant that would have no effect on urethral or bladder smooth muscle.

31. The answer is E.
Theophylline promotes bronchodilation by inhibiting phosphodiesterase and thus raising intracellular levels of cAMP, antagonizing the actions of adenosine, a potent bronchoconstricting agent, and by interfering with Ca^{2+} mobilization. Theophylline also increases mucociliary clearance, inhibits mast cell degranulation, and increases the force of respiratory muscle contraction. Theophylline has a relatively low TI and can cause gastrointestinal (vomiting), CNS (excitement, restlessness, seizures), and cardiac (dysrhythmias) side effects. The latter side effect is seen most often when the drug is administered IV.

32. The answer is B.
The β_2 adrenergic receptor agonists, such as albuterol, increase intracellular cAMP, which results in lower-airway bronchial smooth muscle relaxation. Acetylcholine binds to M3 receptors in large airways, leading to an increase in intracellular Ca^{2+} and bronchial smooth muscle contraction. Muscarinic antagonists block this response and cause large-airway bronchodilation. Muscarinic receptors do not signal by influencing cAMP or cGMP levels in the cell. Muscarinic antagonists lead to a decrease in respiratory airway secretions. Since bronchoconstriction can lead to coughing, bronchodilation by β_2 agonists can decrease cough, but they have no central antitussive action.

33. The answer is C.
Since the loop diuretic furosemide inhibits Na^+ reabsorption in the loop of Henle, large amounts of Na^+ are delivered to the distal tubule where they are exchanged for H^+ and K^+. This increases the urinary excretion of K^+ and H^+, which can lead to serum hypokalemia and metabolic alkalosis, respectively. Furosemide also promotes the excretion of Ca^{2+} and is used to treat serum hypercalcemia. Furosemide is the mainstay of therapy for pulmonary edema associated with congestive heart failure. Its efficacy is related to its ability to institute diuresis quickly and to a mild venodilatory effect that shifts blood away from the pulmonary to the systemic circulation. Although furosemide does not increase glomerular filtration rate in renal failure, it does help to maintain urine output, facilitating the treatment of oliguric renal failure. The retention of organic anions in renal failure can interfere with the tubular secretion of furosemide and may necessitate the use of high doses to promote diuresis. A continuous-rate IV infusion of furosemide is preferred by many clinicians. By decreasing pulmonary capillary pressure, furosemide may help to prevent pulmonary vascular wall damage and the subsequent hemorrhage that occurs during strenuous exercise in horses.

34. The answer is E.
Triamterene inhibits active Na^+ reabsorption in the late distal convoluted tubules. Only the loop diuretics inhibit Na^+ and Cl^- resorption in the loop of Henle. Mannitol is a sugar that is freely filtered at the glomerulus and is poorly reabsorbed. The presence of unabsorbable mannitol in the proximal tubule decreases water reabsorption. Acetazolamide is a carbonic anhydrase inhibitor that blocks proximal tubular resorption of HCO_3^- and Na^+. Hydrochlorothiazide is a thiazide diuretic that blocks Na^+ reabsorption in the distal tubule. Spironolactone is an aldosterone antagonist that blocks Na^+ absorption and K^+ excretion.

35. The answer is C.
Both metophane and halothane will increase intracranial pressure, sensitize the myocardium to dysrhythmias with catecholamines, and reduce cardiac output. Isoflurane has little tendency to cause these adverse effects. Due to its relatively low lipid solubility [minimum alveolar concentration (MAC) = 1.28], isoflurane has a low blood gas partition coefficient, and induction and recovery times are fast. Although the use of nitrous oxide might be safer than isoflurane in this patient, nitrous oxide would not provide the level of anesthesia necessary to perform an exploratory laparotomy.

36. The answer is A.
It is the inactive metabolites generated by metabolism of anesthetic agents by the hepatic CP450 system that are responsible for the observed idiosyncratic hepatotoxic reactions to inhalational anesthetics. The degree of lipid solubility of the agent correlates with the agent's potency. Anesthetic potency is defined by the MAC, which is the concentration of agent that prevents gross purposeful movement in 50% of patients in response to a standardized painful stimulus. Anesthetic potency is inversely related to MAC. The blood gas partition coefficient is an indication of the solubility of the agent in blood. The more soluble the agent in blood, the faster induction and recovery are.

37. The answer is D.
Dissociative anesthetics increase cerebral blood flow and intracranial pressure. They also result in the appearance of epileptiform electrocardiographic patterns in the limbic system and are contraindicated in animals with seizure disorders. The barbiturates are the injectable anesthetics of choice in patients with intracranial disease, as they decrease cerebral blood flow and intracranial pressure. Although the dissociative anesthetics do depress the cardiovascular system, they also activate the sympathetic nervous system, which compensates for the depression. These agents do provide some analgesia, but they actually increase muscle tone and for this reason are commonly combined with a tranquilizer or muscle relaxant such as acepromazine or diazepam.

38. The answer is D.
Although both barbiturates and propofol can cause apnea at induction, it is the barbiturates that cause dose-dependent depression of the brain

stem respiratory center. This property is exploited in the production of solutions for euthanasia. The thiobarbiturates and propofol are highly lipid soluble and rapidly diffuse into the brain, resulting in very fast induction of anesthesia. The drugs then quickly redistribute to more highly perfused tissues such as muscle. This redistribution decreases brain concentrations and thus decreases the level of anesthesia. Further redistribution to more poorly perfused adipose tissue results in recovery from anesthesia. Repeated administration of thiobarbiturates can result in the accumulation of high concentrations in adipose tissue and subsequent release from these stores once anesthetic administration is terminated, which then prolongs recovery times. Final elimination of the barbiturates occurs after extensive hepatic metabolism and excretion of the inactive metabolites in the urine. Propofol undergoes rapid hepatic metabolism and excretion in the urine, and prolonged recovery is not as great a problem with this drug. Neither drug provides any significant analgesia.

39. The answer is C.
Phenothiazines actually have an antiarrhythmic effect and decrease the incidence of catecholamine-associated arrhythmias. This is one of the important reasons why these drugs are used as preanesthetic agents. Other beneficial actions as a preanesthetic agent include a central antiemetic effect [due to inhibition of dopamine neurotransmission in the chemoreceptor trigger zone (CRTZ) and blockage of α adrenergic receptors in the emetic center] and the fact that they decrease the dose of anesthetic necessary to maintain general anesthesia. The preanesthetic benefits of phenothiazines must be weighed against potential adverse effects. The most serious is due to blockage of α_1-mediated peripheral vasoconstriction, which can result in hypotension. This effect is particularly prominent when high sympathetic tone is present such as in hypovolemic or shocky patients. The phenothiazines may also cause sinus bradycardia, hypothermia, sequestration of red blood cells in the spleen, refractory penile prolapse in the stallion, a lowering of the seizure threshold, and occasionally are associated with the development of a rage syndrome. It is important to remember that when they are used as preanesthetics or sedatives for a minor surgical procedure, they have *no* analgesic activity.

40. The answer is D.
Diazepam is a highly lipid-soluble compound and it must be converted to a more polar form before it can be excreted in the urine. The drug undergoes biotransformation in the liver to two active metabolites, desmethyldiazepam and oxazepam. These metabolites undergo conjugation with glucuronide and are excreted in the urine. Diazepam binds to the GABA receptor and facilitates the action of GABA. This results in the influx of large amounts of Cl^-, membrane hyperpolarization, and inhibition of neuronal cell function. It can be used as an anticonvulsant, tranquilizer, and skeletal muscle relaxant. Since it is highly lipid soluble, it reaches the brain quickly after IV administration and is highly effective as a treatment for status epilepticus. In some cats receiving oral diazepam for short periods of time for behavior modification or to relax skeletal muscle components of the urethral sphincter, fulminant hepatic failure has occurred. For this reason, cats receiving diazepam should have liver enzyme activity monitored and the drug should be discontinued at the first sign of an increase in these enzymes.

41. The answer is C.
The peripheral actions of xylazine are related to blockage of α receptors on vascular smooth muscle or secondary to decreased sympathetic tone related to decreased CNS release of norepinephrine. Initial stimulation of postsynaptic α_2 adrenergic receptors on vascular smooth muscle results in vasoconstriction and hypertension. After longer periods, decreases in central norepinephrine release result in activation of presynaptic α_2 receptors and hypotension. Xylazine can actually increase the sensitivity of the myocardium to catecholamine-induced arrhythmias. Since the drug enhances parasympathetic tone to the heart and inhibits sympathetic drive, it can cause sinus bradycardia and heart block. This α_2 adrenergic agonist also decreases cardiac contractility. In most species, respiratory function is maintained, except in ruminants in which marked arterial hypoxemia may occur. This is secondary to the induction of ventilation-perfusion defects and airway constriction. Ruminants are also particularly sensitive to the effects of xylazine and require one-tenth of the dose required for horses.

42. The answer is A.
Xylazine provides good analgesic action in the horse. It is important to note that the sedative effect after IV administration is 1–2 hours while analgesia lasts for only 15–30 min. Xylazine actually promotes gastrointestinal stasis and can cause gastric dilation. Animals under xylazine sedative have an increased sensitivity to auditory stimuli. Xylazine can decrease insulin secretion from the pancreas, leading to an increase in blood glucose levels. Horses do not vomit. Xylazine, however, is a potent emetic agent in the cat.

43. The answer is A.
Binding of opiates to μ receptors mediates analgesia at the level of the brain stem and spinal cord. Brain stem control of the pain response is mediated by enhancing adrenergic neurotransmission in the bulbospinal tract, which in turn blocks spinal cord interneurons that control pain pathways. At the level of the spinal cord, opiates mediate analgesia by inhibiting interneuron neurotransmission by decreasing substance P binding to afferent neurons in the spinothalamic tract. Kappa (κ) receptors mediate analgesia at the level of the spinal cord and are primarily responsible for the dysphoric reaction some animals experience with opiates. This reaction may be mediated in part by inhibition of dopaminergic neurotransmission in the brain. The μ receptors mediate the euphoric feeling that many animals experience under the influence of opiates and are also responsible for the dose-dependent decrease in respiratory drive that accompanies the use of opiates. Opiate binding at both μ and κ receptors actually increases gastrointestinal circular smooth muscle contraction and thus increases transit time. This pharmacological effect forms the basis for the use of opiates in the symptomatic treatment of diarrhea.

44. The answer is D.
Morphine is a full agonist at μ receptors and thus a powerful analgesic. It should be given IM since it has a tendency to cause histamine release from mast cells when given IV. Other disadvantages to the use of morphine are its tendency to cause vomiting and sedation in the dog and the fact that it is a highly controlled schedule II drug. Although fentanyl, which is also a full μ agonist, is 100 times as potent as an analgesic as morphine is, when given by the transdermal route there is a 12–24-hour delay in the onset of analgesic action. Butorphanol is an agonist at κ receptors and a weak antagonist at μ receptors. It is more effective in the control of visceral pain. Naloxone is a pure opioid receptor antagonist and has no analgesic activity.

45. The answer is D.
NSAIDs, by virtue of their ability to inhibit cyclooxygenase activity, prevent the formation of several prostaglandins. Inhibition of thromboxane A_2 production in platelets results in inhibition of platelet aggregation and prolongation of bleeding times. NSAIDs are weak acids and are highly protein bound. They may displace other highly protein-bound drugs, resulting in increased serum levels and duration of action of these drugs. Such interactions may occur with phenytoin, oral anticoagulants, sulfonamides, and sulfonylurea agents. Most of the NSAIDs undergo some degree of hepatic metabolism and then are excreted in the urine. Some glucuronide metabolites may be excreted in the bile, and when these inactive metabolites reach the small intestine, bacterial enzymes may deconjugate them, resulting in generation of the active drug, which then can undergo enterohepatic circulation. This may be the reason why certain NSAIDs, particularly the propionic acid derivatives, are toxic in the dog. All of the NSAIDs are antipyretic.

46. The answer is D.
Although aspirin does inhibit platelet aggregation, this inhibition is irreversible. This differentiation is more than just academic since platelets lack the capacity to synthesize proteins. This means that 1 dose of aspirin can inhibit platelet function for the life span of the platelet (up to 7 days). All of the other NSAIDs are reversible inhibitors of platelet aggregation. All of the NSAIDs have the potential to cause gastric ulceration since

they all inhibit prostaglandin E_2 and I_2 production, which is important in maintaining gastric mucosal blood flow. Naproxen is particularly ulcerogenic in the dog because it undergoes enterohepatic cycling, which increases serum drug concentrations and prolongs the drug's duration of action. All of the NSAIDs are also potential nephrotoxins and should be used cautiously in patients with preexisting renal disease. Renal toxicity is mediated by inhibition of local prostaglandins that normally maintain renal perfusion in the diseased kidney. Foals and ponies are more susceptible to the renal toxic effects of phenylbutazone. Hepatotoxic reactions to NSAIDs have been reported in humans and dogs. Dogs typically develop an acute hepatopathy with carprofen that is reversible if the drug is discontinued promptly once increases in serum transaminases are detected. It appears that this hepatotoxic reaction is idiosyncratic. Acetaminophen is normally metabolized by conjugation to glucuronide. Since cats are deficient in glucuronyl transferase activity, they preferentially metabolize the drug via the hepatic CP450 enzyme system. This leads to generation of a reactive intermediate that must be detoxified by conjugation to glutathione. Since cats also have limited supplies of hepatic glutathione, once this antioxidant is depleted, the reactive intermediate is free to cause oxidant damage. In the cat this toxic reaction is manifested by oxidant damage to red blood cells, which results in methemoglobinemia and cyanosis. With overdosage of acetaminophen in other species, this oxidant damage is manifested as acute hepatic necrosis.

47. The answer is D.
Phenytoin is a membrane-stabilizing agent that interferes with Na^+ and Ca^{2+} conductance across neuronal membranes.

48. The answer is A.
Although the onset of action with IV phenobarbital is delayed 20–30 min, loading doses can be given to achieve steady-state serum concentrations (Css) in a relatively short period of time to achieve long-term seizure control. Oral phenobarbital administration requires up to 2 weeks to obtain Css. Potassium bromide has a very long half-life and oral administration requires 2–3 months to achieve Css. Even with an IV loading dose, potassium bromide would not provide seizure control for several days. Diazepam would be the anticonvulsant of choice to immediately stop seizure activity in status epilepticus, but it must be given IV to achieve this effect. Since the drug is highly lipid soluble, absorption from IM sites is poor and the drug is generally not given by this route. Diazepam's duration of action is short (15–20 min) since it prevents seizures from spreading, but does not prevent the reoccurrence of the epileptogenic foci. A continuous-rate infusion of diazepam can be used if intermittent boluses have been effective in the short-term control of status epilepticus. Pentobarbital IV is highly effective in controlling seizure activity and will do so very quickly. However, this anticonvulsant effect is seen only at doses that induce general anesthesia. For this reason, the use of pentobarbital to control seizures is limited to patients in status epilepticus that fail to respond to diazepam or phenobarbital.

49. The answer is C.
Deprenyl inhibits the enzyme monoamine oxidase, which is responsible for the degradation of biological amines such as dopamine, serotonin, and norepinephrine. This results in an *increase* in the brain concentrations of these neurotransmitters. It is thought that many of the effects of deprenyl in the brain are mediated by an increase in dopamine levels. These effects include an increase in cognitive function and inhibition of the secretion of pituitary hormone secretion. Dopamine is an important neurotransmitter in the prefrontal cortex that plays an important role in cognition. Dopamine also inhibits the release of adrenocorticotropin from the pars intermedia of the pituitary. This latter effect forms the basis for using deprenyl in the treatment of pituitary-dependent hyperadrenocorticism. However, only approximately 15% of the pituitary adenomas responsible for hyperadrenocorticism in the dog arise from the pars intermedia. Therapeutic responses to deprenyl are delayed for up to 30 days once therapy is started. It is important to note that removal of the drug from the body also requires a significant period of time. In general, one should allow 2 weeks before any additional therapy with behavior-modifying drugs such as the tricyclic antidepressants or serotonin uptake inhibitors is initiated.

50. The answer is B.
Buspirone is a serotoninergic receptor agonist that is effective in conditions associated with anxiety. The drug undergoes hepatic metabolism and renal elimination and has no adverse effects on the cardiovascular system. Amitriptyline and imipramine are tricyclic antidepressants that inhibit amine uptake and raise the levels of serotonin and norepinephrine in the brain. They also have significant peripheral anticholinergic and antiadrenergic actions that can lead to hypotension, tachycardia, and prolonged cardiac conduction times. They also have a direct depressant effect on the myocardium. Megestrol acetate is a progestational compound. Several side effects including the induction of diabetes mellitus and mammary hyperplasia/neoplasia limit the use of this drug as first-line therapy for behavior modification in cats.

51. The answer is E.
The use of an antibiotic does not confer any type of pressure to reactivate latent viral infections. The inappropriate use of antibiotics as a "panacea" for all that is wrong promotes the development of acquired bacterial drug resistance. The most important site for the development of bacterial drug resistance is in the gastrointestinal tract where selective pressure leads to the exchange of plasmids among bacteria. These plasmids can carry genetic information that codes for resistance to multiple antibiotic agents. Since many acquired infections in animals, most notably urinary tract infections, are due to host enteric organisms, induction of antibiotic resistance can have devastating consequences. In addition, heavy use of antibiotics in hospitalized patients, especially oral use, can lead to the excretion of multiple drug-resistant bacteria that colonize the hospital. Hospitalized patients then can acquire infections with these organisms by multiple routes. Although most antibiotics are relatively safe, some have direct and some indirect idiosyncratic toxic reactions. For example, aminoglycosides are known renal toxins, while sulfonamides can be associated with serious idiosyncratic immune and hepatotoxic reactions. Some antibiotics kill normal flora and permit the overgrowth of other populations of bacteria. This is particularly true in the horse in which several antibiotics can cause severe life-threatening diarrhea. Antibiotic residues in meat or milk that make their way into the food chain can lead to toxic reactions in susceptible humans.

52. The answer is E.
Subtherapeutic doses of antibiotics given for long periods of time are likely to promote antibiotic drug resistance. The best policy to prevent the emergence of resistant strains of bacteria is to use the narrowest-spectrum antibiotic for the shortest period of time and to base the choice of antibiotic on in vitro sensitivity testing. When empirical antibiotic use must be initiated for economic reasons, one should attempt to obtain material from the infected site for Gram stain. The knowledge of whether the infection is caused by a gram (+) or gram (−) organism coupled with the morphology of the bacteria can greatly assist in the choice of the most appropriate antibiotic. Several antibiotics are illegal to use in food animals owing to the danger of inducing bacterial drug resistance in organisms that eventually reach the human population. Presently, the glycopeptides and fluoroquinolones cannot be used in food animals.

53. The answer is C.
The third-generation cephalosporins have a very good spectrum of action against gram (−) bacteria and 2 third-generation cephalosporins, cefotaxime and ceftazidime, obtain good concentrations in the cerebrospinal fluid. Penicillin G, ampicillin, and first-generation cephalosporins generally do not penetrate the cerebrospinal fluid well, but may gain entry in the presence of meningeal inflammation. Penicillin G has a limited gram (−) spectrum. Although ampicillin and first-generation cephalosporins have a better gram (−) spectrum than penicillin G does, they would still not be considered broad enough in the setting of a life-threatening infection. Vancomycin is a glycopeptide that is effective only against gram (+) organisms.

54. The answer is E.
All of the β-lactam antibiotics are bactericidal. It is important to know whether an antibiotic is bacteriostatic or bactericidal, as bacterial killing will be slower with bacteriostatic drugs. When a bacteriostatic drug is used, the animal's own immune system must rid the body of existing organisms. In general, bactericidal antibiotics are preferred in the presence of life-threatening infection or when treating immunocompromised animals. The β-lactam antibiotics are weak acids that are excreted unchanged in the urine by a combination of glomerular filtration and tubular secretion. Since they obtain high concentrations in the urine, their ability to kill gram (−) pathogens in the urinary bladder is extended. Although rare in animals, life-threatening anaphylactic reactions can occur with oral or parenteral administration of β-lactam antibiotics. Clavulinic acid is an irreversible inhibitor of β-lactamases. The β-lactamases are enzymes produced by bacteria that lyse the β-lactam ring in the antibiotic and destroy its ability to kill bacteria. Although clavulinic acid has no intrinsic antibacterial action, when combined with a β-lactam antibiotic it increases the drug's spectrum of action considerably. All of the β-lactam antibiotics work by binding to penicillin-binding proteins that are involved either in cell wall synthesis or in the control of bacterial cell shape or division.

55. The answer is A.
Although the potentiated sulfonamides may lead to immune-mediated thrombocytopenia, they do not inhibit platelet function. They also have been associated with other immune disorders including immune-mediated hemolytic anemia, glomerulonephritis, arthritis, and dermatitis. All of the sulfonamides can induce dry eye syndrome (keratoconjunctivitis sicca). This may be due to a direct toxicity to the lacrimal glands or secondary to a hypersensitivity reaction. The damage may be irreversible. Two hepatotoxic syndromes, an acute fulminant hepatic necrosis and a milder cholestatic syndrome, have been related to the use of potentiated sulfonamides in dogs. Since the mechanism of action of the potentiated sulfonamides is a 2-hit punch at bacterial folate production (inhibition of dihydropterone synthase and dihydrofolate reductase), long-term therapy can inhibit mammalian folate biosynthesis, leading to folate deficiency.

56. The answer is B.
The fluoroquinolones would be the antibiotic of choice in this case since they have such a good spectrum of action against gram (−) organisms and have no toxic effects on the kidney. Because the fluoroquinolones are highly lipophilic with a low degree of protein binding, they also penetrate tissues well. Since they are excreted in part unchanged in the urine, a high concentration should be obtained in renal tissue. The 2-year-old foal would be at an increased risk for cartilage damage from the fluoroquinolone. Since fluoroquinolones can have less activity against *Streptococcus* species, they would not be the best choice for the cutaneous infection in the dog. There are 2 reasons for not selecting the fluoroquinolone for the cat bite abscess. First, many antibiotics with a much narrower spectrum of action would be effective. Second, some cat bite abscesses are due to anaerobic bacteria for which the fluoroquinolones have poor activity. The use of fluoroquinolones in food animals is illegal.

57. The answer is B.
Several factors increase the nephrotoxicity of aminoglycosides, including the presence of preexisting renal disease, dehydration, fever, very young or old age, and concurrent administration of nephrotoxic drugs. Since NSAIDs should be considered as potential nephrotoxins (by causing renal vasoconstriction), they increase the risk of nephrotoxicity from aminoglycosides. The development of nephrotoxicity correlates with persistently elevated trough concentrations, not peak concentrations. Dosage schedules in which aminoglycosides are given once a day to achieve high peak concentrations are now preferred since they decrease the potential for nephrotoxicity. Aminoglycosides are also more effective in killing bacteria when administered in this manner, since bacterial killing by the aminoglycosides is concentration, not time dependent. Aminoglycoside effectiveness is linked to the magnitude of peak concentration. The aminoglycosides are actively concentrated in the renal proximal tubular cells and in the endolymph and perilymph. Once aminoglycoside administration is stopped, elimination of the drug from these sites can take several days. If one waits for the development of renal azotemia as evidenced by increases in serum urea nitrogen or creatinine to stop the aminoglycosides, renal damage will continue to progress for several days. The best measures of early renal toxicity are the presence of casts in a urine sediment or increased concentrations of the brush border enzyme, γ-glutamyltranspeptidase, in the urine.

58. The answer is C.
Since the aminoglycosides are polar organic bases, their distribution is limited to extracellular fluids. Unless administered intrathecally, they are generally considered ineffective in the treatment of gram (−) meningitis. They are eliminated unchanged in the urine and are not active in acidic environments such as exists in the presence of necrotic material. They are not bioavailable when given orally, but can be given IV, IM, or SC.

59. The answer is A.
Three of the listed antibiotics have activity against rickettsial infections: doxycycline, tetracycline, and chloramphenicol. Tetracyclines are eliminated unchanged in the urine and have been reported to inhibit mammalian protein synthesis, leading to the development of azotemia. Doxycycline is eliminated by secretion into the gastrointestinal tract, with only a small portion undergoing renal or biliary excretion. Chloramphenicol is eliminated by hepatic metabolism and biliary excretion and has no toxic effects on the kidney. The major reason not to use this drug is the fact that owner contact with the drug may be associated with a rare idiosyncratic fatal bone marrow aplasia.

60. The answer is E.
Tetracycline has no activity against β-lactamase–positive gram (+) organisms. The drug of choice among the remaining drugs would be lincomycin. Lincomycin is a narrow-spectrum antibiotic that was developed for the treatment of cutaneous infections. Clindamycin would be effective but has a broader spectrum of action and its use would violate the axiom of striving to use the narrowest spectrum of action that gets the job done. Erythromycin would be a good choice, but many dogs (50%) experience gastrointestinal distress with this medication. Chloramphenicol should not be used because it also would violate the rule of using a narrow-spectrum antibiotic and is associated with a fatal idiosyncratic bone marrow toxicity in humans.

61. The answer is D.
Flucytosine is a fluorinated pyrimidine that fungi deaminate to 5-fluorouracil. Fluorouracil interferes with thymidylate synthase and disrupts fungal RNA and DNA synthesis. Itraconazole interferes with ergosterol synthesis, leading to the accumulation of 14-α-methyl sterols. The methyl sterols impair membrane function, leading to alterations in energy metabolism and growth inhibition.

62. The answer is E.
Although the azole antifungal agents are potential hepatotoxins, this reaction has not been reported with amphotericin B. Normocytic, normochromic anemia occurs in many patients and may be associated with decreased erythropoietin production. Hypokalemia develops owing to renal tubular acidosis. Nephrotoxicity is associated with a direct toxic effect on the renal tubular epithelium and renal vasoconstriction. Fever can be seen during administration and may be controlled by preadministration of antipyretics.

63. The answer is C.
Fluconazole and ketoconazole are azole antifungal agents that are effective against *Cryptococcus* organisms and available for oral administration. Ketoconazole is absorbed variably from the gastrointestinal tract and is highly protein bound and therefore does not penetrate into the CNS. It takes 14–21 days to obtain Css. In cats, anorexia, vomiting, and increases in liver enzyme levels are common side effects. Fluconazole is a water-soluble azole that is absorbed readily from the gastrointestinal tract. It is not highly protein bound and gains entry into the CNS. Effective concentrations are reached in 5–10 days and cats tolerate this medication much better than ketoconazole. Amphotericin B is effective against *Cryptococcus,* but high protein binding limits its distribution to the CNS.

It is also highly nephrotoxic and only available for IV administration. Griseofulvin is primarily active against dermatophytes. Flucytosine is available for oral administration and enters the CNS well, but the occurrence of resistant *Cryptococcus* strains is high. It is almost always used in combination with amphotericin B as the 2 agents are synergistic. It has been reported to cause aberrant behavior and seizures in the cat.

64. The answer is D.
Sulfonamides have no action against *Giardia* species, but are active against *Coccidia* organisms. Metronidazole, furazolidone, and the benzimidazoles, fenbendazole and albendazole, are all effective against *Giardia.* The most commonly prescribed medications in the dog are metronidazole and fenbendazole. The use of albendazole, which is more toxic than the other 2 drugs, is limited to infections resistant to the latter two medications.

65. The answer is A.
Both trimethoprim-sulfadiazine and pyrimethamine work by interfering with protozoal folate synthesis. The sulfonamides work as analogues of *para*-aminobenzoic acid (PABA) and interfere with the conversion of PABA to dihydropteroic acid. Trimethoprim and pyrimethamine work as inhibitors of dihydrofolate reductase, preventing the conversion of dihydrofolate to tetrahydrofolate. Pyrimethamine has a higher selectivity for the protozoan enzyme than does trimethoprim, which is more selective for the bacterial enzyme. Pyrimethamine also crosses the blood-brain barrier better than trimethoprim-sulfadiazine. Monensin is an ionophore coccidiostat that causes fatal cardiac toxicity in the horse. Metronidazole, imidocarb, amprolium, meglumine antimonate, and fenbendazole do not have activity against the causative agent in equine protozoal myelitis, *Sarcocystis.*

66. The answer is D.
Pyrantel acts as a nicotinic cholinergic agonist at the parasite neuromuscular junction, leading to depolarizing block. It is important to understand which antihelminthics might have overlapping mechanisms of action since combinations of antihelminthics typically are used in controlling worm burdens. Since levamisole, organophosphates, and pyrantel work by stimulating cholinergic neurotransmission at the worm's neuromuscular junction, combinations of these drugs may lead to increased host toxicity. The signs of host intoxication with these drugs are salivation, defecation, respiratory distress, muscle tremors, and seizures. The use of piperazine with these 3 drugs would be antagonistic because piperazine is a cholinergic receptor antagonist.

67. The best choice is A.
Although both fenbendazole and organophosphates would be effective against whipworms and hookworms, the much greater toxicity of organophosphates has essentially eliminated their use as antihelminthics in small animals. Pyrantel is not effective against whipworms and piperazine is not effective against whipworms or hookworms.

68. The answer is C.
Ivermectin works by enhancing GABAergic neurotransmission in the parasite. Mammals do have GABA receptors, but unlike nematodes and arthropods in which GABA receptors are located in the peripheral nervous system, GABA receptors in mammals are confined to the CNS. Since ivermectin does not appreciably cross the blood-brain barrier, it has no effect on mammals. In some breeds of dogs, like collies, ivermectin appears to be able to penetrate into the CNS and signs of toxicity as listed in C appear. The heartworm dose of ivermectin, which is 6–12 μg/kg, is safe to use in collies, but the antihelminthic dose, 200 μg/kg, is toxic. Ivermectin is not active against adult heartworms at any dose. It kills the infective stage of the larva at a dose of 6–12 μg/kg in the dog and can be used at 50 μg/kg to kill circulating microfilariae.

69. The answer is B.
Ivermectin has antihelminthic and antiarthropod activity but is ineffective in killing cestodes and trematodes. Praziquantel causes irreversible vacuolization and disintegration of the worm's integument. Albendazole, a benzimidazole, interferes with tubulin function during microtubule assembly and disrupts cell division. Albendazole also inhibits fumerate reductase, an enzyme important in energy generation. Clorsulon is a benzosulfonamide that works by inhibiting 3-phosphoglycerate kinase and phosphoglyceromutase, enzymes that are important for trematode energy production.

70. The answer is D.
Cats are unusually sensitive to many ectoparasital agents. Organophosphates and the pyrethroid permethrin should not be used in cats. The signs of toxicosis with these drugs in cats are similar and include depression, hypersalivation, tremors, vomiting, ataxia, dyspnea, and anorexia. Imidacloprid is safe to use in adult cats but is not recommended for cats under the age of 12 weeks. Lufeneron is safe to use in very young cats, but the drug will not kill the fleas presently on the cat. Lufeneron inhibits insect development by interfering with chitin synthesis. Fleas that ingest lufeneron lay eggs that fail to hatch or die during the first molt. Pyrethrins are natural botanicals derived from the chrysanthemum flower. They have a rapid knockdown effect and kill adult fleas quickly but have little residual action since they are inactivated rapidly by hydrolysis. They would be the safest product to use on this cat. Preparations of pyrethrins containing synergists, however, should be avoided. The synergists prolong the action of the pyrethrin by inhibiting the enzyme responsible for pyrethrin degradation.

71. The answer is E.
Methoprene is an insect growth regulator that mimics juvenile hormone and prevents maturation of the pupa to the adult stage. Insect growth regulators have very low toxicity in mammals and are often combined with other ectoparasital agents that directly kill the adult insects in effective flea control programs. They are not effective against ticks.

72. The answer is C.
The dose-limiting toxicity of vincristine is gastrointestinal, with anorexia, vomiting, and constipation being the most commonly observed side effects. It is important to understand the dose-limiting toxicity of the antineoplastic agents so that combination protocols can be planned in which overlapping toxicity is not seen. Therefore, vincristine could be given safely along with an alkylating agent, such as cyclophosphamide or lomustine, or with the pyrimidine analogue cytosine arabinoside. Since the other dose-limiting toxicity of methotrexate in small animals is severe gastrointestinal toxicity, coadministration with vincristine should be avoided.

73. The answer is D.
Cyclophosphamide, vincristine, and methotrexate do not penetrate into the CNS. Both of the antimetabolites, cytosine arabinoside and 5-fluorouracil, penetrate into the CNS. However, 5-fluorouracil should not be used in cats because it causes severe CNS toxicity.

74. The answer is E.
It is the alkylating agent cyclophosphamide and its structural analogue ifosfamide that are associated with hemorrhagic cystitis. Cystitis is due to chemical irritation of the bladder by acrolein, an inactive metabolite. Rapid IV administration of doxorubicin can result in histamine release, which is accompanied by facial swelling and pruritus. Doxorubicin has been associated with a severe colitis in some dogs that occurs 2–3 days after administration. Myocardial toxicity limits the cumulative dose of doxorubicin that can be administered to any individual. The risk of cardiomyopathy increases above a cumulative dose of 250 mg/m^2 in dogs.

75. The answer is C.
Doxorubicin can be used safely in the cat with similar toxicities as those seen in the dog. Unlike the dog, nephrotoxicity has been reported in cats so renal function should be monitored during therapy. L-Asparaginase is an enzyme that hydrolyzes asparagine to aspartic acid and ammonia. L-Asparaginase is only effective in neoplastic cells that require an exogenous source of asparagine. In general, it is only used to treat lymphoma. Corticosteroids are only cytotoxic for lymphocytes and mast cells and would not be expected to have any effect on carcinoma cells. *Cis*-plati-

num has good activity against carcinomas but causes fatal pulmonary edema in the cat.

76. The answer is A.
Peripheral sensory receptors in the gastrointestinal tract are stimulated in response to serotonin released due to pain, inflammation, distention, or rapid changes in osmolarity. Impulses generated at these receptors are transmitted via sympathetic and vagal nerves to the emetic center. Ondansetron is a serotonin receptor antagonist whose antiemetic action is mediated primarily by inhibition of peripheral sensory receptors. It has some central action since it decreases serotoninergic neurotransmission in the CRTZ. Although neurotransmission in the CRTZ is primarily dopaminergic, serotonergic, cholinergic, histaminergic, and α_2 adrenergic neurons are also present.

77. The answer is A.
Use of xylazine, at doses below those that cause sedation, results in predictable emesis by stimulating α_2 adrenergic receptors in the emetic center and CRTZ. It is much less reliable as an emetic agent in the dog where apomorphine is preferred. Apomorphine is an opioid dopaminergic agent that induces emesis by activating the CRTZ. Hydrogen peroxide can induce emesis in the cat by causing local gastric irritation, but it is unreliable in the cat in which administration of the liquid can be quite difficult. Metoclopramide is an antiemetic agent that inhibits dopaminergic neurotransmission in the CRTZ. Atropine is a peripherally acting muscarinic cholinergic antagonist. It will inhibit gastric motility and by causing gastrointestinal ileus rarely results in vomiting. Anticholinergic agents such as scopolamine and dicyclomine that do penetrate the CNS can be used to control vomiting associated with motion sickness.

78. The answer is B.
The half-life of metoclopramide is 60–90 min so a continuous-rate IV infusion for hospitalized animals is best. When used on an outpatient basis to control emesis, it should be given 20–30 min prior to a meal. The side effects of metoclopramide are related to its antidopaminergic properties in the CNS and include hyperactivity, restlessness, and tremors. These side effects are more common at higher doses and in cats and horses. Metoclopramide has a peripheral prokinetic effect on the proximal gastrointestinal system that is due to sensitization of the smooth muscle to the effects of acetylcholine. Metoclopramide undergoes extensive first-pass metabolism to inactive metabolites after oral dosing. This makes the effective oral dose higher than the parenteral dose. Although metaclopramide is metabolized in the liver to glucuronide and sulfate conjugates, they are excreted primarily in the urine with only about 5% in the bile. About 20%–25% of the drug is excreted unchanged in the urine.

79. The answer is D.
Sucralfate is minimally absorbed and therefore minimally toxic. Local effects include constipation and interference with the absorption of a concurrently administered drug such as tetracycline, fluoroquinolones, cimetidine, and digoxin. Sucralfate dissociates in the stomach into aluminum hydroxide and sucrose octosulfate. It is the latter that binds to negatively charged, damaged gastric epithelium and forms a polymerized matrix. This insoluble polymer entraps pepsin and bile acids and prevents them from further injuring the damaged site. The aluminum hydroxide stimulates the release of local prostaglandins, which promote gastric blood flow.

80. The answer is C.
All 3 H2 antagonists listed, cimetidine, ranitidine, and famotidine, would be equally efficacious in decreasing gastric acidity, but both cimetidine and ranitidine inhibit the hepatic CP450 system. Since dogs with chronic hepatic failure may already have a compromised hepatic CP450 system, it is best to avoid these H2 blockers. Cimetidine also decreases hepatic blood flow. Famotidine does not inhibit CP450 or affect hepatic blood flow and therefore would be the H2 blocker of choice. An added benefit of famotidine is that it need only be given once a day, as opposed to 2 or 3 times a day for ranitidine and cimetidine, respectively. Antacids would be effective in neutralizing gastric acid secretion, but they must be given frequently and it would be impossible for owners to comply. If antacids are not given frequently, acid rebound occurs. Omeprazole is the most effective antisecretory agent by virtue of the fact that it irreversibly inhibits the gastric H^+/K^+ ATPase. Omeprazole, however, is also a potent inhibitor of the hepatic CP450 system. In addition, prolonged administration of omeprazole has been associated with the development of gastric carcinoids in rats. The clinical significance of this in veterinary patients is still not fully appreciated.

81. The answer is C.
Cisapride and metoclopramide are promotility agents in the gastrointestinal tract. Metoclopramide sensitizes smooth muscle in the proximal gastrointestinal tract to the actions of acetylcholine. Cisapride works as a promotility agent in the full length of the gastrointestinal tract. In the proximal gastrointestinal tract, it activates serotonin receptors on myenteric neurons, resulting in enhanced release of acetylcholine. In the colon it stimulates contraction by a noncholinergic mechanism that may involve direct stimulation of serotonin receptors on colonic smooth muscle cells. Erythromycin is a macrolide antibiotic that acts as a motilin agonist. It stimulates contraction of the gastric and small intestinal smooth muscle but has no action on colonic smooth muscle. Loperamide is an opiate that increases circular smooth muscle contraction and decreases longitudinal smooth muscle contraction with the net effect of inhibiting movement through the gastrointestinal tract. Atropine is a muscarinic receptor antagonist that also inhibits gastrointestinal motility.

82. The answer is A.
Contraction of segmental circular smooth muscle in the gastrointestinal tract increases the resistance to intestinal flow while contraction of peristaltic longitudinal smooth muscle moves intestinal contents aborally. Opiates are effective antidiarrheal drugs since they augment circular muscle contraction and inhibit longitudinal muscle contraction, thus increasing resistance to flow. Loperamide is also antisecretory. Loperamide's effects on intestinal motility are due to binding to δ and μ receptors on intestinal smooth muscle. It can be sold as a nonprescription drug since it does not cross the blood-brain barrier and therefore has no abuse potential.

83. The answer is C.
Magnesium sulfate is the only one of the listed medications that can be used as a cathartic. It is poorly absorbed from the intestinal tract and acts as an osmotic particle in the lumen, causing the movement of water into the gut. The increase in luminal volume stretches the intestinal mucosa, activating mechanoreceptors that cause an increase in peristaltic activity. When given in large volume, magnesium sulfate works fairly quickly to promote defecation (within 3 hours), which would be beneficial in this case of acute poisoning. All of the other medications listed will promote defecation, but they will all take more than 24 hours to do so. The docusates, which are detergents, might actually promote absorption of the toxin.

84. The answer is E.
Lactulose works as an osmotic laxative. It escapes absorption in the small intestine and is metabolized by colonic bacteria to lactate, acetate, and formate. These metabolites are not well absorbed and contribute to luminal osmotic pressure drawing water into the colon. These metabolites also acidify the colon and by doing so may promote colonic motility. Lactulose's beneficial action in hepatic encephalopathy is related to its ability to decrease ammonia absorption in the colon. This is due to ion trapping of NH_3 (the decrease in colonic pH promotes protonation of NH_3 to the unabsorbable NH^{4+}), alteration in bacterial metabolism so that less NH_3 is produced, and a reduction in NH_3 absorption due to enhanced colonic transit time.

85. The answer is B.
Somatotropin (growth hormone) is used in dairy cattle to promote milk production. Bovine recombinant growth hormone is formulated as an injectable prolonged release product that is given SC every 14 days. Antidiuretic hormone is used clinically primarily as replacement therapy for diabetes insipidus. Adrenocorticotropin is used in veterinary medicine

primarily to evaluate adrenal functional reserve. Testosterone does have growth-promoting activity in cows but has no effect on milk production and would be contraindicated in a lactating dairy cow. Oxytocin induces contraction of uterine smooth muscle and results in milk letdown at the time of parturition. It is used in veterinary medicine as an aid in parturition.

86. The answer is C.
Synthetic gonadotropin-releasing hormone preparations, buserelin, leuprolid, and histerlin, are given parenterally to induce LH secretion in the cow and subsequent luteinization of follicular cysts in cows. Prostaglandin $F_{2\alpha}$ is used in cows to lyse the corpus luteum for estrous synchronization, to induce abortion or parturition, and to stimulate uterine contraction in cases of retained placenta. Pregnant mare gonadotropin has follicle-stimulating hormone activity. It would stimulate follicle development. Adenocorticotropin causes the release of glucocorticoids from the adrenal and would not have any effect on a follicular cyst.

87. The answer is D.
It is antidiuretic hormone binding to V1 receptors that results in peripheral vascular smooth muscle contraction and hepatic glycogenolysis. Since desmopressin is primarily a V2 agonist, these effects of V1 receptor occupation are not seen. Diabetes insipidus (DI) is a disease associated with lack of antidiuretic hormone secretion from the pituitary. The major clinical sign of DI is severe polyuria and polydipsia (PU/PD) due to an inability of the kidney to produce concentrated urine. Exogenous administration of desmopressin restores urine-concentrating ability and decreases the PU/PD. Desmopressin also stimulates the release of von Willebrand factor and significantly shortens bleeding times in dogs with von Willebrand disease. These hemostatic effects, however, require 5–10 times the dose needed to control PU/PD, and therefore the drug must be given parenterally.

88. The answer is D.
Methimazole inhibits thyroid peroxidase activity in the thyroid gland. This enzyme catalyzes the oxidation of iodine, the coupling of iodotyrosyl groups to form the ether linkage of T_3 and T_4, and iodination of tyrosyl residues in thyroglobulin. Methimazole has no effect on the uptake of iodide or on the secretion of thyroid hormone. It also has no effect on the 5′-deiodinase that catalyzes the peripheral conversion of T_4 to T_3. Idopate, a cholecystographic dye that has been used to treat feline hyperthyroidism, works by inhibiting this 5′-deiodinase. Unlike radioactive iodine, which accumulates in the thyroid gland and is directly cytotoxic, methimazole is not cytotoxic. Therefore, the antithyroid effects of methimazole are readily reversible. This is an important consideration in cats with concurrent chronic renal failure and hyperthyroidism. Since the hyperthyroid state results in an increase in glomerular filtration rates, some cats will experience a worsening of renal failure when their hyperthyroidism is brought under control.

89. The answer is C.
Synthetic T_4 is recommended as first-line replacement therapy for canine hypothyroidism, because some organs, particularly the brain and pituitary, maintain high concentrations of the active hormone, T_3, by extracting T_4 and converting it to T_3. Therefore, T_3 replacement may shortchange these organs. There is no difference in the metabolism and absorption of the 2 forms of thyroid hormone. Very rarely dogs may be unable to absorb T_4 or lack the 5′-deiodinase that converts T_4 to T_3, and in these cases supplementation with T_3 would be warranted. There is no greater danger of inducing a hyperthyroid state when T_3 is given. Indeed, due to the dog's highly efficient capacity to clear thyroid hormone by biliary excretion, the induction of iatrogenic hyperthyroidism in the dog is rare.

90. The answer is A.
Regular insulin is the only insulin preparation that can be given IV or IM. It is also a rapid-acting insulin (onset of action with ½ hour) with a short duration (2–5 hours). Lente, ultralente, and NPH are all suspensions of insulin that must be given SC. Lente and NPH are intermediate-acting insulins with an onset of action of 2–8 hours and a duration of action of 4–14 hours. Ultralente is a long-acting insulin with an onset of action of 4–10 hours and a duration of action of 8–24 hours. These times are only estimates and it is important to remember that the time course of action of any insulin preparation varies among species, among individuals of the same species, and from day to day in any individual. The greatest variation is seen among the products given by the SC route.

91. The answer is E.
The sulfonylureas are drugs used to control type II diabetes mellitus. In this type of diabetes there is a relative lack of insulin, usually due to insulin resistance at the level of the peripheral tissues. The sulfonylureas work primarily by blocking ATP-dependent K^+ channels in the pancreas. This results in an increase in intracellular K^+ in the pancreatic β cells, which causes membrane depolarization and activation of voltage-dependent Ca^{2+} channels. Calcium moves into the β cell through this channel and results in the fusion and exocytosis of granules containing insulin. The sulfonylureas also have minor extrapancreatic effects, including decreasing the hepatic degradation of insulin, increasing the number of peripheral insulin receptors, and decreasing hepatic gluconeogenesis. They have no effect on phosphorylation of the insulin receptor.

92. The answer is D.
Desoxycorticosterone pivalate is a sustained-release preparation of desoxycorticosterone that is used for corticosteroid replacement therapy in hypoadrenocorticism. Mitotane is an adrenolytic drug that causes a relatively selective destruction of the zona fasciculata and the zona reticularis of the adrenal gland and thus destroys corticosteroid-producing tissue. Deprenyl is a monoamine oxidase inhibitor that increases dopamine levels in the brain. Dopamine inhibits adrenocorticotropin hormone secretion from pituitary tumors that are located in the pars intermedia. Unfortunately, only approximately 15% of the pituitary tumors that cause canine hyperadrenocorticism are located in this area. Bromocriptine is a dopamine agonist that is infrequently used in the dog, but is an inexpensive alternative for the treatment of equine hyperadrenocorticism. Ketoconazole inhibits cortisol production in the adrenals by inhibiting 17α-hydrolase activity.

93. The answer is E.
The use of anabolic steroids is not illegal in the horse and there are products approved for use in this species (stanozolol and boldenone). Due to the high abuse potential for anabolic steroids, they are now on the DEA's list of controlled drugs. The beneficial effects of anabolic steroids in treating anemia are related to a stimulation of erythropoietin production as well as an acceleration of heme synthesis and increased red blood cell proliferation. In the setting of adequate protein and calorie intake, anabolic steroids promote the development of muscle mass and reverse catabolism. A major side effect of anabolic steroid use is hepatocellular disease. Hepatic lipidosis, peliosis hepatitis, and hepatic tumors have been reported.

94. The answer is A.
Ursodeoxycholate has no effect on hepatocyte proliferation. Ursodeoxycholate was originally marketed to dissolve cholesterol gallstones in humans. Ursodeoxycholate lowers plasma and bile cholesterol concentrations by inhibiting cholesterol secretion into bile and reducing intestinal reabsorption of cholesterol and hepatic synthesis of cholesterol. When human patients with concurrent gallstones and chronic inflammatory liver disease were treated with ursodeoxycholate to dissolve the stones, their liver condition improved. Although the exact mechanism underlying ursodeoxycholate's hepatoprotective effect is not known, several beneficial effects are known. These include: 1) choleresis, which may help to promote the biliary excretion of potential endogenous hepatotoxins; 2) replacement of the normal hydrophobic and potentially hepatotoxic bile acid pool with a hydrophilic nontoxic bile acid; 3) stabilization of mitochondrial membrane integrity; and 4) a poorly understood immune modulating effect that may be related to interactions with the corticosteroid receptor.

95. The answer is E.
Corticosteroids have no effect on serum electrolytes. Some corticosteroid preparations do contain mild mineralocorticoid activity, but seldom

are changes in serum electrolytes (one would expect hypokalemia) seen. Corticosteroids induce production of a steroid-specific isoenzyme of alkaline phosphatase in the dog. They can also increase the concentration of serum alanine aminotransferase and γ-glutamyltranspeptidase. The presence of increased serum concentrations of hepatic enzymes in dogs on corticosteroid therapy does not mean that hepatobiliary disease is present. However, corticosteroid use does cause the deposition of glycogen in the liver (steroid hepatopathy), which rarely can be associated with hepatic failure. Corticosteroids can enhance erythropoietin production, which results in a mild increase in hematocrit. The mature neutrophilic leukocytosis that accompanies corticosteroid use is due to decreased margination of neutrophils and increased release of neutrophils from the bone marrow. Lymphopenia is due to redistribution away from the peripheral circulation. Corticosteroids are lympholytic to neoplastic lymphocytes, but not normal canine lymphocytes.

96. The answer is A.
Prednisone at 2 mg/kg/day is immune modulating. A dose of 0.125 mg/kg/day is appropriate for physiological replacement in hypoadrenocorticism. A dose of 0.5 mg/kg/day is anti-inflammatory, but not immune modulatory. Since dexamethasone is about 6 times as potent as prednisone, an immune modulatory dose of dexamethasone would be about 0.32 mg/kg/day.

97. The answer is A.
Short-acting preparations of corticosteroids include hydrocortisone, prednisone, and prednisolone. Their duration of action is less than 24 hours so theoretically tapering to alternate-day therapy will allow time for normal functioning of the pituitary-adrenal axis between doses. The use of long-acting repository preparations of corticosteroids should be discouraged as they will result in long-term suppression of the axis. In addition, one cannot "take back" the effect of the corticosteroid should the patient develop a life-threatening complication of corticosteroid therapy such as pancreatitis, laminitis, gastrointestinal ulceration, or sepsis. Alternating the use of several different corticosteroid preparations or using a preparation that lacks mineralocorticoid activity has no effect on the degree of suppression. Corticosteroid preparations that lack mineralocorticoid activity are generally better tolerated for long-term use, especially in the dog. It is the mineralocorticoid activity that results in renal Na^+ and water retention and the development of polyuria and polydipsia.

98. The answer is D.
Azathioprine is generally well tolerated orally and is without adverse effects on the gastrointestinal tract. The most common side effect is bone marrow suppression. Leukopenia is the earliest sign, but anemia and thrombocytopenia also may occur. The suppression is reversible once the drug is discontinued. Both pancreatitis and hepatotoxicity have been reported. Cats are so sensitive to the myelosuppressive effects of azathioprine that the drug is seldom used in this species.

99. The answer is A.
Cyclosporine enters T lymphocytes and binds to a cytoplasmic receptor protein called cyclophilin. The cyclosporin-cyclophilin complex binds to and inhibits calcineurin. Calcineurin is a calcium-dependent phosphatase that normally dephosphorylates regulatory proteins that mediate the movement of a transcription factor to the nucleus. The transcription of the gene encoding interleukin-2 is inhibited. Interleukin-2 is required for the proliferation and clonal expansion of T lymphocytes. Cyclosporine has no effect on DNA synthesis, Fc receptors, or phospholipase A.

100. The answer is A.
Up to 30% of cats on chronic erythropoietin therapy with the human recombinant product develop antibodies that result in resistance to further treatment. In most cases, discontinuation of the drug results in the disappearance of the antibody and resolution of the anemia associated with their presence. In some animals, these antibodies will react with the cat's natural erythropoietin and result in a refractory, transfusion-dependent aplastic anemia. Systemic hypertension is a potential complication of erythropoietin use in humans but is not a serious problem in the cat. Iron depletion will occur with the use of erythropoietin but is easily treated with oral supplementation of iron. Although allergic-type reactions at the SC injection sites have been reported, systemic anaphylactic responses have not.

101. The answer is D.
Heparin binds to the serine protease inhibitor antithrombin III, and increases the activity of this protease 1,000 times. Antithrombin III inhibits activated coagulation factors, particularly thrombin and factor Xa. Administration of heparin SC results in slow sustained release of heparin. Aspirin inhibits platelet aggregation by irreversibly inhibiting platelet production of thromboxane A_2. The use of aspirin in this patient would be contraindicated for 2 reasons. First, the dog already has a defect in platelet number that predisposes him to bleeding. Any further compromise of platelet function may precipitate bleeding that could result in serious cardiac, pulmonary, or neurological damage as well as precipitate the need for red blood cell transfusion. Second, the combination of aspirin and corticosteroids is likely to precipitate gastrointestinal ulceration. Streptokinase is a fibrinolytic agent used to dissolve clots already present. Warfarin interferes with vitamin K–dependent activation of coagulation factors II, VII, IX, and X and the anticoagulants proteins C and S. It takes several days of oral therapy with warfarin before coagulation factors are sufficiently depleted and the anticoagulation effects are seen.

REFERENCES

Basic Pharmacology

Adams H, ed. Veterinary pharmacology and therapeutics, 7th ed. Ames, Iowa: Iowa State University Press, 1995.

Ahrens FA. Pharmacology. Philadelphia: Williams & Wilkins, 1996.

Boothe DM. Drug therapy in cats: mechanisms and avoidance of adverse drug reactions. J Am Vet Med Assoc 1990;196:1297–1305.

Goodman A, Rall TW, Nies AS, Taylor P, eds. The pharmacological basis of therapeutics, 8th ed. New York: McGraw Hill, 1990.

Griffiths JP. Drug interactions. Vet Clin North Am Small Anim Pract 1988; 18:1243–1266.

Martinez M. Use of pharmacology in veterinary medicine: article I: non-compartmental methods of drug characterization: statistical moment theory. J Am Vet Med Assoc 1998;213:984–999.

Martinez M. Use of pharmacology in veterinary medicine: article II: volume, clearance and half-life. J Am Vet Med Assoc 1998;213:1122–1127.

Martinez M. Use of pharmacokinetics in veterinary medicine: article III: physiochemical properties of pharmaceuticals. J Am Vet Med Assoc 1998;213:1274–1277.

Neff-Davis C. Therapeutic drug monitoring. Vet Clin North Am Small Anim Pract 1988;18:1287–1308.

Papich M. 1995. Incompatible critical care drug combinations. In: Bonagura JD, Kirk RW, eds. Current veterinary therapy XII. Philadelphia: WB Saunders, 1995:194–199.

Riviere JE. Veterinary clinical pharmacokinetics. Part 1: fundamental concepts. Comp Cont Ed Small Anim Pract 1988;10:24–30.

Riviere JE. Veterinary clinical pharmacokinetics. Part II: modeling. Comp Cont Ed Small Anim Pract 1988;10:314–327.

Autonomic Pharmacology

Cunningham JE. Neurophysiology. In: Cunningham JE, ed. Textbook of veterinary physiology, 2nd ed. Philadelphia: WB Saunders, 1997:33–124.

Guyton A, Hall JE. The autonomic nervous system. In: Guyton A, Hall JE, eds. Textbook of medical physiology. Philadelphia: WB Saunders, 1996:769–780.

O'Brien D. Autonomic nervous system: function and dysfunction. Semin Vet Med Surg (Small Anim) 1990;5:1–80.

Robershaw D. Visceromotor (autonomic control). In: Swenson MJ, Reece WO, eds. Dukes physiology of domestic animals, 11th ed. Ithaca, NY: Comstock Publishing Associates, 1993:875–885.

Local Anesthetics/Neuromuscular Blockers

Cullen LK. Muscle relaxants and neuromuscular blocking agents. In: Thurman JC, Tranquilli WJ, Benson GJ, eds. Lumb and Jones: Veterinary anesthesia. Philadelphia: Williams & Wilkins, 1996:337–360.

Hildebrand S. Neuromuscular blocking agents. Vet Clin North Am Small Anim Pract 1992;22:341–350.

Hildebrand S. Neuromuscular blocking agents in equine anesthesia. Vet Clin North Am Equine Pract 1990;6:587–606.

Heavner JE. Local anesthetics. In: Thurmon JC, Tranquilli WJ, Benson GJ, eds. Lumb and Jones: Veterinary anesthesia, 3rd ed. Philadelphia: Williams & Wilkins, 1996:330–336.

Hunter JM. New neuromuscular blocking agents. N Engl J Med 1995; 332:1691–1699.

Pascoe P. Local and regional anesthesia and analgesia. Semin Vet Med Surg (Small Anim) 1997;12:94–105.

Cardiology

Adams HR. New perspectives in cardiology: pharmacodynamic classification of anti-arrhythmic drugs. J Am Vet Med Assoc 1986;189:525–532.

Cooke KL, Synder PS. Calcium channel blockers in veterinary medicine. J Vet Intern Med 1998;12:123–131.

Fox PR, Sisson DD. Angiotensin converting enzyme inhibitors. In: Bonagura JD, Kirk RW, eds. Current veterinary therapy XII. Philadelphia: WB Saunders, 1995:786–791.

Keene BW. Therapy of heart failure. In: Bonagura JD, Kirk RW, eds. Current veterinary therapy XII. Philadelphia: WB Saunders, 1995:780–786.

Knight D. Efficacy of inotropic support of the failing heart. Vet Clin North Am Small Anim Pract 1991;21:879–903.

Novotny MJ, Adams HR. New perspectives in cardiology: recent advances in anti-arrhythmic drug therapy. J Am Vet Med Assoc 1986; 189:533–539.

Roth AL. Use of angiotensin converting enzyme inhibitors in dogs with congestive heart failure. Comp Cont Ed Pract Vet 1993;15:1240–1244.

Schlesinger DP, Rubin SI. Potential adverse effect of angiotensin converting enzyme inhibitors in the treatment of congestive heart failure. Comp Cont Ed Pract Vet 1994;16:275–283.

Sisson D. Evidence for and against the efficacy of afterload reducers for management of heart failure in dogs. Vet Clin North Am Small Anim Pract 1991;21:945–954.

Strickland KN. Advances in anti-arrhythmic therapy. Vet Clin North Am Small Anim Pract 1998;28:1515–1530.

Synder PS, Atkins CE. Current use and hazards of the digitalis glycosides. In: Bonagura JD, Kirk RW, eds. Current veterinary therapy XI. Philadelphia: WB Saunders, 1992:689–693.

Ocular Pharmacology

Davidson M. Ocular therapeutics. In: Bonagura JD, Kirk RW, eds. Current veterinary therapy XI. Philadelphia: WB Saunders, 1992:1047–1060.

Grahn BH, Wolfer J. Therapeutics. In: Peiffer PL, Petersen-Jones SM, eds. Small animal ophthalmology: a problem oriented approach. Philadelphia: WB Saunders, 1997:27–39.

Olivero DK. Ophthalmic products, topical. In: Plumb DC, ed. Veterinary drug handbook, 3rd ed. Ames, Iowa: Iowa University Press, 1999: 655–674.

Slatter D. Ocular pharmacology and therapeutics. In: Fundamentals of veterinary opththamology. Philadelphia: WB Saunders, 1990:32–67.

Micturition

Barsanti JA, Coates JR, Bartges JW, et al. Detrusor-sphincter dysynergia. Vet Clin North Am Small Anim Pract 1996;26:327–338.

Lane IF. Disorders of micturition. In: Osborne CA, Finco DR, eds. Canine and feline nephrology and urology. Philadelphia: Williams & Wilkins, 1995:693–717.
Lane IF. Pharmacologic management of feline lower urinary tract disorders. Vet Clin North Am Small Anim Pract 1996;26:515–533.
O'Brien DP. Disorders of the urogenital system. Semin Vet Med Surg (Small Anim) 1990;5:57–66.
Page SW. Diethylstilbestrol—clinical pharmacology and alternatives in small animal practice. Aust Vet J 1991;7:226–230.

Respiratory Drugs

Corcoran BM, Foster DJ, Fuentes VL. Feline asthma syndrome: a retrospective study of the clinical presentation in 29 cats. J Small Anim Pract 1995;11:481–488.
Dixon PM, Railton DI, McGorum BC, Tothill S. Equine pulmonary disease: a case control study of 300 referred cases. Part 4: treatments and re-examination findings. Equine Vet J 1995;27:436–439.
Dye JA. Feline bronchopulmonary disease. Vet Clin North Am Small Anim Pract 1992;22:1187–1201.
McKiernan BC. Current uses and hazards of bronchodilator therapy. In: Bonagura JD, Kirk RW, eds. Current veterinary therapy XI. Philadelphia: WB Saunders, 1992:660–668.
Viel L. Small airway disease as a vanguard for chronic obstructive pulmonary disease. Vet Clin North Am Equine Pract 1997;13:549–560.

Diuretics

Brater DC. Diuretic therapy. N Engl J Med 1998;339:387–395.
Hinchcliff KW, Mitter LA. Furosemide, butetanide and ethacrynic acid. Vet Clin North Am Equine Pract 1993;9:511–522.
Rose DB. Clinical physiology of acid-base and electrolyte disturbances, 3rd ed. New York: McGraw-Hill Book, 1990. New York.

Inhalant Anesthetics

Haskins SC. Inhalational anesthetics. Vet Clin North Am Small Anim Pract 1992;22:297–307.
Steffey L. Inhalation anesthetics. In: Thurmon JC, Tranquilli WJ, Benson GJ, eds. Lumb and Jones: Veterinary anesthesia, 3rd ed. Philadelphia: Williams & Wilkins, 1996:297–329.

Injectable Anesthetics

Hall LW, Clarke KW. General pharmacology of intravenous anesthetic agents. In: Veterinary anesthesia. London: Balliere Tindal, 1991:80–97.
Haskins S. Injectable anesthetics. Vet Clin North Am Small Anim Pract 1992;22:245–260.
Mama K. New drugs in feline anesthesia. Comp Cont Ed Pract Vet 1998; 20:125–139.
Muir WW, Mason D. Side effects of etomidate in dogs. J Am Vet Med Assoc 1989;194:1430–1434.
Robinson EP, Sanderson SL, Machon RG. Propofol: a new sedative-hypnotic agent. In: Bonagura JD, Kirk RW, eds. Current veterinary therapy XII. Philadelphia: WB Saunders, 1995:77–81.

Tranquilizers and Sedatives

Cullen LK. Medetomidine sedation in dogs and cats: a review of its pharmacology, antagonism and dose. Br Vet J 1996;5:519–535.
Gross ME, Tranquilli J. Use of α_2 adrenergic receptor antagonists. J Am Vet Med Assoc 1989;195:378–381.
Klide AM. Precautions when using α_2 agonists as anesthetics or anesthetic adjuvants. Vet Clin North Am Small Anim Pract 1992;22:294–296.
Tranquilli WJ, Benson GJ. Advantages and guidelines for using α_2 agonists as anesthetic adjuvants. Vet Clin North Am Small Anim Pract 1992;22:289–293.

Opioids

Hellyer P. Management of acute surgical pain. Semin Vet Med Surg (Small Anim) 1997;12:106–144.
Hellyer PW, Gaynor JS. Acute post-surgical pain management in dogs and cats. Comp Cont Ed Pract Vet 1998;20:140–154.
Hosgood G. Pharmacologic features of butorphanol in dogs and cats. J Am Vet Med Assoc 1990;196:135–136.
Kamerling SC. Narcotics and local anesthetics. Vet Clin North Am Small Anim Pract 1993;9:605–620.
Kyles AE. Transdermal fentanyl. Comp Cont Ed Pract Vet 1998;21:721–726.

Nonsteroidal Anti-inflammatory Agents

Budsberg SC, Johnston SA, Schwarz PD, et al. Efficacy of etodolac for the treatment of osteoarthritis of the hip joints in dogs. J Am Vet Med Assoc 1999;214:206–210.
Conlon PD. Nonsteroidal drugs used in the treatment of inflammation. Vet Clin North Am Small Anim Pract 1988;18:1115–1130.
Hulse D. Treatment methods for pain in the osteoarthritic patient. Vet Clin North Am Small Anim Pract 1998;28:361–370.
Kallings P. Nonsteroidal anti-inflammatory drugs. Vet Clin North Am Equine Pract 1993;9:523–540.
MacPhail CM, Lappin MR, Meyer D, et al. Hepatotoxicity associated with the administration of carprofen in 21 dogs. J Am Vet Med Assoc 1998; 212:1895–1901.
Papich M. Principles of analgesic drug therapy. Semin Vet Med Surg (Small Anim) 1997;12:80–93.
Short CE. Equine pain: use of NSAIDs and analgesics for its prevention and control. Equine Pract 1995;17:12–22.

Anticonvulsant Therapy

Boothe D. Anticonvulsant therapy in small animals. Vet Clin North Am Small Anim Pract 1998;28:411–448.
Brown SA. Anti-convulsant therapy in small animals. Vet Clin North Am Small Anim Pract 1988;18:1197–1240.
Trepainer LA. Use of bromide as an anticonvulsant for dogs with epilepsy. J Am Vet Med Assoc 1995;207:163–166.

Behavioral Pharmacology

Chew DJ, Buffington T, Kendell MS, et al. Amitriptyline treatment for severe recurrent idiopathic cystitis in cats. J Am Vet Med Assoc 1998; 213:1282–1286.
Dodman NH, Shuster L, eds. Psychopharmacology of animal behavior disorders. Malden, Mass.: Blackwell Science, 1998.
Marder AR. Psychotropic drugs and behavioral therapy. Vet Clin North Am Small Anim Pract 1981;21:329–342.
Overall K. Pharmacologic treatment for behavioral problems. Vet Clin North Am Small Anim Pract 1997;27:637–663.

Antibiotics

Boothe DM. Drug therapy in cats: a therapeutic approach. J Am Vet Med Assoc 1990;196:1659–1669.
Cockerill FR, Edson R. Trimethoprim-sulfamethoxazole. Mayo Clin Proc 1987;62:921–929.
Donowitz GR, Mandell GL. Beta lactam antibiotics (part 1). N Engl J Med 1988;318:419–426.
Donowitz GR, Mendell GL. Beta lactam antibiotics (part 2). N Engl J Med 1988;318:490–498.
Edson RS, Terrell CL. The aminoglycosides. Mayo Clin Proc 1987;62: 916–920.
Gold HS, Moellering RC. Anti-microbial drug resistance. N Engl J Med 1996;335:1445–1454.
Greene C, ed. Infectious diseases of the dog and cat, 2nd ed. Philadelphia: WB Saunders, 1998.
Harari J, Lincoln J. Pharmacology of clindamycin in dogs and cats. J Am Vet Med Assoc 1989;195:124–125.
Papich M. The beta lactam antibiotics: clinical pharmacology and recent developments. Comp Cont Ed Pract Vet 1987;9:68–74.
Papich M. Anti-bacterial drug therapy—focus on new drugs. Vet Clin North Am Small Anim Pract 1998;28:215–232.
Prescott JF, Baggot JD. Antimicrobial therapy in veterinary medicine, 2nd ed. Ames, Iowa: Iowa State University Press, 1993.

Riond JL, Riviere JE. Effects of tetracyclines on the kidney in cattle and dogs. J Am Vet Med Assoc 1989;195:995–997.
Settepani J. The hazard of chloramphenicol use in food animals. J Am Vet Med Assoc 1984;184:930–931.
Thompson R. Cephalosporin, carbapenem, and monolactam antibiotics. Mayo Clin Proc 1987;62:821–834.
Walker RC, Wright AJ. The quinolones. Mayo Clin Proc 1987;62:1007–1012.
Washington J, Wilson W. Erythromycin: a microbial and clinical perspective after 30 years of clinical use. Mayo Clin Proc 1985;60:189–203.
Whitten T, Gaon D. Principles of anti-microbial therapy. Vet Clin North Am Small Anim Pract 1998;28:197–214.
Wilson WR, Cockerell FR. Tetracycline, chlortetracycline, erythromycin and clindamycin. Mayo Clin Proc 1987;62:906–915.
Wright AJ, Wilkowske CJ. The penicillins. Mayo Clin Proc 1987;62:806–820.

Antifungal Agents

Greene CE, Watson ADJ. Antifungal chemotherapy. In: Greene CE, ed. Infectious diseases of the dog and cat, 2nd ed. Philadelphia: WB Saunders, 1998:357–361.
Legendre A. Anti-mycotic drug therapy. In: Bonagura JD, Kirk RW, eds. Current veterinary therapy XII. Philadelphia: WB Saunders, 1995:327–334.

Antiprotozoal Agents

Barr SC, Jamross GF, Hornbuckle W, et al. Use of paromomycin for treatment of cryptosporidiosis in a cat. J Am Vet Med Assoc 1994;205: 1742–1743.
Dubey J. Intestinal protozoal infections. Vet Clin North Am Small Anim Pract 1993;23:37–55.
Greene C, Watson D. Antiprotozoal chemotherapy. In: Greene C, ed. Infectious diseases of the dog and cat, 2nd ed. Philadelphia: WB Saunders, 1998:441–444.
Haberkorn A. Chemotherapy of human and animal coccidioses: state and perspectives. Parasitol Res 1996;82:193–199.
McDouglad L. Coccidiosis prevention and treatment. Poultry Dig 1994; 53:20–21.

Antiparasitic Drugs

Beasley VR, ed. Toxicology of selected pesticides, drugs and chemicals. Vet Clin North Am Small Anim Pract 1990;20:283–564.
Bennett D. Clinical pharmacology of ivermectin. J Am Vet Med Assoc 1986;189:100–103.
Craig TM. Developing successful internal parasite control programs. Comp Cont Ed Pract Vet 1997;19:S112–S119.
Hansen SR, Stemme K, Villar D, Buck W. Pyrethrins and pyrethroids in dogs and cats. Comp Cont Ed Pract Vet 1994;16:707–712.
Jacobs, D ed. Selamectin—A novel endectocide for dogs and cats. Veterinary Parasitology 2000;91:161–405.
Lynn RC. Anti-parasitic drugs. In: Bowman DD, Lynn RC, eds. Georgi's parasitology for veterinarians, 6th ed. Philadelphia: WB Saunders, 1996:247–292.
Martin RJ. Neuromuscular transmission in nematode parasites and anti-nematodal drug action. Pharmacology and Therapeutics 1993;58:13–50.
McKellar QA. Developments in pharmacokinetics and pharmacodynamics of anti-helminthic drugs. J Vet Pharmacol Ther 1997;20:10–19.
Paradis M. Ivermectins in small animal dermatology. Part 1: pharmacology and toxicology. Comp Cont Ed Pract Vet 1998;20:193–198.
Sangster N. Pharmacology of anti-helminthic resistance. Parasitology 1996;113:S201–S216.
Shoop WL, Mrosik H, Fisher MH. Structure and activity of avermectins and milbemycins in animal health. Vet Parasitol 1995;59:139–156.

Antineoplastic Agents

Couto CG. Management of the complications of cancer chemotherapy. Vet Clin North Am Small Anim Pract 1990;20:1037–1053.
Dobson JM, Gorman NT. Cancer chemotherapy in small animal practice. UK: London Blackwell Science, 1993.
Golden DL, Langston VC. Uses of vincristine and vinblastine in dogs and cats. J Am Vet Med Assoc 1988;193:114–117.
Hahn KA, Richardson RC, eds. Cancer chemotherapy: a veterinary handbook. Malvern, PA: Williams & Wilkins, 1995.
Helfand SC. Principles and application of chemotherapy. Vet Clin North Am Small Anim Pract 1990;20:987–1013.
Oglivie GK, Moore AS. Managing the veterinary cancer patient. Trenton, NJ: Vet Learning System, 1995.

Gastrointestinal Drugs

Andrews FM, MacAllister C, Jenkins CC, Blackford JT. Omeprazole: a promising anti-ulcer drug in horses. Comp Cont Ed Pract Vet 1996; 18:1288–1230.
Boeckh A. Misoprostol. Comp Cont Ed Pract Vet 1999;24:66–67.
Dart AJ, Hodgson D. Role of prokinetic drugs for treatment of post-operative ileus in the horse. Aust Vet J 1998;76:25–31.
Dowling PM. Therapy of gastrointestinal ulcers. Can Vet J 1995;36:276–277.
Guilford WG, Center SA, Strombeck DR, et al, eds. Strombeck's small animal gastroenterology, 3rd ed. Philadelphia: WB Saunders, 1996.
Hall J, Washabau R. Gastrointestinal prokinetic therapy: dopaminergic antagonist drugs. Comp Cont Ed Pract Vet 1997;19:214–220.
Hall J, Washabau R. Gastrointestinal prokinetic therapy: motilin like drugs. Comp Cont Ed Pract Vet 1997;19:281–287.
Hall J, Washabau R. Gastrointestinal prokinetic therapy: serotinergic drugs. Comp Cont Ed Pract Vet 1997;19:473–480.
Hall J, Washabau R. Gastrointestinal prokinetic therapy: acetylcholinesterase inhibitors. Comp Cont Ed Pract Vet 1997;19:615–621.
Matz M. Gastrointestinal ulcer therapy. In: Bonagura JD, Kirk RW, eds. Current veterinary therapy XII. Philadelphia: WB Saunders, 1995:706–710.
Papich MG. Anti-ulcer therapy. Vet Clin North Am Small Anim Pract 1993; 23:497–512.
Washabau RJ, Elie MS. Anti-emetic therapy. In: Bonagura JD, Kirk RW, eds. Current veterinary therapy XII. Philadelphia: WB Saunders, 1995: 679–684.

Endocrine

Bruyette D. Alternatives for the treatment of hyperadrenocorticism in dogs and cats. In: Bonagura JD, Kirk RW, eds. Current veterinary therapy XII. Philadelphia: WB Saunders, 1995:221–245.
Bruyette D, Ruehl WW, Entriken T, et al. Management of canine pituitary dependent hyperadrenocorticism with L-deprenyl. Vet Clin North Am Small Anim Pract 1997;27:273–286.
Cowan LA, McLaughlin R, Toll P, et al. Effect of stanozolol on body composition, nitrogen balance and food consumption in castrated dogs with chronic renal failure. J Am Vet Med Assoc 1997;211:719–722.
Feldman E, Nelson R. Use of ketoconazole for control of canine hyperadrenocorticism. In: Bonagura JD, Kirk RW, eds. Current veterinary therapy XI. Philadelphia: WB Saunders, 1992:349–352.
Feldman EC, Nelson R. Canine and feline endocrinology and reproduction, 2nd ed. Philadelphia: WB Saunders, 1993.
Greco D. Oral hypoglycemic agents for noninsulin dependent diabetes in the cat. Semin Vet Med Surg (Small Anim) 1997;12:259–262.
Kintzer P. Considerations in the treatment of feline hyperthyroidism. Vet Clin North Am Small Anim Pract 1994;24:577–586.
Kintzer P, Peterson M. Mitotane (o'p'-DDD) treatment of hyperadrenocorticism in dogs. In: Bonagura JD, Kirk RW, eds. Current veterinary therapy XII. Philadelphia: WB Saunders, 1995:416–420.
Lean I, Weaver LD, Galland JC, et al. Bovine somatotropin: biological implications. Comp Cont Ed Pract Vet 1989;11:1168–1173.
Lynn RC, Feldman E, Nelson R. Efficacy of microcrystalline desoxycorticosterone pivalate for treatment of hypoadrenocorticism in dogs. J Am Vet Med Assoc 1993;202:392–396.
McCabe MD, Feldman EC, Lynn RC, Kass PH. Subcutaneous administration of desoxycorticosterone pivalate for the treatment of canine hypoadrenocorticism. J Am Anim Hosp Assoc 1995;31:151–155.

McKinnon AO, Voss JL. Equine reproduction. Philadelphia: Lea & Febiger, 1993.
Nelson RW. Diagnosis and treatment of canine hypothyroidism. Vet Q 1996;18:S29–S31.
Nichols R, Hohenhaus A. Use of vasopressin analogue desmopressin for polyuria and bleeding disorders. J Am Vet Med Assoc 1994;205:168–173.
Purswell B. Pharmaceuticals used in canine reproduction. Semin Vet Med Surg (Small Anim) 1994;9:54–60.
Snow DH. Anabolic steroids. Vet Clin North Am Equine Pract 1993;9: 563–576.
Webster CRL. Bile acids: what's new. Semin Vet Med Surg (Small Anim) 1997;12:2–9.

Corticosteroids

Behrend E, Kemppainen R. Glucocorticoid therapy: pharmacology, indications and complications. Vet Clin North Am Small Anim Pract 1997; 27:187–213.
Cohn LA. The influence of corticosteroids on host defense mechanisms. J Vet Intern Med 1991;5:95–104.
Harkins JD, Carney JM, Tobin T. Clinical use and characteristics of corticosteroids. Vet Clin North Am Equine Pract 1993;9:543–555.
Waddell LS, Drobatz KJ, Otto CM. Corticosteroids in hypovolemic shock. Comp Cont Ed Pract Vet 1998;20:571–585.

Immunosuppressive Drugs

Arndt H, Palitzsch K, Grisham M, Granger D. Metronidazole inhibits leukocyte-endothelial cell adhesion in rat mesenteric venules. Gastroenterology 1994;106:1271–1276.
Cohn L. Glucocorticoids as immunosuppressive agents. Semin Vet Med Surg (Small Anim) 1997;12:150–156.
Dowling PM. Immunosuppressive drug therapy. Can Vet J 1995;36:781–783.
Elizondo G, Ostrosky-Wegman P. Effects of metronidazole and its metabolites on histamine immunosuppression activity. Life Sci 1996;59: 285–297.
Miller E. The use of danazol in the therapy of immune-mediated disease of dogs. Semin Vet Med Surg (Small Anim) 1997;12:167–169.
Miller E. The use of cytotoxic agents in the treatment of immune-mediated disease of dogs and cats. Semin Vet Med Surg (Small Anim) 1997; 12:157–160.
Scott-Moncrieff CR, Reagen WJ. Human intravenous immunoglobulin therapy. Semin Vet Med Surg (Small Anim) 1997;12:178–185.
Vaden S. Cyclosporine and tacrolimus. Semin Vet Med Surg (Small Anim) 1997;12:161–166.
Van Kampen K. Immunotherapy and cytokines. Semin Vet Med Surg (Small Anim) 1997;12:186–192.

Drugs Acting on Blood

Boothe D. Drugs acting on blood and/or blood forming organs. Vet Clin North Am Small Anim Pract 1998;28:329–360.
Cowgill LD, James KM, Levy JK, et al. Use of recombinant human erythropoietin for management of anemia in dogs and cats with chronic renal failure. J Am Vet Med Assoc 1998;212:521–528.
Ramsey CC, Burney DP, Macintire D, Finn-Bodner S. Use of streptokinase in 4 dogs with thrombosis. J Am Vet Med Assoc 1996;209:780–785.
Weitz J. Low molecular weight heparins. N Engl J Med 1997;337:688–699.

INDEX